Containment of High-Level Radioactive and Hazardous Solid Wastes with Clay Barriers

Spon Research

Spon Research publishes a stream of advanced books for built environment researchers and professionals from one of the world's leading publishers. The ISSN for the Spon Research programme is ISSN 1940–7653 and the ISSN for the Spon Research E-book programme is ISSN 1940–8005.

Published

Free-Standing Tension Structures
From tensegrity systems to cable-strut systems
978–0–415–33595–9
W. B. Bing

Performance-Based Optimization of Structures
Theory and applications
978–0–415–33594–2
Q. Q. Liang

Microstructure of Smectite Clays & Engineering Performance
978–0–415–36863–6
R. Pusch and R. Yong

Procurement in the Construction Industry
The impact and cost of alternative market and supply processes
978–0–415–39560–1
W. Hughes, P. M. Hillebrandt, D. Greenwood and W. Kwawu

Communication in Construction Teams
978–0–415–36619–9
S. Emmitt and C. Gorse

Concurrent Engineering in Construction Projects
978–0–415–39488–8
C. Anumba, J. Kamara and A.-F. Cutting-Decelle

People and Culture in Construction
978–0–415–34870–6
A. Dainty, S. Green and B. Bagilhole

Very Large Floating Structures
978–0–415–41953–6
C. M. Wang, E. Watanabe and T. Utsunomiya

Tropical Urban Heat Islands
Climate, buildings and greenery
978–0–415–41104–2
N. H. Wong and C. Yu

Innovation in Small Construction Firms
978–0–415–39390–4
P. Barrett, M. Sexton and A. Lee

Construction Supply Chain Economics
978–0–415–40971–1
K. London

Employee Resourcing in the Construction Industry
978–0–415–37163–6
A. Raiden, A. Dainty and R. Neale

Managing Knowledge in the Construction Industry
978–0–415–46344–7
A. Styhre

Collaborative Information Management in Construction
978–0–415–48422–0
G. Shen, A. Baldwin and P. Brandon

Containment of High-Level Radioactive and Hazardous Solid Wastes with Clay Barriers
978–0–415–45820–7
R. N. Yong, R. Pusch and M. Nakano

Forthcoming

Organisational Culture in the Construction Industry
978–0–415–42594–0
V. Coffey

Relational Contracting for Construction Excellence
Principles, practices and partnering
978–0–415–46669–1
A. Chan, D. Chan and J. Yeung

Performance Improvement in Construction Management
978–0–415–54598–3
B. Atkin and J. Borgbrant

Containment of High-Level Radioactive and Hazardous Solid Wastes with Clay Barriers

Raymond N. Yong,
Roland Pusch and
Masashi Nakano

Routledge
Taylor & Francis Group

LONDON AND NEW YORK

First published 2010 by Spon Press

2 Park Square, Milton Park, Abingdon, Oxon OX14 4RN
711 Third Avenue, New York, NY 10017, USA

Routledge is an imprint of the Taylor & Francis Group, an informa business

First issued in paperback 2017

Copyright © 2010 Raymond N. Yong, Roland Pusch and Masashi Nakano

Typeset in Gill Sans and Sabon by Prepress Projects Ltd, Perth, UK

British Library Cataloguing in Publication Data
A catalogue record for this book is available from the British Library

Library of Congress Cataloging-in-Publication Data
Yong, R.N. (Raymond Nen)
Containment of high level radioactive and hazardous solid wastes with
clay barriers / Raymond N. Yong, Roland Pusch, and Masashi Nakano
p. cm.
Includes bibliographical references and index.
1. Radioactive waste disposal in the ground. 2. Clay–Permeability. 3.
Engineered barrier systems (Waste disposal). 4. Clay soils. 5. Hazardous
waste sites. I. Pusch, Roland. II. Nakano, Masashi, 1937– III. Title.
TD898.2.Y66 2009
621.48'38–dc22
2009012840

ISBN13: 978-0-415-45820-7 (hbk)
ISBN13: 978-1-138-11536-1 (pbk)

Contents

About the authors

Raymond N. Yong has authored and co-authored eight other textbooks and over 500 refereed papers in various journals in the field of Geoenvironmental Engineering. He is a fellow of the Royal Society (Canada), and a Chevalier de l'Ordre National du Québec. He and his students were amongst the early researchers in Geoenvironmental Engineering who were engaged in research on the physico-chemical properties and behaviour of clays, and their use in buffer/barriers for the containment and isolation of high-level radioactive waste (HLW) and hazardous solid waste (HSW). He is currently engaged in research on issues in geoenvironmental sustainability and environmental soil behaviour.

Roland Pusch has made significant contributions in research on the microstructure of clays and their impact on the properties and performance. He has been very active, nationally and internationally, in pioneering work on clay buffers and underground repository systems for HLW containment in association with the Swedish Nuclear Fuel and Waste Management Company (SKB) Stockholm, and the European Commission. He has long been active in European Union projects relating to HLW and HSW containment and isolation. He is currently the Scientific Head and Managing Director of Geodevelopment International AB, Sweden, and is working on issues of long-term stability of clay buffers in HLW repositories.

Masashi Nakano has worked and educated many students as a professor of the University of Tokyo and has published many significant papers on mass transport in soils in the field of soil physics. He has served as a counsellor for the Japan Atomic Energy Agency and its predecessors, on clay barrier systems for radioactive wastes disposal, for 30 years – since the inception of plans for radioactive waste disposal in Japan. He is currently a member of the Science Council of Japan, and he is now working on such issues in clay science as ion adsorption on soils and mineral corrosion by microorganisms.

Preface

The subject material in this book concerns the nature and performance of clay as an engineered material in the multi-barrier system proposed for high-level radioactive waste (HLW) containment. The use of engineered multi-barrier systems that include clay as one of the component barriers is not unique in itself. This type of barrier system is now being used in many jurisdictions for secure landfill disposal of hazardous solid wastes (HSWs). Containment and disposal of low- and medium-level radioactive wastes also use technologies that are similar to those practised in HSW landfill disposal schemes. Isolation of HLW canisters and HSW using multi-barrier systems brings with it challenges of heat and chemical attacks on the clay component, and other degradative forces that are associated with abiotic and biotic reactions. These types of forces and reactions are particularly important, as the barrier systems are required to function according to design expectations over time spans of hundreds of thousands of years in the case of HLW containment/isolation.

The nature, function and performance of clays have been studied for a long time in the various fields of science and engineering. These include the various branches of soil science, soil engineering and mechanics, geology, etc. In order to utilize clay as an engineered material in a multi-barrier system, knowledge gained in all the fields of soil science, soil engineering and geology is needed in as much as these disciplines deal directly with the environment within which HLW and HSW are contained and isolated. This book collects and synthesizes the fundamental physical, chemical, mechanical, geological and biological knowledge of clay that is required for the design of the multi-barrier systems. In addition, the material contained in this book provides the latest knowledge gained from research and full-scale studies of HLW containment and isolation in deep underground repositories, together with various methods and procedures that are necessary to measure the parameters required for analysing different phenomena in the clay barrier.

Chapter 1 provides an outline of the multi-barrier system and clay barrier performance in relation to the system. The nature of the threats posed by radioactive wastes, from low- to medium-level and especially high-level radioactive wastes, is described in Chapter 2. The length of time needed for radioactivity to diminish, coupled with the slow dissipative nature of the heat from the spent fuel rods, are two of the principal features that characterize the HLW problem. The containment schemes discussed in this chapter for HLW and HSW include the multi-barrier systems that are used as repositories to isolate HLW canisters and the landfills for containment of HSW. Chapter 3 discusses the nature of clays, and especially those clays that are most suited for the clay barrier component in the multi-barrier system. The importance of the structure of the clays and their capability to physically and chemically attenuate pressures and chemicals are discussed. Chapters 4 and 5 elaborate on the chemical and physical reactions that occur in clays when they are under environmental regimes characterized by their repository and/or waste containment schemes. The forces and fluxes associated with thermal, hydraulic, mechanical, chemical and biological processes are discussed in Chapter 6. The importance of obtaining a better knowledge of these processes is not only with respect to the nature of the processes, but also with respect to the outcome of these processes on the properties and performance of the clay barrier component. How do these processes and factors impact on the clay?

The longer-term aspects of the impact of the processes described in Chapter 6 are discussed in Chapter 7. Although much has yet to be fully studied and determined, we can identify the many aspects of the long-term clay evolution problem that will pose difficulties for those who use clays as engineered clay barriers. Clay alteration and transformation under abiotic and biotic circumstances constitute issues that impact severely on clay performance. Some of the kinds of information required for us to examine short- and long-term clay performance may be found in the field and mock-up experiments that are described in Chapter 8. The kinds of information obtained with full-scale experiments are invaluable for validation of the predictive and performance models discussed in Chapter 9. Given the long time span requirement for secure service life of the clay barrier, one has no recourse except to rely on valid relevant and robust models to check on design performance of the constructed facility. That being said, there is still the requirement for determination of the safety of the facility – in this case, the engineered clay barrier of a multi-barrier system; this is discussed in Chapter 10. Safety assessment is a critical process that is essentially a measure of risk acceptable for the system under consideration – a balance of reliability and risk.

It is our hope that the material contained in this book will be useful for those responsible for regulatory oversight of HLW and HSW containment/isolation, and especially for those charged with (a) planning, design and construction of multi-barrier systems for containment and isolation of high-level radioactive wastes and (b) design and construction of safe and secure landfills for the containment of HSWs. When properly used, the clay barrier component in a multi-barrier containment system is an effective long-term protective barrier.

We would like to express our gratitude to our colleagues working in research and development on the isolation/containment schemes for HLW canisters and HSW secure landfills for their discussions and input. We are grateful to the Swedish company Swedish Nuclear Fuel and Waste Management Company (SKB) Stockholm for giving us access to its database and other information, particularly through the former head of the Äspö Hard Rock Laboratory, Christer Svemar. Thanks are extended to the Geological department, Greifswald University, Germany, for valuable contributions by Jörn Kasbohm, and to Ragn-Sell Waste Management AB, Stockholm, for constructive discussions with the former head of its R&D office, Anders Kihl, and the numerous research scientists/engineers at Atomic Energy of Canada Ltd (AECL) and the Japan Atomic Energy Agency (JAEA).

Raymond N. Yong, North Saanich, Canada
Roland Pusch, Lund, Sweden
Masashi Nakano, Tokyo, Japan

Chapter 1

Introduction

1.1 Containment and isolation

1.1.1 Hazardous and dangerous wastes

A basic requirement in the design and construction of waste containment facilities in the land environment is to contain and isolate waste substances from the surrounding environment in a manner akin to a tightly sealed garbage bag embedded in the ground. The objective is to deny exposure of the contained waste substances and their reaction products to humans, other biotic receptors, and the environment – most especially if these waste substances and their leachates pose health and environmental threats. When waste substances and their reaction and/or dissolution products – such as waste streams and leachates – are threats to public health and the environment, they are generally identified as *dangerous wastes*. A detailed discussion of the nature and required management of these wastes is given in Chapter 2.

Because of some very significant differences in (a) the kinds of dangers or hazards presented by certain kinds of dangerous wastes and (b) the manner in which these wastes should be managed and disposed, it has been found necessary to separate dangerous wastes into two major categories: *radioactive wastes* and *hazardous solid wastes (HSWs)*. We do not include hazardous liquid wastes (e.g. organic and volatile chemicals) in our discussions, as they constitute liquid streams and must be detoxified before discharge. The exceptions are those liquid wastes that are housed in leak-proof containers (drums and similar containers), which are then disposed of in landfills in a manner similar to the disposal of HSWs. As will be discussed in detail in Chapter 2, it is the high-level radioactive waste (HLW) that demands the strictest and most demanding attention to containment of dangerous wastes.

Technically speaking, there is no strict scientific definition of HSWs. This is because of the very wide variety of (a) ways or manners in which the hazard presented by the waste material or product can be expressed, for example fire hazard, health hazard, radiological hazard, and (b) sources, actions, materials, substances, etc., that can be responsible for generating the hazard in question. It is because of the dangers posed by hazardous wastes that they are considered to be *dangerous wastes*. Even although *radioactive wastes* are hazardous (dangerous) as well and fall within the definition/classification of hazardous wastes, the guidelines formulated by most countries for characterizing and disposal of dangerous wastes have relegated these wastes into a separate and distinct category – to a very large extent because of the radiation hazard posed by these wastes. Accordingly, regulations and governmental oversight management of HSWs and radioactive wastes in most countries are administered by agencies with primary focus on one or the other type of dangerous waste – with perhaps some overlapping, for example in the case of health issues. It is not unusual for two or more agencies to have jurisdictional oversight on the use, management and disposal of these wastes.

The HSWs considered in this book are those that have been classified as such by the various governmental agencies and jurisdictions tasked with environmental protection and public health in many different countries. By and large, HSWs are defined as wastes that have properties which are inherently harmful to the state of the environment and harmful to the health of biotic receptors. Most jurisdictions have detailed lists of HSWs, classified according to characteristics such as properties, type, source, industry, or as *non-specific*. In the case of *non-specific* HSW, any one of the following critical characteristics will qualify a waste as a hazardous waste:

- *ignitability* – potential for fire hazard during storage, transport or disposal;
- *corrosivity* – potential for corrosion of materials in contact with the waste, resulting in damage to the environment and human health;
- *reactivity* – potential for adverse chemical reactions;
- *toxicity* – as determined from some well-established toxicity tests on the leachate or reaction products obtained from the waste in question.

The radioactive wastes that are of primary concern in this book are those generated in the nuclear fuel cycle, as, for example, obtained as spent nuclear fuel irradiated in nuclear reactors. These wastes are commonly known as high-level nuclear wastes or high-level radioactive wastes (HLWs) – to be distinguished from intermediate-level radioactive wastes (ILWs) and low-level radioactive wastes (LLWs). The high level of radioactivity results from the waste product obtained in the reprocessing of spent fuel to extract uranium and plutonium. Subsequent drying of the liquid waste residue obtained in the reprocessing phase does nothing to alleviate the high source level of radioactivity associated with the spent fuel. A detailed discussion of these wastes is provided in Chapter 2. Low-level radioactive wastes (LLWs) are obtained from a variety of sources, for example reactor decommissioning, discarded medical equipment, protective clothing exposed to neutron radiation and discarded laboratory material that is associated with

X-radiology, etc., with low concentrations of beta and gamma contamination. Disposal of these kinds of wastes may be handled in the manner used to contain the general class of hazardous wastes. Intermediate-level radioactive wastes (ILWs) refer to wastes that are derived from nuclear power reactor operations maintenance and from operations involved in reprocessing. They contain concentrations of gamma and beta contamination that are higher than LLWs, with half-lives of radionuclides not exceeding about 30 years. As opposed to disposal of LLWs, encapsulation techniques are used as part of the canister containment procedure for ILWs. The nature of these two types of wastes, i.e. HLWs and hazardous wastes (HWs), will be described and discussed in Chapter 2.

For the purposes of discussion in this chapter, because HLW qualifies as a hazardous waste, and in as much as common accepted usage of the term *hazardous solid wastes (HSWs)* distinguishes this from radioactive wastes, the term *dangerous waste* will be used when general reference to both of these classes of wastes is required. Exposure to, and contact with *dangerous wastes* and their products (radioactive and other hazardous wastes) will be harmful to not only the environment and human health, but also land and aquatic animals and other organisms.

1.1.2   *Regulatory concerns, attitudes and strategies*

The different ways of categorizing waste materials are:

1   by the medium in which they are released, for example air, water or on/ in land;
2   in accordance with their physical characteristics, for example gas, liquid or solid;
3   by their origin or generating source, for example mining, municipal or industrial;
4   by the type of risk they pose, for example hazardous, non-hazardous, toxic or non-toxic or radioactive.

Although the first three methods of categorization can be very useful, they require another step in characterization to determine whether the waste materials pose a health and/or environmental threat. Category 4 is the favoured method for categorizing waste materials by many regulatory agencies. This is because of the dangers posed to public health and/or the environment by (a) the actual wastes themselves; (b) the waste products; and/or (c) the waste streams produced. The term *regulated wastes* is used to identify the category 4 type of classification. Although all types of wastes discharged into the environment require some level of tracking by the appropriate regulatory agency, the category 4 type of regulated waste demands the highest level of tracking. The minimal requirement for tracking these kinds of wastes is an integrated tracking system that tracks the wastes from the generating source to final disposal. In many instances, further tracking beyond final disposal is required to ensure that the containment system fulfils its task, i.e. until the risks from exposure to the waste substances, their products and waste streams do not pose any undue threat to the biotic receptors and the environment. Chapter 2 provides a detailed discussion of the classification protocols used.

High-level nuclear wastes and HWs are regulated wastes, i.e. wastes that fall within the jurisdiction of regulatory agencies and require a high level of *tracking response sequence* as indicated previously. It is not uncommon for many governments to separate the responsibilities for regulating and monitoring the management of these wastes, with one regulatory agency being responsible for low- and high-level nuclear wastes, and another separate agency being responsible for HWs. To a large extent, this is because of some differences in the nature of the dangers posed by the two different types of dangerous wastes; it is also because of the different levels and types of complexities faced in the containment and isolation techniques required – especially in the containment/isolation of HLW, as will be evident in the Chapter 2 discussion. The radioactive nature of the waste has generated a whole host of international conventions and cooperative agreements (e.g. Euratom Treaty Article 37, EC Directive 85/337/EEC (as amended by 97/11/EC) on Environmental Assessment, International Convention on the Safety of Spent Fuel and Radioactive Waste Management).

The regulatory attitudes adopted by the various regulatory agencies responsible for waste management and disposal play a very important part in the determination of the nature, role and effectiveness of contaminant facilities. In general, regulatory attitudes can be broadly grouped into two types:

- command and control (CC);
- performance assessment (PA).

The earlier CC attitudes of many regulatory agencies for management and disposal of all kinds of wastes have resulted in the development of land disposal techniques and practices that have a common base – secure containment to prevent escape of contaminants. Most containment structures consist of at least two lines of containment defence: a *primary impermeable liner* that acts as a container, and a *secondary back-up impermeable liner* that is separated from the first by a separation or buffer layer, which will generally house a 'leak'-detection system (see Figure 1.1 and Chapter 2). A third line of defence can be incorporated into the containment design scheme using the surrounding rocks and subsoil system as an attenuating barrier.

Even although specified and mandated requirements for a containment scheme may be implicit in the CC regulatory framework, responsibility for the proper design, construction and operation of the containment facility rests with the stakeholder. Adherence to the mandated requirements does not relieve one of the responsibility for design/construction/operation *faults and failure-to-perform*. The PA approach adopted by many regulatory agencies today recognizes that design and construction techniques will vary, depending on site and waste specificities. The difference between the PA approach and the CC attitude is in the articulated requirements for the constructed facility. The PA approach requires the constructed facility to perform according to specified performance standards, particularly with respect to containment control, prevention of contaminant escape and contaminant contact with biotic receptors and the environment. Detailed discussions of PAs and the approaches adopted are provided under 'Performance assessment', later, and Chapters 9 and 10 provide detailed discussions of performance assessments (PAs) and the approaches adopted.

1.1.3 *Multiple containment barriers*

Because of the dangers posed by fugitive contaminants, techniques developed to securely contain the dangerous wastes described in the previous section most often adopt a *multi-barrier* approach. Figure 1.1 shows a simple diagram of the main items constituting this type of approach. By and large, land disposal/containment is undertaken in deep strata for HLW and ILW, and in shallow underground containment facilities for LLW and HSW. In principle, all of these containment techniques require multiple barriers as shown in the diagram. These consist of:

* A *primary or first barrier*, consisting of a container system that could be an impermeable rigid canister or membrane or some similar system designed to totally contain the dangerous substance, thus isolating it from the surroundings. The analogy of a leak-proof garbage bag has been used to describe this form of containment.
* A *second barrier* or second line of defence against fugitive contaminants. For HSW landfills, this generally consists of a double membrane or some similar system. A leak collection and leak detection barrier is generally used to separate the first and second membranes in the double-membrane system. The *leak* substance can be contaminants carried as an escaping leachate (from the first barrier, which acts as the first line of defence) or fugitive radionuclides or other kinds of contaminants (see Chapter 2). For HLW canisters in boreholes located in rock, a clay buffer is used as the cushion and separation layer between the canisters and the host rock. This clay buffer functions as second barrier system, i.e. second to the canister housing the spent nuclear fuel.
* An *engineered clay barrier system* consisting of in-place or transported clay material underlying the double-membrane system in HSW landfills. The type and properties of the material used and the manner

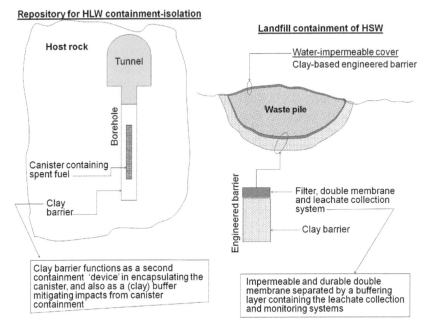

Repository for HLW containment-isolation

Host rock

Tunnel

Borehole

Canister containing spent fuel

Clay barrier

Clay barrier functions as a second containment 'device' in encapsulating the canister, and also as a (clay) buffer mitigating impacts from canister containment

Landfill containment of HSW

Water-impermeable cover
Clay-based engineered barrier

Waste pile

Engineered barrier

Filter, double membrane and leachate collection system

Clay barrier

Impermeable and durable double membrane separated by a buffering layer containing the leachate collection and monitoring systems

Figure 1.1 Examples of containment systems for HLW (left) and HSW (right). For the HLW containment system, the example illustrates an underground repository-type situation, in which the canister containing the spent nuclear fuel is in a borehole surrounded by a smectite buffer as a second containment system. For the HSW landfill shown, an impermeable durable double-liner membrane system constitutes the first and second barrier system overlying an engineered clay barrier.

of placement are all designed to provide this engineered clay barrier with the capability of impeding the transport of contaminants through the barrier, assuming that both the first and second barriers have been breached. For HLW containment, the host rock constitutes the third member of the multi-barrier system.

These containment objectives, requirements and implementation are discussed in detail in Chapter 2 together with the issues and factors impacting on the integrity of the total containment barrier system and the various processes and conditions that contribute to the degradation of this system.

1.2   Engineered clay barrier systems

1.2.1   Barriers and buffers

The designs and types of materials used in the construction of the first and second containment barriers as lines of defence against fugitive contaminants will vary somewhat depending on the kind of dangerous wastes to be contained. For high-level nuclear wastes, the first containment barrier consists of well-constructed triple-walled (or more) rigid canisters or their equivalent. The details of the first and second barrier systems for HSW are well discussed in various other textbooks dedicated to these particular subjects. Prudent engineering practice requires one to consider the possibility that at some point, sooner or later, the first and second containment barriers will lose their integrity. This may take many years, decades, centuries or even many millennia. When this happens, the engineered clay barrier system will be required to prevent fugitive contaminants from escaping into the environment.

The terms *clay barrier* and *clay buffer* have often been used interchangeably to mean the same thing. Strictly speaking, a clay buffer is a clay barrier with buffering properties that are chemical, mechanical, electrical, etc. For example, the clay barrier encapsulating the canister shown in the left-hand diagram in Figure 1.1 is quite often referred to as a clay buffer, as one of its primary responsibilities is to impede the transport of fugitive radioactive nuclides and other contaminants, as will be discussed in detail in later chapters. The term *clay barrier* will be most often used in the discussions in this book, and when specific reference to buffering capabilities is intended, the term *clay buffer* will be used.

The focus of this book is directed towards the clay barrier system, which constitutes the second or third containment system commonly identified as the engineered clay barrier in a multi-barrier system. The clay used to construct the engineered clay barrier system must demonstrate a high capability for contaminant sorption and impedance. This means that the compositional features of the clay and its properties, together with the placement technology, must provide these capabilities. Considerable research and prototype studies conducted have resulted in the development of clay barrier placement technology to obtain the optimum functional capability of the engineered clay barrier system (Yong, 2001; Yong and Mulligan, 2004; Pusch and Yong, 2006).

For clay to meet the barrier functional requirements, it must not allow any fugitive contaminant to reach harmful contact with biotic receptors and receiving waters. The clays chosen for engineered clay barriers are mostly phyllosilicates with high cation exchange capacities and specific surface areas (SSAs). This means denying transport of fugitive contaminants to positions where they can harm the environment and directly or indirectly cause health hazards to biotic receptors. The significant issues are (a) to completely eliminate contaminant transport to the targets of concern such as receiving environmental bodies and biotic receptors, and/or (b) to reduce the concentrations and especially the toxicity of the contaminants such that the level and/or the toxicity of the contaminants do not pose any health and environmental hazard to the targets of concern. This is discussed further in Chapter 10.

1.2.2   Long-term clay barrier performance expectations

The minimum required full-functional lifespan design capability of clay barriers varies in accord with the kinds of waste being contained. For example, for high-level radioactive waste (HLW) containment (left-hand diagram in Figure 1.1), regulatory requirements in most countries expect disposal of high-level long-lived radioactive wastes in secure containment facilities. These must be designed and constructed to contain the wastes and errant discharges for a long period of time – generally for at least 100,000 years. The reason for this demanding requirement is that not only are the fission products and transuranic wastes highly radioactive (see Chapter 2), but the amount of heat generated in the canister containing the spent fuel rods will be active over a long period of time. Although the time period for containment effectiveness for HSWs is not as long as the HLW case, the prime requirement for *containment until the waste no longer poses a hazard* remains the same. This means that the clay barrier must be expected to perform according to design standards over the entire specified or required lifespan, i.e. the time required when escaping contaminants are threats to biotic receptors and the environment.

The input parameters used for design performance of clay barriers include the properties and characteristics of both the waste to be contained and the clay soil itself. Laboratory testing and assessment techniques are used to obtain the characteristics and properties of the clay, and design performance expectations are generally obtained using analytical computer modelling procedures. Distinction must be made with respect to short- and long-term performance of clay barriers. One needs to determine whether time, internal and external reactions, and environmental surroundings will produce changes in the nature and values of the parameters used in the original design performance calculations. The essential points for barrier performance consideration are:

- *contained waste system*: dissolution reactions and products; interactions of reaction products and discharges with complete engineered barrier system, i.e. first, second and third barriers;
- *engineered barrier system*: durability and integrity of the canister or first and second barrier membranes; alterations and transformation of the clay in the clay barrier due to (a) interaction with fugitive contaminants and waste leachate and (b) geochemical and biogeochemical processes.

Considerable continuing research and reporting have been undertaken on contained wastes and interactions between waste reaction products and the durability of double-membrane liners and canisters or over-packs. Changes in the nature of the clay in clay barriers due to its geochemical and biogeo-chemical properties, and especially in the properties and characteristics of the *changed clay*, have not been well perceived or understood. Present intuitive understanding of physical, chemical and biological weathering of clays suggests very strongly that changes in the nature and properties of the clay due to long-term processes and activities in the clay substrate will impact on the long-term performance of the clay barrier. Safety requirements demand that detrimental changes to the barrier clay, and hence degradation of the functional capability of the clay barrier, should not occur during the design life of the clay barrier. The question of when changes in the clay will occur depends on many factors in the immediate environment and, obviously, on whether escape of contaminants occurs.

Changes in the nature of the clay occur because of (a) alteration and transformation of the clay minerals and other soil fractions that constitute the clay as a result of interactions with fugitive contaminants, and also through various kinds of geochemical processes; (b) the introduction of new fractions or removal of particular clay fractions as, for example, through processes associated with biomineralization, precipitation and dissolution; and (c) changes in the chemical, thermal, pressure and hydraulic environment due to factors external to the system under consideration. The various elements constituting these changes will occur with different intensities and over different time periods and intervals – depending on the factors or forces instigating the changes. Thus, for example, because bioweathering processes involve a two-step process: (1) its initiation time will be later than the ordinary chemical weathering process because the first bioweathering step involves processes associated with biological activities, and (2) its active weathering time period will probably be longer and/or later than the normal chemical weathering process.

It follows that because of changes in the nature of the clay, the characteristics and properties of the clay will also change. The principal properties and/or features that impact directly on clay barrier design performance include:

- clay macro- and microstructure;
- transmission properties for such forces/fluxes as fluid, thermal, gaseous, vapour, radiant, diffusive, etc.;
- mechanical properties, such as strength, compressibility, creep and consolidation;
- clay–water characteristics, accumulative and chemical buffering properties.

1.3 Clay barrier evolution

1.3.1 Intensity and influence of phenomena in barrier

Changes in clay properties and structure occur continuously in nature and especially in engineered clay barriers in short- and long-term design life periods. The changes are the results of various actions and activities: thermal, hydraulic, mechanical, chemical and biological. In the short term these changes will be in the unsteady state, graduating towards the pseudo-steady state in the long term. For example, in HLW deep geological containment (left-hand diagram in Figure 1.1), the boundary conditions given for clay surfaces at the interface between over-packs and clay as well as between rocks and clay also will change continuously with time. The various actions and activities will probably cause alteration of clay minerals and hence changes in their properties and performance. Figure 1.2 shows the various types of changes occurring in the clay barrier due to actions and activities of the different driving forces. The designations *R side* and *OP side* refer to the location of the driving force or activity – host rock and over-pack (HLW canister) respectively. Although long-term *evolution* is commonly thought to be associated with the process of development of living organisms, i.e. *organic evolution*, the term is used in this book in a broader context. We use the term *evolution* with respect to a clay or clay barrier to mean the gradual development of the clay into a transformed or altered form. The emphasis is on *gradual development*. Both intrinsic and extrinsic factors contribute to the processes that result in this gradual development. In a sense, it is possible to call this gradual development a *maturation* process, i.e. maturation of a clay barrier involves the evolution of the clay that constitutes the clay barrier.

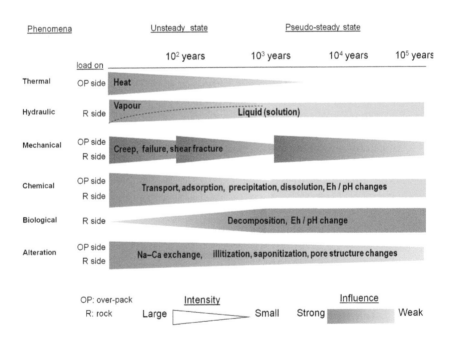

Figure 1.2 Time chart showing the intensity and influence of phenomena in the long term. The term *over-pack* is often used to designate the HLW canister. Eh, redox potential.

The external and internal processes and clay–aqueous and interparticle reactions involved in the maturation process include abiotic chemical reactions and biotic activities leading to biogeochemical reactions. As evolution of the clay in the clay barrier is a slow process, it follows that the change in clay barrier properties and characteristics will be correspondingly slow. Whether or not these slow changes are noticeable with the use of monitoring systems that are used to capture information on barrier integrity and performance remains problematic. To a large extent, this is because the interpretation of barrier performance using transmissivity or transport-based models has yet to be fully proven. Chapter 5 provides a detailed discussion on these processes and reactions, but only on the main elements of the various processes. The mechanisms contributing to the evolution of the clay in the barrier in the maturation process are listed later in this section.

1.3.2   Chemical processes and reactions

The complex aqueous chemical regime in a clay soil barrier is the result of the presence of naturally occurring salts and fugitive inorganic and organic contaminants. The different kinds of reactions between these salts, contaminants and the surface reactive groups in the clay make for a very complex thermodynamic system, especially so when we include biologically mediated chemical processes and reactions.

- Acid–base reactions are protolytic. They involve the transfer of protons between proton donors and proton acceptors in a process that is called *protolysis*.
- Hydrolysis is in essence an acid–base reaction with particular reference to the reaction of the H^+ and OH^- ions of water.
- Oxidation–reduction (redox) reactions involve the transfer of electrons between electron donors and acceptors. Generally speaking, as the transfer of electrons in a redox reaction is accompanied by a proton transfer, there is a link between redox reactions and acid–base reactions.

1.3.3   Microorganisms

The various types of microorganisms found in soils include those that fall within the Whittaker (1969) five-kingdom classification. They include protozoa, fungi, algae, viruses, bacteria and worms (such as nematodes, flatworms, roundworms, etc.). Microorganisms are classified by their energy or carbon requirements. For example, chemoorganotrophs utilize organic substrates for energy, whereas heterotrophs use them as a carbon source. Autotrophs, on the other hand, use carbon dioxide instead of organic substrates as their carbon source.

Factors favourable for optimal performance of microorganisms include nutrient availability, temperature, oxygen, moisture and osmotic pressure. Psychrophiles grow well in the low-temperature range, whereas mesophiles fare well in the mid-temperature range of about 10–45°C. Thermophiles are found in the high-temperature range – even at temperatures of about 100°C. The availability of moisture is essential for microorganism survival. Biotic redox conditions in the clay soil are important factors in producing activities from different microorganisms.

1.4 Performance assessment

1.4.1 Pathways and exposure to contaminants

One of the primary concerns in land containment and isolation of HLW and HSW is prevention of contaminant escape into the immediate surroundings. Improper design of containment facilities or failure of such facilities could result in contamination of the immediate surroundings by fugitive contaminants. Contamination of the ground by the products of dangerous waste substances, for example leachates and pollutants, provides the pathways from these substances to direct and indirect contact with groundwater, rock and soil materials as well as the various kinds of biotic receptors through water, air and dust, as shown in Figure 1.3. The dangerous waste substances gradually accumulate in water, ground surface and biota. The source–receptor pathway (SRP) diagram shows a region of a ground subsurface with fugitive contaminants and pollutants, i.e. contaminants and pollutants that have escaped confinement and are no longer isolated from the immediate environment, as the source for health and environmental threats. The term *contaminants* is used to denote substances that are not native to the specific region, location, site or material under consideration. No distinction is made as to whether the contaminants are environment/health-threat substances or otherwise.

The term *pollutants* refers to contaminants that have been designated by governmental regulatory and health agencies as toxicants, toxic chemicals or compounds, and are threats to the environment and public health. As even non-health-threatening contaminants can become a threat to human health by virtue of increased levels of concentration of the substance under consideration, the discussions in this book will use the term *contaminants* in the general sense to include both pollutants and non-health-threatening contaminants.

Regulations require HLW and HSW containment facilities to be constructed in a way that meets specifications and standards designed to protect public health and the environment. The issuance of standards and/or criteria for *environmentally safe management of waste* requires one to consider not only the nature of the health and environmental threats posed by the waste

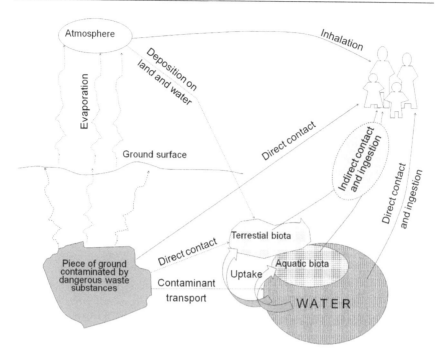

Figure 1.3 Source–receptor pathway (SRP) showing the routes to biotic receptors. Contaminants carried into the atmosphere by evaporative processes and deposited on land and receiving waters allow for uptake by terrestrial and aquatic biota. The diagram shows that human contact with the contaminants (top right-hand corner) can be direct and indirect, as shown by the solid arrows.

material, and the problem of level of exposure, as, for example, *acceptable daily intake (ADI)* of the toxicant, but also the technical requirements for the waste containment system. In all instances, the contained waste is required to be completely isolated and protected in such a way that there will be no opportunity for contaminants to escape. In practice this is not always easily achieved, especially in the case of HLW, for which designed containment facilities are required to protect biotic receptors against contact with fugitive contaminants for a period of at least 100,000 years. This recognizes the reality that long-term chemical processes and biological activities may degrade the containment structures, resulting in the escape of contaminants into the immediate environment. Imposing a requirement of zero contact or non-lethal contact with biotic receptors for least 100,000 years forces one to construct secondary lines of defence designed to impede the transport of contaminants to the biotic receptors (Figure 1.3). Chapter 2 addresses the HLW and HSW containment objectives, requirements, concerns and present-day concepts in greater detail.

1.5   Concluding remarks

Containment of dangerous wastes such as high-level radioactive waste (HLW) and hazardous solid waste (HSW) in the land environment requires containment structures that include multi-barrier containment systems. In almost all cases, the final containment system consists of a clay barrier that is designed to impede or completely deny the transport of contaminants and other kinds of waste streams to locations where contact can be harmful to the environment and to biotic receptors. The nature and properties of the clay used to construct the clay barrier, frequently referred to as the engineered clay barrier or clay buffer, will change with time because of the clay–water reactions brought about by intrinsic and extrinsic factors.

Design and construction of clay barriers using knowledge of the properties and characteristics of clays from laboratory and other kinds of tests do not necessarily pay attention to the evolution of the clay during the required design lifespan of the clay barrier. Little is known about, and little attention has been paid to, the magnitude of changes in the transitory (i.e. changing) properties of the clay as time progresses. Also, little is known of the processes and mechanisms involved in causing these changes and whether these transitory properties will degrade the design performance of the clay barrier. However, one fact is eminently clear. The clay barrier will evolve with time. Information and knowledge of the transitory properties and characteristics are needed if we are to produce the proper design and construction standards for a long-lasting and durable containment system, of which the clay barrier is an integral and vital component, and in many cases the last line of defence.

The remaining chapters will (a) provide background information on the nature of high-level radioactive and hazardous wastes in order to demonstrate why the present-day secure containment systems are needed; (b) offer a detailed discussion of the various elements of the clay barrier, beginning with a discussion of the nature of clay and proceeding onwards to a description of the various factors, forces and processes that are involved in clay barrier evolution; and (c) extend the knowledge gained from the basic fundamentals to present-day practice, and discuss the significant issues remaining in the maturation process that may lead ultimately to untimely and disastrous consequences.

Chapter 2

Radioactive and hazardous solid waste isolation

2.1 Introduction

In Chapter 1 it was indicated that radioactive and hazardous solid wastes fall into a class that can be called *dangerous wastes*, based on the harm that these wastes and their reaction products can render to the environment and the health of biotic receptors. The nature of the threats from these dangerous wastes needs to be determined in order that secure containment systems can be constructed to isolate these wastes and their reaction products from contact with biotic receptors. The discussions in this chapter will examine (a) the nature of these dangerous wastes; (b) the nature and form of the threats presented by these wastes; and (c) the basic elements of containment systems, with particular focus on the clay barrier method that is used in these systems.

2.2 Radioactive wastes

2.2.1 *Radioactivity and radioactive decay*

Because of the type of hazard that radioactive wastes present, it is useful to briefly revisit the basic essentials that define the nature of the hazard known generally as *radiation hazard*. The process of decay of a radioactive atom into a stable atom is called *transmutation*. The emission of radiation, in the form of α or β particles (with or without γ rays) in the spontaneous radioactive decay of an unstable atom into a stable atom is defined as *radioactivity*. Unstable atoms are radioactive, as they will decay, and in doing so will emit radioactivity from the disintegrating nucleus. The energy associated with radioactive decay or transmutation of atoms is generally large enough to qualify as ionizing radiation; α and β particles constitute *direct ionizing radiation*, whereas γ rays constitute *indirect ionizing radiation*.

An α particle is a helium (He) nucleus, i.e. it consists of two neutrons and two protons. Radioactive decay involving emission or ejection of α particles is called α decay or α radioactivity. In the spontaneous α decay of a ^{239}Pu nucleus, for example, one obtains ^{235}U with the discharge of α particles and γ rays. The α decay of ^{238}U gives one ^{234}Th and the β decay of ^{234}Th will give us ^{234}Pa (protactinium-234). Alpha particles are very short range and are not known to penetrate the outer layer of human skin or thick paper. Ingestion of matter containing α particles, however, can be lethal.

As with the discharge of α particles, emission of β particles is called β radioactivity. As β particles can be either negative (negatron) or positive (positron), we will have either β$^-$ decay or β$^+$ decay. β$^-$ particle discharge or emission occurs when there is an excess of neutrons in the unstable atom's nucleus. Negative β particles are similar to electrons. In the case when there is an excess of protons in the unstable atom's nucleus, β$^+$ particle emission will occur if discharge of α particles is not possible. Release of γ rays will generally accompany β$^+$ decay. Although they are considerably smaller in size than α particles, the penetrating power of β particles is larger. They are known to be impeded by glass or aluminium sheets. On the other hand, as γ rays are essentially high-energy photons with no charge or mass, they have the ability to travel large distances and can penetrate concrete, and are only impeded by several lead sheets of some significant thicknesses.

As the decay of radioactive atoms is a spontaneous process and decay of atoms of a similar nature is a probabilistic process, one obtains the following for $N(t)$, the time-dependent number of not yet decayed atoms:

$$\frac{dN}{dt} = -\lambda N \tag{2.1}$$

where λ represents the characteristic decay constant. For the initial condition where $N = N_0$ at time $t = 0$, it can be shown that τ, the mean lifetime of the atoms present, will be obtained as:

$$\tau = <t> = \int_0^\infty te^{-\lambda t}\,dt = \frac{1}{\lambda} \tag{2.2}$$

with the corresponding half-life T_{50} (i.e. time to reach one-half of the total radioactivity of the element) given as:

$$T_{50} = <t> = \tau\ln 2 \tag{2.3}$$

Figure 2.1 shows the exponential decay of an element with time. The three different $e^{-\lambda t}$ decay curves represent three different elements with their correspondingly different T_{50} times. If the solid curve representing the decay of element B is the $\lambda = 1$ decay curve (i.e. e^{-t} curve), values for the characteristic decay constant λ will be < 1 for decay curves above this element B curve, and λ will be > 1 for decay curves below the element B curve. The range of half-lives for the various nuclides can be from milliseconds or even microseconds

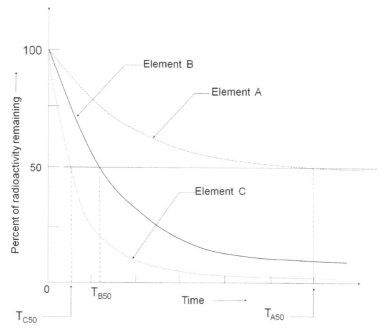

Figure 2.1 Exponential decay curves showing percentage of radioactivity remaining in an element in relation to time. Note that the different decay curves represent different elements with differing decay times with corresponding differences in the T_{50} times to reach their half-lives. Note too that the percentage of radioactivity remaining continues indefinitely and will never reach zero.

to hundreds, thousands and even millions of years. For example, the half-life (T_{50}) of uranium-226 (i.e. ^{226}U) is 0.5 s – compared with ^{238}U and ^{235}U, which have half-lives of 4.47×10^9 years and 7×10^8 years respectively.

2.2.2 Ionizing radiation

We have stated in the previous subsection that α and β particles are direct ionizing radiation whereas γ rays are indirect ionizing radiation. Ionizing radiation refers to radiation with energy that is sufficient or greater than that necessary to remove bound electrons from the orbit of an atom as a result of interaction with the atom. Ionizing radiation is harmful to living organisms because it has the capability of changing the molecular structure of the bio-logical cells that make up living organisms. In that respect, it is well known that it is a cause of cancer in humans. As α and β particles carry charges, they are considered as direct ionizing radiation because of their capability for coulombic interaction with the atoms. On the other hand, γ rays are electromagnetic in nature, and they are considered to be indirect ionizing radiation as their interactions with the atoms do not involve coulombic forces.

The extent of damage to biological cells from ionizing radiation is a function of (a) the exposure route, i.e. direct dermal contact leading to dermal absorption, ingestion and inhalation; (b) the radiation dosage received or absorbed by the subject in question; and (c) the frequency of exposure or length of time of exposure. In humans, the likelihood of contracting radiation-induced or radiation-related diseases, because of attack on the genetic DNA in the cell nucleus, is a function of radiation dosage received either over a specific time period or in periodic frequencies, i.e. acute as opposed to chronic exposure.

2.2.3   Sources of radioactive wastes

Although radioactive wastes are hazardous wastes (i.e. HSWs), accepted regulatory practices in most countries have separated them from the hazardous solid wastes (HSWs) classification into a separate category because of the radiation hazards presented by these wastes. Radioactive wastes contain radioactive isotopes that emit α and β particles and γ rays with resulting ionizing radiation that can be harmful to human health and the environment. Radioactive wastes are commonly defined as materials that (a) serve no further useful purpose and (b) contain concentrations of radioactive isotopes (i.e. radionuclides) exceeding the threshold limit considered as safe to human health by regulatory and health authorities. They are generated from a variety of sources ranging from non-nuclear fuel cycle activities, such as processing of uranium from mining operations at the one end to the waste products obtained in the various processes in the nuclear fuel cycle at the other end. Besides military operations involving production of nuclear weapons and the nuclear fuel cycle, sources of radioactive wastes include industries using radioactive substances and also hospitals and other medical facilities, such as government and university research laboratories.

Some of the key reasons for distinguishing between radioactive wastes generated by non-nuclear fuel cycle and nuclear fuel cycle processes are (a) the nature of the waste; (b) the principal medium in which they are released; and (c) the level of radioactivity and hence the risk they present to the various receiving bodies. Non-nuclear fuel cycle types of radioactive wastes include those facilities or organizations using radioactive substances in measurement and detection devices such as scintillating counters, and for treatment purposes such as in nuclear medicine. Other sources include materials, tools and protective clothing that are used in conjunction with uranium extraction/processing, and in facilities and organizations utilizing radioactive substances. Radioactive wastes generated in the nuclear fuel cycle include spent nuclear fuel and those pieces of equipment, tools, protective clothing, and waste discharge associated with decommissioning activities (see

Figure 2.2). The level of radioactivity of spent fuel is considerably larger than the levels associated with decommissioning and those obtained in the non-nuclear fuel cycle.

2.2.4 Classification of radioactive wastes

Classification or categorization of radioactivity wastes is necessary for at least two very important reasons:

- to call attention to the level of radioactivity in the waste and hence the extent of risk involved when one is exposed to the radiation hazard;
- to indicate and/or provide the necessary requirements for protection of biological cells and the environment from the radiation hazards – depending on the level of risk.

As previously discussed in section 1.2.1 with regard to the many ways of categorizing wastes, the same can be said for radioactive wastes. There is consensus among many countries and regulatory agencies that whereas no theoretical scientific base for categorizing or classifying the different kinds of radioactive wastes exists, classification schemes adopted by stakeholders should be based on criteria that include consideration of (a) their radioactive contents; (b) the type of activity generating the waste or the source of the radioactive waste; and (c) the types of radioactive isotopes and the length of time these isotopes will remain hazardous (i.e. the half-lives of the radioactive isotopes). The levels of radioactivity of the wastes and the health hazards they present bear some relationship to the source generating the waste. The benefit of classifying radioactive wastes on the basis of health hazards lies in the class-specific robustness of the containment–isolation schemes that are required for safe and secure disposal of the radioactive wastes. Regulatory requirements for management of radioactive wastes for containment and isolation or final disposal in most stakeholder countries are based, by and large, upon the class of radioactive waste involved.

The International Atomic Energy Agency uses five categories for its classification scheme (IAEA, 2007). These are:

- *Exempted waste (EW).* 'Activity levels at or below national clearance levels which are based on an annual dose to members of the public of less than 0.01 mSv'. The SI unit of dose equivalent, sievert (Sv), is used in place of the older REM (Röntgen equivalent in man). 1 Sv = 100 REM.
- *Low- and intermediate-level waste (LILW).* 'Activity levels above clearance levels and thermal power below about 2 kW/m^3'.
- *Short-lived LILW (SL-LILW).* 'Restricted long-lived radionuclide concentrations (limitation of long-lived alpha emitting radionuclides to 4000 Bq/g in individual waste packages and to an overall average

of 400 Bq/g per waste package)'. The SI unit of radioactivity, the becquerel (Bq), defined as a unit of radioactivity equal to one unit of nuclear transition or disintegration, is used in favour of the previous conventional curie (Ci) expression for the unit of activity of a radioactive material. 1 Ci = 37 GBq.

- *Long-lived LILW (LL-LILW)*. 'Long-lived radionuclide concentrations exceeding limitations for short-lived LILW'. The same techniques for disposal of HLW are generally recommended for this class of waste.
- *High-level waste (HLW)*. 'Thermal power above about 2 kW/m³ and long-lived radionuclide concentrations exceeding limitations for short-lived LILW'. Geological containment systems are recommended for disposal–isolation of this class of wastes.

The UK uses a slightly different classification scheme. It is a combination of threshold levels for the lower level wastes and containment/disposal requirements for the higher levels of radioactive wastes. The Department for Environment, Food and Rural Affairs (DEFRA, 2001) considers:

- The IAEA-exempted waste to be *very low-level waste (VLLW)*, i.e. 'waste of activity less than 400 kBq/0.1 m³ β/γ activity or single items of less than 40 kBq to be VLLW' – the main generators for wastes that qualify under this categorization are in the non-nuclear fuel cycle sector.
- *Low-level wastes (LLWs)* to be 'radioactive materials other than those not suitable for suitable for ordinary refuse disposal but not exceeding 4 GBq/tonne of α or 12 GBq/tonne of β/γ activity'. These are primarily solid wastes such as contaminated equipment, tools, protective clothing, laboratory supplies and materials.
- *Intermediate level wastes (ILWs)* to be 'wastes with radioactivity levels exceeding the upper boundaries for LLW, but which do not need heating to be taken into account in the design of storage or disposal facilities'. The types of wastes that qualify as ILW include those obtained in the operation and maintenance of nuclear power plants, and reprocessing of spent fuel.
- *High-level wastes (HLWs)* to be 'wastes in which the temperature may rise significantly as a result of their radioactivity' – the principal generators for these wastes are the processes within the nuclear fuel cycle, with spent nuclear fuel as the major constituent.

The United States Nuclear Regulatory Commission (US NRC, 2002), which regulates storage and disposal of all radioactive wastes generated in the US, uses two broad classifications for radioactive wastes. These are *low-level radioactive waste (LLW)* and *high-level radioactive waste (HLW)*. The LLW class of wastes include those classed as VLLW and LLW by IAEA and DEFRA, and the HLW classification is in common with most regulatory agencies. Emphasis on methods of containment and isolation is common with almost all jurisdictions that have responsibility and control of management of radioactive wastes.

Figure 2.2 shows the various components (in the shaded ellipse) that constitute the nuclear fuel cycle. There are some that contend that the *conversion and enrichment* component in the shaded ellipse does not belong in the nuclear fuel cycle. It is argued that this component should be included in the *front-end processes*, together with *mining and milling*, as it precedes the all-important *fuel production* fuel source for the reactor. We have included this component in the nuclear fuel cycle because it is an integral part of the process that is necessary to obtain the nuclear fuel for the reactor.

Figure 2.2 shows that radioactive wastes will be produced at all stages of the nuclear fuel cycle and also in the front-end processes. The requirements for handling and disposal of LLW and SL-LILW (using the IAEA classification scheme) from the various generating sources in the nuclear fuel cycle are well articulated in the regulations that are issued by the various regulatory authorities (see next subsection). In terms of the nuclear fuel cycle, these

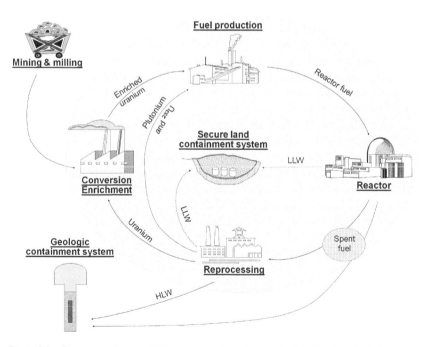

Figure 2.2 Components constituting the nuclear fuel cycle (within the shaded ellipse). The 'mining and milling' component is a front-end process and the 'geological containment system' is the containment scheme recommended by many stakeholder countries as a prime candidate for implementation of HLW containment (see later in chapter).

wastes are by-products generally obtained in the course of operation of the reactor and the other facilities in the nuclear fuel cycle. The prime requisite before final disposal in secure landfills is the containment of LL-LILW in secure containers or in solidified form for placement in the landfill. In the case of HLW, secure triple-barrier canisters are generally used for placement and isolation in deep underground repositories, generally referred to as deep *geological disposal.*

2.2.5   Class-associated radioactive waste containment/ disposal requirements

Requirements for containment or disposal of radioactive wastes take into account the level of radioactivity present in the generated waste, i.e. the waste to be managed and disposed. The sets of reasoning or criteria used to classify these wastes – detailed in the previous section – generally include consideration of containment or disposal scenarios that will provide the necessary protection against the level of radiation hazard associated with the particular class of radioactive waste. For example, IAEA (2007) indicates 'no radiological restrictions' in its disposal options for LILW and SL-LILW. The disposal options indicate a requirement for 'near surface or deep underground facility'. In general, although no shielding or cooling is required in the disposal scheme, some form of treatment and conditioning is generally required before final disposal. For example, solid combustible wastes are incinerated and the collected ashes are mixed with a binder material (concrete or bituminous material) for disposal in deep underground or regular hazardous waste (HW) facilities.

For LL-LILW and HLW, a 'deep underground disposal facility' is the indicated disposal option. Different jurisdictions and different regulatory agencies have similar views on the containment and disposal of their radioactive wastes – all of which are designed to provide protection of public health and the environment from the radiation hazards presented by the different classes of radioactive wastes. For the discussion in this book, the disposal option for LILW and SL-LILW will be included in the disposal scheme for HSWs, as they are essentially similar. Discussion of HLW requirements for disposal is given separately, in the next section.

2.2.6   High-level radioactive waste

Practically all regulatory bodies concerned with regulating the disposal of HLW require HLW to be contained and isolated in deep underground disposal facilities. Strictly speaking, HLWs include both spent nuclear fuel and various highly radioactive materials such as transuranic radionuclides obtained in the reprocessing of spent nuclear fuel. In general, transuranic wastes are solid radioactive elements obtained from the irradiation of uranium and thorium in the reactors. They contain isotopes that have atomic numbers higher than uranium, such as ^{239}Pu (plutonium), ^{243}Am (americium) and ^{237}Np (neptunium), and are generally α radioactive. The half-lives for ^{239}Pu, ^{243}Am and ^{237}Np are 24.1×10^3, 7.38×10^3 and 2.14×10^6 years respectively.

The proportions of the various constituents in the spent fuel will vary depending on the type of reactor used, for example pressurized water reactors (PWRs), boiling water reactors (BWRs) or liquid metal fast-breeder reactors (LMFBRs). About 95 per cent of spent fuel is ^{238}U and about 1 per cent (of spent fuel) is ^{235}U. The remaining 4 per cent consists of plutonium, activation and fission products, and transuranics. Two particular hazard problems in spent nuclear fuel distinguish themselves from the types hazards encountered in the disposal of HSWs generated by other types of industries. These are (a) long half-lives of highly radioactive isotopes and (b) high thermal heat generated by the spent nuclear fuel. These two problems are in part responsible not only for the separation of HLW from HSW classification, but also for the very stringent requirements for final disposal of these wastes. Figure 2.3 shows the level of radioactivity of a nuclear fuel canister containing spent fuel rods a function of time of removal of the fuel rods from the reactor.

The information reported in Figure 2.3 is for the decay in the level

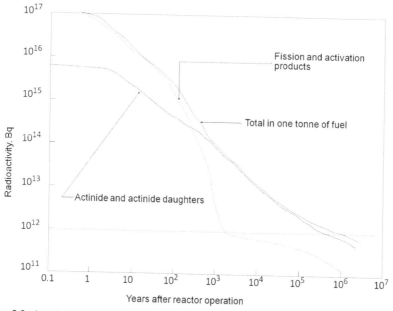

Figure 2.3 Level of radioactivity of spent nuclear fuel with a burn-up of 38 MW d/kg U in relation to time after reactor operation (adapted from Hedin, 1997).

radioactivity of spent fuel rods with a burn-up of 38 MW d/kg (megawatt days per kilogram). For about the first 100 years, fission and activation products are responsible for the major portion of the total radioactivity. Beyond this time period, actinides and actinide daughters contribute the greater proportion of radioactivity to the total radioactivity. The horizontal line shown at the 10^{12} Bq level refers to the level of radioactivity of 8 tonnes of natural uranium with uranium daughters. The information shown in Figure 2.3 indicates that the level of radioactivity of the spent fuel rods will decrease to the level of natural uranium in a period of about one million years for the type of nuclear fuel used to provide the information portrayed in the figure.

The heat generated by the decay heat of the spent fuel rods in the canisters is shown in Figure 2.4. As will be seen, the heat generated by the actinides and actinite daughters persist for almost one million years. The transuranics are not known to produce as much heat as the fission products, mainly because of the longer half-lives of the transuranics. The shorter half-lives of most of the fission products and the higher rate of decay are responsible for the higher heat generation.

Activation products are obtained in the fission process when neutrons interact with different elements present, mainly as impurities, in the nuclear fuel. They are essentially products of neutron capture. Although ^{14}C and ^{36}Cl are found as activation products, most of these products are metals, such as

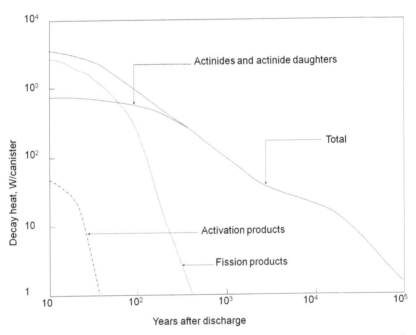

Figure 2.4 Reduction in decay heat in the spent fuel canisters in relation to time after removal from operation (adapted from SKB, 1999).

Co (cobalt), Nb (niobium) and Ni (nickel), which would become radioactive. Although a large number of these are short lived, some have half-lives that stretch into thousands of years. For example, the half-lives of ^{54}Co and ^{60}Co are 0.193 s and 5.3 years, respectively, whereas the half-lives of ^{94}Nb, ^{95}Nb, ^{56}Ni and ^{59}Ni are 2×10^4 years, 3.5 days, 6 days and 8×10^4 years respectively.

Fission products are obtained in the fission process as nuclei formed by the fission of higher mass elements. They constitute a collection of numerous kinds of radioisotopes that include ^{3}H (tritium) with β^- radiation and a half-life of 12.3 years; ^{85}Kr (krypton) with β^- and γ radiation and a half-life of 10.7 years; ^{135}Cs (caesium) with β^- radiation and a half-life of 2.3×10^6 years; and ^{129}I (iodine) with β^- and γ radiation and a half-life of 2.6×10^7 years. As with the case of activation products, fission products have both short and long half-lives – as demonstrated for example by ^{129}I and ^{131}I; although both of these emit β^- and γ radiation, the half-life of ^{131}I is 8 days compared with the half-life of 2.6×10^7 years for ^{129}I.

Actinides are chemical elements with atomic numbers that range from 89 (actinium, Ac) to 103 (lawrencium, Lr). The absorption by ^{238}U of some of the neutrons produced when ^{235}U is fissioned in the reactor produces not only plutonium, but also various kinds of actinides. These are essentially the heavier elements that are obtained in the fission process in the reactor – as opposed to the lighter elements produced (fission products) when ^{235}U is fissioned.

Transuranics are transuranium elements with atomic numbers greater than 92 (uranium). Except for Np (neptunium, atomic number 93) and Pu (plutonium, atomic number 94), the others are generally considered to be synthetic elements, i.e. produced in the course of reactions in reactors or particle accelerators. Table 2.1 gives a short sample of the various kinds of radionuclides obtained in the course of operations in the nuclear fuel cycle.

2.3   Hazardous solid wastes

2.3.1   *Characterization and classification of hazardous solid waste*

Hazardous wastes include both solid and liquid forms of wastes, and the term *hazardous wastes* (HWs) is used in general discussions of these wastes. The term *hazardous solid waste* (HSW) is used specifically to indicate that the hazardous waste under discussion is in a solid form. Hazardous wastes are regulated wastes, and as such, are subject to regulatory control, i.e. treatment, storage, and final disposal or discharge of these wastes is regulated by the appropriate regulatory agencies. In this book, we are concerned with hazardous solid wastes (HSWs) as, in most cases, their final resting place is in landfills. HSWs have properties that are inherently harmful to the state of the environment and harmful to the health of biotic receptors. The US

Table 2.1 Representative radioactive nuclides shown in order of their half-lives in each group

Nuclides	Decay	Half-life
Activation and fission products		
Cobalt-62	β^-, γ	1.51 months
Niobium-95	β^-, γ	3.5 days
Iodine-131	β^-, γ	8.04 days
Cerium-141	β^-, γ	32.5 days
Strontium-89	β^-, γ	50.52 days
Polonium-210	α, γ	138.4 days
Promethium-147	β^-, γ	2.62 years
Cobalt-60	β^-, γ	5.27 years
Krypton-85	β^-, γ	10.72 years
Strontium-90	β^-	29 years
Caesium-137	β^-, γ	30.17 years
Nickel-63	β^-	100 years
Technetium-99	β^-, γ	2.13×10^5 years
Zirconium-93	β^-, γ	1.5×10^6 years
Caesium-135	β^-	2.3×10^6 years
Iodine-129	β^-, γ	1.59×10^7 years
Actinides and transuranics		
Protactinium-233	β^-, γ	27 days
Thorium-228	α, γ	1.91 years
Curium-244	α, γ	18.11 years
Plutonium-238	α, γ	87.74 years
Americium-241	α, γ	432.2 years
Plutonium-240	α, γ	6.5×10^3 years
Plutonium-239	α, γ	2.41×10^4 years
Protactinium-231	α, γ	3.28×10^4 years
Curium-248	α	3.39×10^5 years
Neptunium-237	α, γ	2.14×10^6 years
Curium-247	α, γ	1.56×10^7 years
Plutonium-244	α	8.3×10^7 years
Thorium-232	α, γ	1.4×10^{10} years

Resource Conservation Recovery Act (1976), for example, defines a *hazard-ous waste* as

> a solid waste or combination of solid wastes, which because of its quantity, concentration, or physical, chemical or infectious characteristics may – (a) cause, or significantly contribute to an increase in mortality or an increase in serious irreversible, or incapacitating reversible, illness; or (b) pose a substantial present or potential hazard to human health or the

environment when improperly treated, stored, transported, or disposed of, or otherwise managed.

As we have noted in Chapter 1, most jurisdictions have detailed lists of HWs, classified according to characteristics, properties, type, source, industry, and in general, a *non-specific* category. The United States Environment Protection Agency (US EPA), for example, has compiled four hazardous waste lists identified as the F-, K-, P- and U-lists. The F-list covers non-specific source wastes, whereas the K-list covers source-specific wastes and the P- and U-lists include discarded commercial chemical products such as pesticides and pharmaceutical products. The items detailed in the lists can be found in Code of Federal Regulations (CFR) Title 40, Part 261. A short summary of these lists is as follows:

- The *F-list*, which covers non-specific source wastes, consists of 39 classes that include spent halogenated and non-halogenated solvents, sludges, plating bath residues, process wastes, wastewaters and discarded unused waste formulations of organic chemicals, residues and leachates.
- The *K-list*, which attends to source-specific wastes, consists of a detailed listing of wastes sources grouped into categories that include wood preservation, inorganic pigments, organic chemicals, inorganic chemicals, explosives, petroleum refining, metal industries that include iron, steel, copper, lead, zinc, aluminium and ferroalloys, veterinary pharmaceuticals, ink formulation and coking. The categories of organic and inorganic chemicals contain 54 and 6 listings of source-specific wastes, respectively – for example 'distillation side cuts from the production of acetaldehyde from ethylene' (K10), 'distillation bottoms from the production of 1,1,1-trichloroethane' (K095) and 'chlorinated hydrocarbon waste from the purification step of the diaphragm cell process using graphite anodes in chlorine production' (K073).
- The *P- and U-lists* of discarded chemicals and products include hundreds of specific compounds and items, such as 'arsenic oxide As_2O_5' (P011), '1,3-dithiolane-2-carboxaldehyde, 2,4-dimethyl-,O-[(methylamino)-carbonyl]oxime' (P185) and 'benzamide,3,5-dichloro-N-(1,1-dimethyl-2-propynyl)' (U192).

In addition to the lists, the US EPA also states that a waste material is considered to be a hazardous waste if it possesses any of the following characteristics:

- *ignitability* – potential for fire hazard during storage, transport, or disposal;
- *corrosivity* – potential for corrosion of materials in contact with the waste, resulting in damage to the environment and human health;
- *reactivity* – potential for adverse chemical reactions;

- *toxicity* – as identified from the Toxicity Characteristics Leaching Procedure (TCLP), with the procedure and requirements detailed in CFR Title 40, Part 261.

The type of HSWs that find their way into landfills most often fits the fourth characteristic, i.e. toxicity. The aim of the TCLP is to determine whether the leachates generated by the solid waste in question qualifies as a hazardous (liquid) waste because of the presence of chemical compounds or heavy metals that are identified in the lists provided in CFR Title 40, Part 261. The zero headspace extractor (ZHE) to be used with the TCLP is shown in Figure 2.5. The pressure exerted on the piston can be either from a gas pressure (shown in the figure) or via a mechanical pressure with a piston rod attached to the piston within the chamber. The specifications call for a piston pressure of 344.7 kPa to be exerted on the sample for 1 hour in order to produce the extract that is harvested at the top. The reagent-grade chemicals required for the extraction are determined and specified according to whether one is performing a volatile or non-volatile extraction procedure.

The regulatory limits for the extracts obtained with the ZHE are shown in Table 2.2. The compounds listed in Table 2.2 are included in all of the US EPA lists that have been described previously. The TCLP provides one with an assessment of the leachability of the solid waste tested, which together with an analysis of the chemical compounds in the leachate will tell us whether the leachate will cause harm to the environment and human health. This is

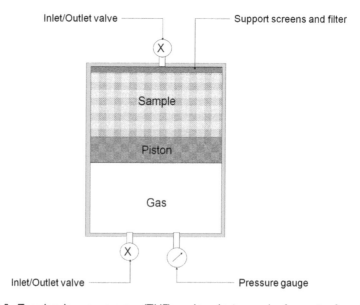

Figure 2.5 Zero headspace extractor (ZHE) used to obtain samples for testing for volatile analytes (see Table 2.2) using EPA TCLP, as prescribed in CFR Title 40, Part 261.

Table 2.2 TCLP compounds and regulatory limits in extract

Compound	Chemical symbol/formula	Level (mg/L)
Arsenic	As	5.0
Barium	Ba	100.0
Benzene	C_6H_6	0.5
Cadmium	Cd	1.0
Carbon tetrachloride	CCl_4	0.5
Chlordane	$C_{10}H_6Cl_8$	0.03
Chlorobenzene	C_6H_5Cl	100.0
Chloroform	$CHCl_3$	6.0
Chromium	Cr	5.0
m-Cresol	C_7H_8O	200.0
o-Cresol	C_7H_8O	200.0
p-Cresol	C_7H_8O	200.0
1,4-Dichlorobenzene	$C_6H_4Cl_2$	7.5
1,2-Dichloroethane	$C_2H_4Cl_2$	0.5
1,1-Dichloroethene	$C_2H_2Cl_2$	0.7
2,4-Dinitrotoluene	$C_7H_6N_2O_4$	0.13
Endrin	$C_{12}H_8Cl_6O$	0.2
Heptachlorepoxide	$C_{10}H_5Cl_7O$	0.008
Hexachlorobenzene	C_6Cl_6	0.13
Hexachlorobutadiene	C_4Cl_6	0.5
Hexachloroethane	C_2Cl_6	3.0
Lead	Pb	5.0
Lindane	$C_6H_6Cl_6$	0.4
Mercury	Hg	0.2
Methoxychlor	$C_{16}H_{15}Cl_3O_2$	10.0
Methyl ethyl ketone	C_4H_8O	200.0
Nitrobenzene	$C_6H_5NO_2$	2.0
Pentachlorophenol	C_6HCl_5O	100.0
Pyridine	C_5H_5N	5.0
Selenium	Se	1.0
Silver	Ag	5.0
Tetrachloroethene	C_2Cl_4	0.7
Toxaphene	$C_{10}H_{10}Cl_8$	0.5
Trichloroethene	C_2HCl_3	0.5
2,4,5-Trichlorophenol	$C_6H_3Cl_3O$	400.0
2,4,6-Trichlorophenol	$C_6H_3Cl_3O$	2.0
2,4,5-TP (Silvex)	$C_9H_7Cl_3O_3$	1.0
Vinyl chloride	C_2H_3Cl	0.2

of utmost importance in the management of landfills containing the type of waste tested, and particularly in the management of leachates generated from the landfill wastes. The environmental mobility of pollutants in the leachate, i.e. contaminants deemed to be harmful to the environment and human health by regulatory agencies, is of considerable concern to regulatory authorities. The listing of the pollutants, as for example in the K- to U-lists, provides only one aspect of the 'environment and health protection' issue. Figure 1.1 has shown that the pathways by which the effects of these pollutants reach their targets is the other aspect of the issue at hand. The environmental mobility of pollutants is an important consideration in the assessment of the threat of pollutants to the environment and human health. This subject is addressed in detail in Chapter 4. The TCLP test is meant to provide one with an assessment of the mobility of inorganic and organic pollutants, for example in leachates described in the K-list.

For the Member States of the European Community, the European Council Directive 91/689/EEC, otherwise known as the *Hazardous Waste Directive (HWD)* provides the categories 'or generic types of hazardous waste listed according to their nature or the activity which generated them' and a detailed listing of the hazardous wastes that could be in liquid, semi-liquid or solid forms. This listing has been revised according to Commission Decision 2000/532/EC, which in turn was amended by Commission Decisions 2001/118/EC, 2001/119/EC, and Council Decision 2001/573/EC. The result is a revised catalogue of hazardous wastes known as the *European Waste Catalogue EWC-2002*. In this new revised catalogue, there are 14 main groups with one subgroup listed, as shown in Table 2.3.

The harmonized list of wastes developed in accord with *Article 1(a) of*

Table 2.3 Hazardous properties [adapted from Environment Agency (UK), 2003]

H1	Explosive
H2	Oxidizing
H3A	Highly flammable
H3B	Flammable
H4	Irritant
H5	Harmful
H6	Toxic
H7	Carcinogenic
H8	Corrosive
H9	Infectious
H10	Toxic for reproduction
H11	Mutagenic
H12	Toxic gaseous reaction products[a]
H13	Reaction and dissolution products[b]
H14	Ecotoxic

Notes

a Toxic or very toxic gases reaction products released when substances come in contact with water, air or an acid.

b Reaction and dissolution products, such as leachates, obtained after disposal, which possess any of the characteristics listed.

Directive 75/442/EEC on waste and *Article 1(4) of Directive 91/689/EEC on hazardous waste* contains both hazardous and non-hazardous wastes lists. By and large, the classification scheme for this harmonized list of wastes uses a combination of source-specific and waste-type procedure that has considerable similarities to the schemes used, in one form or another, in many jurisdictions. The first 12 and the last four categories (lists) are source specific, ranging from wastes generated in the mining industries through to wastes obtained from metal treatment and shaping, and, finally, to municipal wastes. The remaining four categories are waste types ranging from oily wastes to wastes that are not considered in the total listing.

2.3.2   Pollutants, contaminants and fate

In Chapter 1, we referred to *pollutants* as contaminants that are considered to be potential threats to human health and the environment. These health hazard contaminants (pollutants) are both naturally occurring substances (such as arsenic and iron) and various kinds of chemicals produced by mankind (such as chlorinated organics). Most, if not all, of these kinds of substances or compounds are found in many hazardous and toxic substances lists issued by various governments and regulatory agencies in almost all countries of the world. We will use the term *pollutant* to emphasize the contamination problem under consideration, and also when we mean to address known health hazard contaminants (specifically or in general). We will continue to use the term *contaminant* when we deal with the collection of both contaminants and pollutants, and also when we consider the general theories of contaminant–soil interactions. The ultimate nature and distribution of contaminants in the subsurface environment is generally referred to as the *fate* of contaminants. It is important to have proper knowledge of the fate of contaminants in subsurface soils, and especially in the design and construction of clay barriers as integral components in HLW and HSW containment schemes. The TCLP test described above (under 'Characterization and classification of HSW') is designed to provide information on environmental mobility of both organic and inorganic pollutants – as part of the information needed to determine the fate of pollutants/contaminants in transport into and through the clay barriers.

The fate of contaminants depends on the various interaction mechanisms established between contaminants and clay fractions, and also between contaminants and other dissolved solutes present in the porewater that constitutes the clay–water system. The general interactions and processes contributing to the fate of contaminants will be described in detail in the next chapter. For the discussion in this chapter, we turn our attention to the characterization of the fate of contaminants generally, which is referred to as *fate description*:

- *persistence* – includes contaminant recalcitrance, degradative and/or intermediate products, and partitioning;
- *accumulation* – basically describes the processes that are involved in the removal of the contaminant solutes from solution, for example adsorption, retention, precipitation and complexation;
- *transport* – accounts for the environmental mobility of the contaminants and includes partitioning, distribution and speciation;
- *disappearance* – should include the final disappearance of the contaminants; in some instances the elimination of pollutant toxicity or threat to human health and the environment of the contaminant (even although it may still be present in the substrate) has been classified under this grouping, i.e. disappearance of the threat posed by the pollutant.

The term *persistence and fate* is often used when one encounters contaminants in the subsurface soil environment in conjunction with pollutants and contaminants detected in the substrate. The *fate* of a contaminant refers to the final outcome or *state* of a contaminant found in the subsurface environment or with reference to a specific site or location. The term *fate* is most often used in studies on contaminant transport when concern is directed towards whether a contaminant will be retained (accumulated), attenuated within the domain of interest or transported (mobile) within the domain of interest.

A contaminant in the domain of interest is said to be *persistent* if it remains in the domain of interest or in the subsurface environment in its original form or in a transformed state that poses an immediate or potential threat to human health and the environment. One could define a *persistent organic chemical pollutant (POP)* as an organic chemical pollutant that is resistant to conversion by biotic and/or abiotic transformation processes. Strictly speaking, we can consider *persistence* to be part of *fate*. An organic chemical is said to be a *recalcitrant chemical or compound* or labelled as a *persistent organic chemical or compound* when the original chemical that has been transformed in the substrate persists as a threat to the environment and human health. An example of this can found in the persistence of certain pesticides in the ground. *Persistence* is most often used in conjunction with organic chemicals, when one is concerned with not only the presence of such chemicals, but also the state the organic chemicals found in the subsurface environment. This refers to the fact that the chemical may or may not retain its original chemical composition because of transformation reactions. Most organic chemicals do not retain their original composition over time in the subsurface environment because of the various abiotic and biologically mediated chemical reactions. It is not uncommon to find *intermediate products* along the transformation path of an organic chemical. The reductive dehalogenation of tetrachloroethylene or perchloroethylene (PCE) is a very good example. Progressive degradation of the compound through removal and substitution of the associated chlorines with hydrogen will form intermediate products. However, because of the associated changes in the water solubility and partitioning of the intermediate and final products, these products can be more toxic than the original pollutant.

2.3.3   Hazardous solid waste pollutants

The term *pollutants* is used deliberately instead of *contaminants* in this section to emphasize the seriousness of the issues concerning HW containment – to protect the environment and public health from contact with or exposure to the pollutants associated with HWs. The pollutants of concern fall into two groups: heavy metals and organic chemicals. The discussion in this section follows the treatment given by Yong (2001).

Heavy metals

Elements with atomic numbers higher than 38 (strontium, Sr) are classified as *heavy metals*. That being said, it is not uncommon to find the term *heavy metals* used to include elements with atomic numbers greater than 20 (e.g. calcium, Ca). Common practice includes heavy metals with atomic numbers found in the lower right-hand portion of the periodic table, i.e. the *d*-block of the periodic table. In all, there are 38 elements that can be grouped into three convenient groups of atomic numbers as follows:

*   from atomic numbers 22 to 34 – Ti, V, Cr, Mn, Fe, Co, Ni, Cu, Zn, Ga, Ge, As and Se;
*   from 40 to 52 – Zr, Nb, Mo, Tc, Ru, Rh, Pd, Ag, Cd, In, Sn, Sb and Te,
*   from 72 to 83 – Hf, Ta, W, Re, Os, Ir, Pt, Gu, Hg, Tl, Pb and Bi.

Except for Zn and the metals in group III to group V, the metals in the three groups are *transition metals*. This is because these are elements with at least one ion with a partially *d* subshell. Almost all the properties of these transition elements are related to their electronic structures and the relative energy levels of the orbitals available for their electrons.

The heavy metals commonly found in leachates originating from landfills are lead (Pb), cadmium (Cd), copper (Cu), chromium, (Cr), nickel (Ni), iron (Fe), mercury (Hg) and zinc (Zn). The heavy metal ions such as Cu^{2+}, Cr^{2+}, etc. (M^{n+} ions) are generally coordinated to six water molecules as they do not readily exist in aqueous solutions or in the porewater of a clay–water system as individual metal ions. They exist as $M(H_2O)_x^{n+}$ in their hydrated form. As the M^{n+} coordination with water is in the form of bonding with inorganic anions, replacement of water as the ligand for M^{n+} can occur if the candidate ligand, generally an electron donor, can replace the water molecules bonded to the M^{n+}.

Some of the more common inorganic ligands that will form complexes with metals include CO_3^{2-}, SO_4^{2-}, Cl^-, NO_3^-, OH^-, SiO_3^-, CN^-, F^- and PO_4^{3-}. In addition to anionic-type ligands, metal complexes can be formed with molecules with lone pairs of electrons, e.g. NH_3 and PH_3. Examples of these kinds of complexes are $Co(NH_3)_6^{3+}$, where the NH_3 is a Lewis base and a neutral ligand, and $Fe(CN)_6^{4-}$, where CN^- is also a Lewis base and an anionic ligand. As all metal ions M^{nx} are Lewis acids, it follows that heavy metals are bonded with Lewis bases, as Lewis acids can accept and share electron pairs donated by Lewis bases. Although Lewis bases are also Brønsted bases, it does not follow that Lewis acids are necessarily Brønsted acids, as Lewis acids include substances that are not proton donors. It is important to note that the use of the Lewis acid–base concept permits us to treat metal–ligand bonding as acid–base reactions. Complexes formed between soil organic compounds and metal ions are generally chelated complexes. These naturally occurring organic compounds are humic and fulvic acids, and amino acids.

Some of the heavy metals can exist in the porewater in more than one oxidation state, depending on the pH and redox potential of the porewater in the microenvironment. As an example, Se can occur as SeO_3^{2-} with a valence of +4, and as SeO_4^{2-} with a valence of +6. Similarly, we have two possible valence states for Cu in the porewater, i.e. valences of +1 and +2 for CuCl and CuS respectively. Cr has more than one ionic form for each of its valence states: CrO_4^{2-} and $Cr_2O_7^{2-}$ for the valence state of +6, and Cr^{3+} and $Cr(OH)_3$ for the +3 valence state. The same holds true for Fe. We have Fe^{2+} and FeS for the +2 valence state, and Fe^{3+} and $Fe(OH)_3$ for the +3 valence state.

Variability in oxidation states is a characteristic of transition elements (i.e. transition metals). Many of these elements have one oxidation state that is most stable, for example the most stable state for Fe is Fe(III) and Co(II) and Ni(II) for cobalt and nickel respectively. Much of this is a function of the electronic configuration in the d orbitals. Unpaired electrons which constitute one-half of the sets in d orbitals are very stable. This explains why Fe(II) can be easily oxidized to Fe(III) and why the oxidation of Co(II) to Co(III) and Ni(II) to Ni(III) cannot be as easily accomplished. The loss of an additional electron to either Co(II) or Ni(II) still does not provide for one half unpaired electron sets in the d orbitals. This does not mean to say that Co(III) does not readily exist. The complex ion $[Co(NH_3)_6]^{3+}$ has Co at an oxidation state of +3.

Organic chemical compounds

The organic chemicals and compounds found in the ground can be (a) point-source pollutants that are the contained in leachates generated from the wastes; (b) point-source pollutants obtained as a result of direct illegal dumping of these chemicals; or (c) non-point source pollutants such as those associated with the use of soil amendments in the agroindustry. These organic chemicals have origins in various chemical industrial processes and as commercial substances for use in various forms with derived products for commercial use, which include organic solvents, paints, pesticides, oils, gasoline, creasotes and greases. It is possible to find hundreds of thousands of organic chemical compounds registered in the various chemical abstracts services that are available. The more common organic chemical pollutants can be grouped into two main groups as follows:

- *Hydrocarbons*, including the petroleum hydrocarbons (PHCs), the various alkanes and alkenes, and aromatic hydrocarbons such as benzene, multicyclic aromatic hydrocarbons (MAHs), e.g. naphthalene; and polycyclic aromatic hydrocarbons (PAHs), e.g. benzo-pyrene.
- *Organohalide compounds*, of which the chlorinated hydrocarbons are perhaps the best known. These include TCE (trichloroethylene), carbon tetrachloride, vinyl chloride, hexachlorobutadiene, polychlorinated biphenyls (PCBs) and polybrominated biphenyls (PBBs).

In addition to the two main groups, there is a case to be made with respect to the formation of a group that could include oxygen-containing organic compounds, such as phenol and methanol, and nitrogen-containing organic compounds such as trinitrotoluene (TNT). A well-accepted classification is the NAPL (non-aqueous phase liquid) scheme, which breaks the NAPLs down into the light NAPLs (identified as LNAPLs) and the dense ones (DNAPLS). The LNAPLs are considered to be lighter than water, and the DNAPLs heavier than water. Because LNAPL is lighter than water, it stays above the water table. On the other hand, as the DNAPL is denser than water, it will sink through the water table and will come to rest at the impermeable bottom (bedrock). Some typical LNAPLs include gasoline, heating oil, kerosene and aviation gas. DNAPLs include the organohalide and oxygen-containing organic compounds such as 1,1,1-trichloroethane, creosote, carbon tetrachloride, pentachlorophenols, dichlorobenzenes and tetrachloroethylene.

Municipal solid wastes, waste products and liquid waste streams

Municipal solid wastes (MSWs) are also classified as regulated wastes in most jurisdictions. Although they do not by themselves fall into the category of hazardous wastes, their leachates contain substances that qualify under the CFR lists previously discussed. We can typify, in general, the kinds of waste products and waste streams generated by the various groupings of industries. A sample of some of the major kinds of waste products produced, together with the kinds of pollutants found in municipal solid waste leachates, can be seen as follows:

- *Laboratories*: acids, bases, heavy metals, inorganics, ignitable wastes and solvents.
- *Printing and other print industries*: acids, bases, heavy metals, inorganic wastes, solvents, ink sludges and spent plating.
- *Agroindustries*: pesticides, fertilizers, fungicides, solvents, inorganics such as nitrates and phosphates, herbicides, etc.
- *Metal manufacture and fabrication*: acids, bases, solvents, cyanide wastes, reactives, heavy metals, ignitable wastes, spent plating wastes, etc.
- *Leachates from municipal solid wastes*: arsenic, heavy metals, chlorinated hydrocarbons, cyanides, selenium and miscellaneous organics.

A large proportion of solid waste substances are not hazardous to humans, i.e. they do not threaten the health of biotic receptors. It is their leachates, however, which are of concern – as proven, for example, by the last item listed above, which details the kinds of pollutants found in MSW leachates. There are two types of leachates in any waste pile: (a) the *primary leachate*, which essentially comes from the liquid part of the waste pile, and (b) the *secondary leachate*, which is composed of water percolating into the waste pile in combination with the dissolution products of the waste pile itself. Common practice assumes that the primary and secondary leachates will combine themselves in one fashion or another. There are two options available for the collected leachate: (a) treatment of the collected leachate in conformance with regulatory discharge standards or (b) re-introduction of the collected leachate into the waste pile as part of a biochemical reactor treatment scheme for the waste pile in the landfill.

Leachates or waste liquids may be water or organic liquids. These leachates or waste liquids may be classified into four groups as follows:

1   *Aqueous inorganic, with water as the liquid (solvent) phase*: brines, electroplating wastes, metal etching wastes, caustic rinse solutions, salts, acids, bases, dissolved metals, etc.

2   *Aqueous organic, with water as the liquid (solvent) phase*: wood-preserving wastes, water-based dye water, rinse water from pesticide containers, ethylene glycol production, polar and charged organic chemicals, etc.

3   *Organic, with organic liquid as the solvent phase*: oil-based paint waste, pesticide and fertilizer-manufacturing wastes, spent motor oil, spent cleaning solvents, spent solvents, refining and reprocessing wastes, etc.

4   *Sludge, with organic liquid or water as the solvent phase*: separator sludge, storage tank bottoms, treatment plant sludge, filterable solids from all kinds of production plants, etc.

2.4   High-level radioactive waste containment and disposal

2.4.1   Containment and disposal

We have used the terms *containment* and *disposal* previously without defining them because, up until now, they have been used in the general sense as defined in standard dictionaries. We now need to provide greater clarity in the use of these terms – in the context of high-level radioactive wastes (HLWs) and hazardous solid wastes (HSWs). If we take the case of HLW, for example, the term *containment* is used to describe the combined system comprising a spent fuel canister, clay buffer or barrier, and host rock (i.e. rock mass within which the canister–buffer system is placed). In the case of surface burial of HSW, the term *containment* is used to denote the landfill double-liner system together with the underlying engineered clay barrier. The term *disposal* is used when subsequent or potential recovery of high-value items from the HLW or HSW is not considered in the final scheme of events. For our purpose, as the decision of possible recovery of high-value items is a decision that is made in conjunction with economic, environmental, engineering and many other kinds of analyses, we will use the term *disposal* when we mean final resting place for the waste under consideration.

The requirements for containment of radioactive wastes, from LLW to LILW and HLW, are detailed in the various documents dealing with regulatory control of the management of these kinds of wastes, which are issued by the various Member States and governments. By and large, the options for management and disposal of these wastes in all these documents are classified as follows: (a) deep disposal in repositories sited in massive clay formations at depths of at least 250 m for HLW containment; (b) deep geological repository containment in crystalline or argillaceous rock and isolation at depths of more than 1000 m for HLW containment; and (c) secure landfill or surface burial with secure containment systems for those treated and conditioned LLW and LILW that qualify as HW disposal. In all cases, there are strict requirements for monitoring systems to provide early warning and tracking of fugitive radionuclides and pollutants.

The discussions on the design or specification of the required containment element or hardware such as canisters and concrete- and bituminous-embedded systems are not within the range of this book. These subjects deserve detailed treatment and are contained in books and manuals that are specifically designed to address these subjects. The subjects of concern in this book are the clay buffers and clay barriers that form integral parts of the multi-barrier systems in deep geological containment, surface burial and entombment schemes. This section deals with the general schemes for HLW containment–disposal, and with the principal issues or impacts on the clay buffer and barrier that form part of the containment system. The specific details of these will be discussed in the next two chapters.

2.4.2 High-level radioactive waste deep geological repository disposal

Figures 2.3 and 2.4 show that there are two very significant factors that require attention in any engineering scheme that is designed to contain spent fuel canisters. These are (a) long-lived emitted radiation and (b) long-lived heat emission, both of which last for more than 100,000 years. The deep underground disposal requirement articulated in most regulatory documents in response to these two demanding factors reflects the requirement that the travel times for the most harmful fugitive radionuclides (obtained as a result of a breach in the canisters) to reach ground surface would exceed their half-lives.

Figure 2.6 provides the essence of the deep geological containment or disposal scheme, showing a repository that consists of a tunnel with boreholes containing spent fuel canisters (bottom right-hand corner). The canisters (top right-hand corner of the figure), which are designed to provide the proper shielding from radiation, emit decay heat from the spent fuel rods that will persist for hundreds of thousands of years, albeit at a diminishing intensity. The clay barrier surrounding the canisters is called a *clay buffer*,

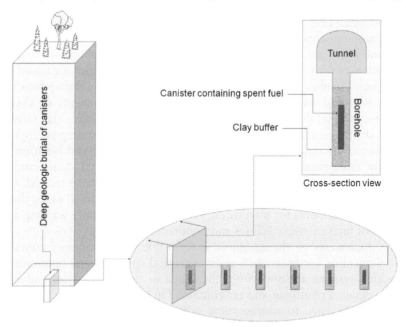

Figure 2.6 Deep geological burial of spent fuel canisters. The 'exploded' view in the bottom right-hand corner shows a tunnel with regularly spaced boreholes with embedded canisters, as shown in the cross-section view at the top right-hand view. The clay buffer is most often a smectitic clay, and the tunnel is generally in-filled with a clay that could be mixed with other fine-grained and/or coarse-grained soils.

as its role is to provide a secure buffer between the canisters and the host rock. In the example shown in the figure, the clay or clay–soil mixture in the tunnel forms the *clay barrier*.

More examples of deep geological repository containment of HLW are shown in Figure 2.7. Although there is no fixed rule as to what the configurations or placement schemes should be, there is consensus on the type of repository that would provide secure protection from breeched canisters that allow fugitive radionuclides to escape into the environment, resulting in harmful impacts to both the environment and biotic receptors. The single over-riding rule for all the configurations revolves around the time taken for fugitive radionuclides to reach positions of actual threat to the various receptors. In the configurations shown in Figure 2.7, smectitic clay buffer is most often used to embed the spent fuel canisters for all borehole schemes. Other kinds of swelling clay can be used in place of smectites, provided that they manifest closely similar (or better) properties and characteristics. The reasons for using swelling clays are discussed in the next subsection.

The clay buffer that surrounds the canister in any single borehole is expected to provide a cushion support for the canister while offering a

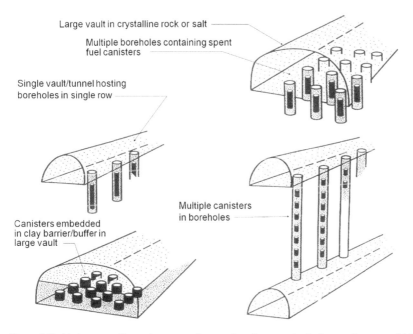

Figure 2.7 Various configurations or schemes for deep geological containment of HLW canisters. Smectitic clay is used as the buffer in the boreholes. The multiple canisters in the long vertical boreholes containing multiple containers in a stacked row, shown in the bottom right-hand corner, can also be configured as horizontal boreholes (adapted from Pusch, 1994).

buffering capability against the decay heat emitted from the canister. In addition, the buffer is expected to function as an attenuating barrier for transport of corrosion products and fugitive radionuclides. The effects of all of these – heat, corrosion products and radionuclides – on the clay buffer, together with (a) the impact of the interface features between the clay buffer and the host rock and (b) the wetting of the clay buffer from the ingress of groundwater, on the design performance characteristics of the clay buffer will be discussed in detail in the chapters to follow. These chapters will discuss the elements contributing to the considerable changes in the nature and properties of the clay buffer – not only from these actions, but also from long-term processes that include oxidation–reduction reactions and biologically mediated chemical reactions. For now, we will consider only the main actions on the clay buffer coming from the canister and from the surrounding host rock (Figure 2.8).

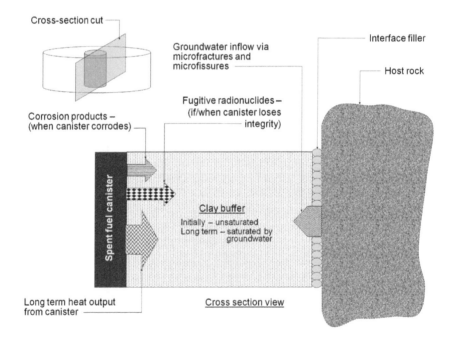

Figure 2.8 Impact of spent fuel canister on the clay buffer surrounding the canister. The primary role of the buffer is to act as a physical and chemical buffer between the hot canister and the host medium (host rock shown in this example). The competent rock which is the most likely candidate for the host medium may have microfissures and microfractures that will permit inflow of groundwater into the initially unsaturated clay buffer.

2.4.3 High-level radioactive waste containment – clay buffer/barrier functional requirements

The use of swelling clays for buffer and barrier systems in the containment and isolation of hazardous and non-hazardous waste materials has been promoted widely in the past two decades, and especially with respect to the containment of HLW, largely because of the Swedish research and development work in response to requirements for safe isolation of HLW. Smectitic clays, which are swelling clays, offer the best features that fit the buffer and barrier requirements for containment of HLW repositories and HSW landfills. The major attributes of this family of clays (i.e. smectitic clays) are (a) low hydraulic conductivity; (b) volume expansion (swelling) upon water uptake; (c) ability to seal internal cracks, joints and openings in tightly confined situations; (d) development of a more homogeneous macrostructure as a result of the tightly constrained swelling process; and (e) high capability for partitioning of inorganic and organic contaminants.

The spent fuel canister placed in a borehole (Figures 2.6 and 2.7) requires such a clay. It should tightly surround and cushion the canister so that it will be secure in the borehole. Ideally, it needs to maintain its design functional capabilities for at least 100,000 years, i.e. the properties of the surrounding clay soil should remain constant or improve (through maturation processes) as time progresses – for the 100,000-year period. The clay buffer must:

- keep the emplaced canister physically stable to avoid damage to it from displacements and movements initiated not only by the weight of the canister itself, but also from external forces; a supporting cushion-type platform is needed, i.e. a canister embedment scheme in which the embedding material (i.e. the smectitic clay) acts as a competent platform and also as a surrounding cushion;
- protect the embedded canister from processes initiated by the host rock environment that would threaten or harm the integrity of the canister; the detrimental processes initiated in the clay barrier and against the integrity of the canister by water availability from the surrounding host medium (rock) must be denied, at best, or at least minimized;
- deny or mitigate transport of corrosion products and fugitive radionuclides to the surrounding host medium (rock), which requires the smectitic clay to have mechanical, physico-chemical and chemical buffering capabilities; the ultimate aim is to allow only diffusive transport of the corrosion products and radionuclides in the smectitic clay, in addition to promote partitioning of these solutes.

The requirement for knowledge of the changes that might occur in the nature and properties of the smectitic clay over a period of tens of thousands of years is a very demanding requirement to fulfil. Nevertheless, this is necessary because the many long-term processes provoked by abiotic and biotically mediated acid–base and oxidation–reduction reactions will result in various kinds of alterations and transformations of the clay minerals. The effect of all of these on the nature of the clay soil and its properties can be profound and may even be detrimental to the design performance characteristics of the clay buffer and barrier. Chapters 3 and 4 provide detailed discussions of the various aspects of the nature and properties of clays that allow them to function as competent clay buffers and barriers.

Considering only the external inputs and factors acting on the clay buffer, the processes initiated by the actions of the canister and the groundwater ingress shown in Figure 2.8 include (a) conductive and advective heat flow; (b) hydration and water uptake, and diffusive and advective fluid flow; and (c) transport of corrosive products and radionuclides in the clay. These are shown in Figure 2.9 for a three-phase (solid, liquid and gaseous) representation of the clay buffer. The resulting interactions and processes in the clay buffer due to these actions are discussed in detail in Chapters 4 and 5. How these various processes are manifested in the clay buffer, how the responses arising from the interactions between the clay particles, water, solutes, temperature, resultant swelling pressures developed, and how the

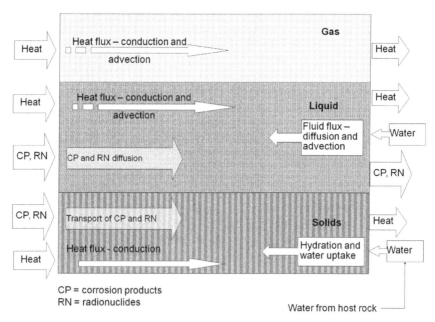

CP = corrosion products
RN = radionuclides

Water from host rock

Figure 2.9 Three-phase (solids, liquid and gaseous) clay buffer, showing major processes initiated by the actions shown in Figure 2.8. Only the processes acting on the clay buffer are shown. The processes provoked in the clay buffer such as osmotic swelling, volume change, partitioning of solutes, evaporation and condensation and diffusive vapour flow are not shown, as these are processes that will be discussed in detail in Chapters 4 and 5.

chemical reactions, partitioning, biotransformations, biomineralization, etc. are integrated into the total performance of the buffer over a protracted time period, constitute the principal discussion topics in the various chapters of this book.

The problems of particular concern centre on how one can determine system performance over the 100,000-year life of the repository. By system performance, we mean the performance of the canister, clay buffer and the surrounding host medium. Analytical computer modelling tools have been developed to provide predictive capability. However, these tools are only as good as their capability to appreciate or perceive the various interactions and coupled processes that contribute to the total system performance. To date, these have not met with the level of credibility required for confident acceptance as analytical or predictive tools. In large measure, this is because the necessary background information and knowledge of chemical reactions and contributions from microbial activities to the transitory nature and properties of the clay buffer over the required regulatory lifespan of 100,000 years have yet to be fully obtained.

2.5 Hazardous solid waste containment and disposal – landfills

2.5.1 Garbage bag analogy

As we have previously indicated, HSWs are regulated wastes, and, as such, containment and disposal of these wastes have mandated regulatory requirements. In compliance with these requirements, designs and construction of HSW landfills have used double-membrane techniques in combination with engineered clay barriers as a multi-barrier system to contain non-hazardous and hazardous wastes. Both non-hazardous and hazardous waste landfill barrier–liner systems have one primary aim, i.e. to contain and secure the wastes contained within the barriers such that (a) the dissolution products obtained from the waste pile will not escape from their containment systems and/or (b), if escape does occur, the contaminants represented in these products must be attenuated to the extent that their presence in the subsurface will not threaten the health of the environment and biotic receptors. The more general term *contaminants* will be used to identify the dissolution products issuing forth from the waste pile. These contaminants comprise toxicants and non-toxic substances.

Design attitudes with respect to landfill-engineered barrier–liner systems are guided, to a very large extent, by the regulatory requirements of the country or region, site specificities and economics. The *dry garbage bag contents* principle adopted in construction of, and in operating, waste landfills, shown in the generic form in Figure 2.10, satisfies the primary aim. It is acknowledged, however, that any containment–protection system designed to be totally impermeable to liquids will probably not remain impermeable forever, i.e. the dry garbage bag contents cannot remain dry forever. Accordingly, prudent engineering practice requires one to seek designs and methods that would deal with the outcome of leakage of leachates through multi-barrier systems.

2.5.2   Engineered multi-barrier systems

The engineered cover and engineered bottom barrier systems of secure and even medium-secure landfills consist of layers of clay, impermeable geomembranes, filter systems, leachate collections systems, detection systems, and other protective devices or measures designed to meet the regulatory requirements governing containment of the waste in the landfill. These are called *engineered multi-barrier* or *engineered barrier–liner systems*. The aim of the top engineered cover system is to deny entry of water into the waste pile – to maintain a 'dry' waste condition. This follows the school of thought that argues that, by limiting the amount of water entering the waste pile, one limits the generation of leachates. By this strategy, one minimizes potential

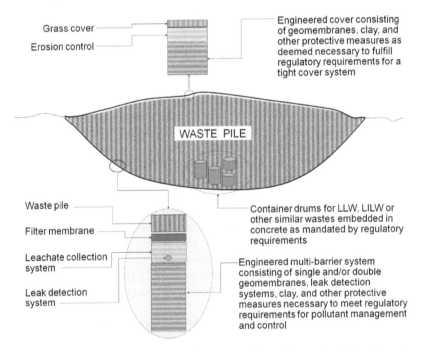

Figure 2.10   General 'sealed (dry) garbage bag contents' type of waste landfill configuration for non-hazardous (e.g. MSWs) and HSWs. Note that the landfill configurations with HSW sealed containers, such as concrete and bituminous blocks embedding HW (e.g. LLW and LILW), may be surface or deep subsurface land disposal systems, depending on regulatory requirements.

leachate escape from the waste pile. Depending on the strategy devised for the life status of the waste pile, the top cover can be designed to permit air entry while denying water entry, to allow for an oxic environment in the waste pile. The side- and bottom-engineered barrier systems are designed to contain and trap leachates so that these may be removed and treated before discharge into acceptable receptors – or recycled into the waste pile to create the biochemical reactor effect. Of course, if the top cover of the landfill performs as expected, and if no liquids are present in the waste pile, no leachates will be developed, and the side- and bottom-engineered barrier systems will only need to provide containment while denying entry of subsurface water.

Saturation of the top, side and bottom engineered clay layers in a landfill is a critical issue depending on the strategy or type of waste material being contained, except for the cases in which bioreactor simulation is desired. If, and when, the engineered clay cover system is saturated, it is no longer capable of protection against water entry into the waste pile from the top, unless an impermeable geomembrane is used. Conservative design procedures anticipate that, at some stage, imperfections, construction practice and degradation will all combine to produce eventual leaks into the waste pile. Leachates escaping from the landfill can occur only after penetrating the geomembrane(s) and the engineered clay barrier or liner. Full saturation of the clay material would facilitate transport of contaminants to the underlying subsoil. An important question is therefore 'How quickly does the clay layer fully saturate, or more exactly how does the wetting process take place?' If it is sufficiently slow it may mean that, for the top liner system, little or no water reaches the waste in several hundred years and, for the side and bottom liner system, escape of fugitive contaminants to the subsoil will not be facilitated as long as the clay material remains partly saturated or dry.

Figure 2.11 provides a comparison of the basic elements constituting typical bottom-engineered multi-barrier systems of non-hazardous and HSW landfills. The choices that need to be made with respect to types of geomembranes, leachate collection systems (LCSs), leak detection systems (LDSs), and especially the kinds and thickness of clay material and thickness of clay layer, are most critical. The LCS shown in the figure, for example, consists of graded granular material in the layer, and a leachate collection pipe system (white oval in the diagram). The leak detection system (LDS), shown in the right-hand side of the figure, consists of a layer of graded granular materials and a leak detection unit (grey oval). There is flexibility on the choice and use of technology for leachate collection. However, the system chosen must satisfy mandated requirements regarding the maximum leachate head permitted in the system (generally < 0.5 m). This essentially dictates the leachate removal programme. Correspondingly, the leak detection system is governed by permeability constraints on the granular material in the layer (generally > 0.5 m) and a minimum permeability coefficient.

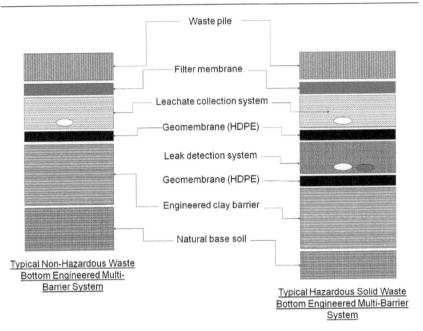

Figure 2.11 Typical elements in landfill waste bottom-engineered barrier systems. Left, non-hazardous solid wastes; right, HSWs. Note that the white ovals in the leachate collection system are the collection pipes, and the grey oval in the leak detection system is the leak detection unit (adapted from Yong, 2004). HDPE, high-density polyethylene.

The basic elements for the multi-barrier systems shown previously as Figure 2.11 can now be seen in greater detail in Figure 2.12 for a typical engineered multi-barrier system for HSW and MSW landfills, using either single- or double-membrane liner system. The differences for the two depicted systems depend upon the security requirements of the waste to be contained. It is important to note that the thickness of the engineered clay layer overlying the natural foundation base must be determined on the basis of 'what is required to fully attenuate contaminants entering the engineered clay layer'. Although this may not always produce thickness requirements that will be easily met, it is important to obtain design calculations for the thickness of this layer based on the ideal requirement for full attenuation of contaminants. In any event, the minimum thickness of the engineered clay layer must not be less than 1 m. All of the other specifications of materials and dimensions of the various layers that constitute the multi-barrier engineered system must be determined on the basis of the capabilities of materials and placement methods to meet the regulatory and/or performance requirements.

Geomembranes, geosynthetic clay liners (GCLs), geonets and geotextiles are some of the geosynthetics used in the liner–barrier systems that

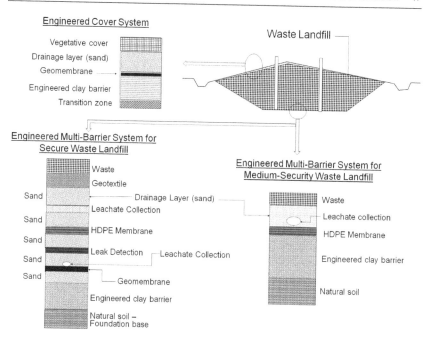

Figure 2.12 Typical engineered multi-barrier systems for secure waste (e.g. HSW) and medium-security (e.g. MSW) landfills. Note that the thickness of the engineered clay layer overlying the natural soil foundation base should be (a) thick enough to perform satisfactory attenuation of contaminants escaping from the landfill and (b) at least 1 m thick.

line landfills. Detailed material and performance specifications required for geomembranes used in the liner–barrier system, for example, include specifications or minimum values on thickness, tensile strength, durability, texture, gas and vapour transmission rate, and solvent vapour transmission rate, etc. Typical types of geomembranes include high-density polyethylene (HDPE), very flexible polyethylene (VFPE) and polyvinyl chloride (PVC).

It is useful to note that there are detailed specifications and requirements covering specific situations and the use of various types of barrier systems, both as design standards and regulatory requirements. Except for the clay, all of the other elements in the engineered multi-barrier system are manufactured items that can be produced to meet properly articulated specifications. The nature of the clay, the methods for preparation of the clay barrier itself, the required properties and characteristics of the clay and the performance expected of the clay barrier are all subjects that need careful evaluation. For example, as in the case of the United States Resource Conservation and Recovery Act (US RCRA) legislation, the European Commission has recently defined the technical characteristics of waste landfills in the Council Directive 1999/31/EC. It specifies the following permeability (Darcy) coefficient

k and clay layer thickness requirements for top, side and bottom liners as follows:

- landfill for hazardous waste: $k < 10^{-9}$ m/s, clay layer thickness > 5 m;
- landfill for non-hazardous waste: $k < 10^{-9}$ m/s, clay layer thickness > 1 m.

Chapters 3 and 4 will develop these subjects in greater detail. The types of materials and designs for top cover liner and side and bottom multi-barrier systems vary considerably with regard to (a) different countries' regulatory requirements; (b) different kinds of wastes to be contained in the landfill; (c) site specificities, such as hydrogeological setting and subsurface features; (d) available engineering and technological capabilities; and (e) economics.

Although regulations, criteria and standards (i.e. regulatory attitude) adopted by the regulatory body responsible for waste management and disposal (e.g. Ministry of Environment, Pollution Control Board, Environmental Management Bureau, Department of Environment, Environment Protection Agency, European Commission) will essentially dictate the type of landfill and liner or barrier system to be used, the actual design and construction of the landfill are the responsibility of the stakeholder. In the US for example, MSWs and HSWs are regulated under the Resources Conservation Recovery Act (RCRA), including the Hazardous and Solid Waste Amendments (HSWA) to RCRA. As stated previously, Part 261 of Title 40 of the Code of Federal Regulations (CFR Title 40, Part 261) provides the definitions for hazardous waste. Legislation applicable to MSW is contained in Subtitle D of RCRA and the regulations governing MSW landfills are covered in CFR Title 40, Part 258.

2.5.3   Hazardous solid waste containment – engineered clay barrier functional requirements

The properties and characteristics of clays that are used to construct clay barriers that line or surround the landfill are discussed in Chapter 3, and the interactions between these clays and contaminants etc. that result in the attenuation capability of these clays are discussed in Chapter 4. For this section, we want to focus our attention of the functional requirements for these engineered clay barriers, for example those illustrated in the previous figures.

As engineered barrier systems are essential components in the design and construction of waste landfills, so long as the barriers function according to design requirements, escape of harmful leachates into the subsurface environment will not occur. Leachates generated in the landfill will be captured and treated before discharge in the proper receiving facilities. Failure of the geomembrane or similar synthetic impermeable material to contain waste leaches within the confines of the landfill can occur as a result of any or all of the following factors: fabrication imperfections, ageing processes, construction activities, acid–base reactions and biological activities. When failure or breach of the membranes occurs, leachates must travel through the engineered clay barrier to the immediate surrounding environment. If this should occur, the role of the clay barrier is to attenuate contaminant concentrations to the extent that the concentrations will not harmfully impact the environment and biotic receptors. To meet its design function the engineered clay must attenuate the contaminants in leachate transport through the barrier (see Chapter 4).

The attenuation of contaminants and/or pollutants in clays arises from the assimilative processes of these soils. Attenuation of contaminants in clays refers to the reduction of concentrations of contaminants and/or pollutants during transport in clays and other soils in general. This can be accomplished by (a) dilution of the leachate because of mixing with (infusion of) uncontaminated groundwater; (b) interactions and reactions between contaminants and clay solids (particles) that result in partitioning of the contaminants between the clay solids and porewater; and (c) transformations that serve to reduce the toxicity threat posed by the original contaminants/pollutants in the leachate. Short of overwhelming dilution with groundwater, it is generally acknowledged that partitioning is by far the more significant factor in attenuation of contaminants.

Contaminant attenuation – dilution, retardation and retention

The details relating to partitioning processes are discussed in Chapter 4. For now, we want to focus on attenuation of contaminants as determined by temporary and permanent partitioning of contaminants. Figure 2.13 depicts various contaminant concentration–time (CC–T) pulses. The input CC–T is shown at the contaminant concentration ordinate as a uniform pulse, i.e. constant contaminant concentration input over a specified time period. With the condition that the input contaminant concentration remains constant over a specified time period, one obtains the vertical rectangle shown on the ordinate. The area of the rectangle represents the total contaminant load input. The distance away from the contaminant source is shown as the abscissa. The three separate contaminant attenuation mechanisms (dilution, retardation and retention) are all assumed to be solely responsible for contaminant attenuation.

The figure shows the different shapes of pulses as one progresses away from the contaminating source, ranging from horizontal–rectangular to different kinds of bell-shaped pulses. The shapes and areas contained with the shapes are reflective of the kinds of mechanism responsible for contaminant attenuation. The effect of contaminant attenuation due to dilution is shown as the horizontal rectangle resting on the abscissa. For purposes of illustration, this rectangle has been shown as a constant area in accord with

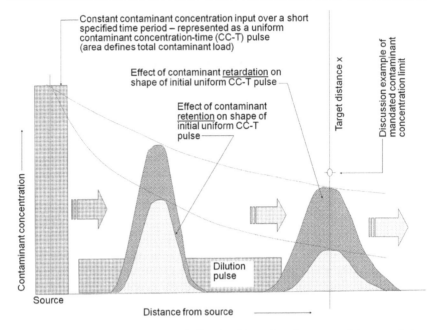

Figure 2.13 Progress of the input CC–T pulse. Note that the areas of the retardation CC–T pulses are constant and that they are equal to the original contaminant pulse load. The areas of the retention CC–T pulses decrease as one progresses further from the contaminant source. Assuming a fixed-volume input of dilution water, the area of the dilution CC–T pulse (shown as the horizontal bar at the bottom) remains constant, as the total amount of contaminants is unchanged.

the assumption of a fixed-volume dilution load. Should dilution water be continuously available, it follows that the size of the rectangular area should increase, reflecting decreasing contaminant concentration due to dilution. In other words the area of the horizontal rectangle will be larger than the vertical rectangle, as dilution has effectively reduced the concentration of contaminants and hence can reach the point at which they will satisfy regulatory requirements that dictate a limit on contaminant concentration delivery (*trigger limits*) at control points (see point x in Figure 2.13). This leads one to believe the dilution of leachates bearing contaminants is a good strategy in managing control of escaping leachates from a landfill. In the past, the 'slogan of the day' for control of contaminant plumes (of leachates) used to be: 'dilution is the solution'. This slogan is now no longer used, for obvious reasons, and also for reasons that are complicated. The one most obvious reason is that, although dilution will reduce contaminant concentration in the leachate, the total contaminant load will nevertheless be delivered ultimately to the target.

Similarly, as retardation mechanisms are thought to rely on temporary reversible sorption processes and on physical constraints to retard movement of the contaminants, the areas in the retardation pulse shapes shown in Figure 2.13 are equal to the original rectangular shape shown on the left. If the contaminants are retained by the clay solids by means of sorption mechanisms (Chapters 3 and 4) that are more or less permanent, retention of the pollutants occurs. When such occurs, contaminant attenuation is achieved via retention mechanisms that are more or less permanent, depending on the nature of the partitioning processes (interactions between the contaminants and the soil solids). As transport of the pulse continues away from the source, the decreasing areas under the CC–T retention bell-shaped curves indicate that total contaminant load is smaller than the original input load.

The likelihood of only one mechanism being solely responsible for attenuation of contaminants in transport in the soil is quite remote. In all probability, all of the various processes or mechanisms will participate – to varying degrees – in the attenuation of contaminants. It is obvious, however, that the greatest benefit obtained from the attenuating capacity of clays is when irreversible sorption, for example retention, is the primary mechanism of attenuation. For such to be obtained, it is necessary to have an understanding of the various partitioning processes, and the role of soil and clay composition and structure, in interactions with the contaminants. To illustrate how these mechanisms affect contaminant attenuation, we consider a target distance x away from the contaminant source as a hypothetical regulatory decision point. The concentration of contaminants at any one instance of time t passing point x is given by the height of the CC–T pulse passing the point. If regulatory or design requirements mandate a trigger limit at this control target distance x, the diagram in Figure 2.13 shows that this mandate can be met with the decreasing concentration heights of the dilution pulse and CC–T bell-shaped pulses. However, as stated previously, although this criterion is an important one, it says nothing about the total load delivered to this point over some time interval. Another kind of control requirement could specify that the total contaminant load delivered to point x or reaching a distance x away from the source over a specified time interval should not exceed some critical limit. It is this control requirement that will place a demand on attenuation mechanisms that require permanent partitioning of the contaminants during transport through the soil substratum. This limiting contaminant load is important when the point of delivery (target or receptor) of contaminants is a water resource or an environmentally sensitive piece of land. For contaminant load-limiting requirements, attenuation of contaminants by retention mechanisms will satisfy the requirements – until such time when the retention capacity of the clay barrier is reached. This is addressed in detail in Chapter 4. When this occurs, retardation and dilution mechanisms will be the operative attenuation operators.

2.6   Evolution of clay barrier/buffer

As we will see in the next chapter, clays and other kinds of soils are living dynamic systems. It is more usual to consider soils as *soil–water systems* and, correspondingly, clays as *clay–water* systems. They are the habitat for a whole host of microorganisms. Because of the chemical nature of the clay fractions and their formational characteristics, and because of microbiological activities, clay is continually subjected to changes in its nature, properties and characteristics. Evolution of engineered clay barriers and buffers due to chemical reactions and microbial activities can be expected during the operating life of the containment facility using such barriers and buffers.

The question that is asked is: What is the impact of evolution on the design function of engineered clay buffers and barriers? The importance of knowledge-based information contained in the answers to this very critical question cannot be overstated. The various processes contributing to the evolution of clays and the effect of the transitory nature and properties of these clays as clay buffers and barriers are discussed in Chapters 6 and 7. One thing is absolutely clear: the nature of the clays used to construct engineered clay barriers and/or buffers will evolve with time, resulting in corresponding changes in the properties and characteristics of the clays. It follows that the performance of these clays as they evolve is not likely to be identical to their initial performance, i.e. performance in the short term.

2.7   Concluding remarks

The differences that make it necessary to distinguish between radioactive wastes and HWs also make it necessary for regulatory authorities to require deep geological disposal for HLW and strict engineered multi-barrier systems for HSW landfills. The purpose of clays used as buffers or barriers in the containment of HLW and HSW differs – to the extent that one serves as a physical, hydraulic, thermal, mechanical and physico-chemical buffer (HLW), whereas the other serves primarily as a hydraulic, physical and contaminant attenuation barrier (HSW). Evolution of the clays used as engineered buffers or barriers is inevitable over the design operating life of the containment facilities. It is eminently clear that unless one can foresee what 'evolution' would do or, more specifically, what the effect of the processes responsible for the evolution of the clays would be, a proper evaluation of the long-term performance of the containment systems will not be available. The aim of the remaining chapters is to provide the background details and information necessary to determine or anticipate 'what evolution would do to the clays'.

Chapter 3

Nature of clays

3.1 Clays and soil classification

3.1.1 Particle size classification

Clays belong in the overall family of soils derived from the weathering of rocks, and are either weathered in place or transported by various transport agents (such as glacial activity, wind, water, anthropogenic activities) to various locations. In general, one used to say that clays consist of three supposedly separate phases: a solid phase that consists of various minerals, an aqueous phase that contains dissolved solutes, and a gaseous phase. However, we now know that this is not exactly true – not in the light of our greater understanding of the nature of interlayer water in certain kinds of clays.

The inorganic crystalline fractions consist primarily of the following: (a) oxides and hydrous oxides of iron, aluminium and silicon; (b) carbonates, sulphates, phosphates and sulphides; and (c) primary and secondary minerals. It is common in soil and geotechnical engineering practice to categorize soils into groups according to particle size ranges, as, for example, gravels, sands, silts and clays. This is because if we consider soils to consist of discrete particles, particle size distribution is a significant physical property. It tells us about the stability of the discrete particle soil system through our understanding of (a) packing or arrangement of particles and (b) particle contact mechanics. In short, the physical behaviour of the discrete soil particle system responds to gravitational forces.

The following average particle diameter sizes provide an example of this kind of soil classification scheme: gravels $> 2\,mm >$ sands $> 0.06\,mm >$ silts $> 0.002\,mm >$ clays.

There are, however, several shortcomings to this method of soil classification when we need to determine the kinds of clays needed to fulfil the required functions of engineered clay buffers and barriers. Size classification schemes discriminate only along the lines of particle size. According to the commonly used size classification schemes, soils with effective particle diameters less than 2 microns (0.002 mm or 2 μm) are called clays. This can be misleading as it does not pay attention to the many significant characteristics that distinguish between *clays* and soil materials that are clay sized. The term *clay-sized soil* should be used to describe a soil for which discrimination of soil type is based solely on particle size distinction.

Strictly speaking, *clays* consist of clay-sized particles, sometimes referred to as *clay particles* and *clay minerals*. Clay minerals are alumino-silicates, the structure of which can range from highly crystalline to amorphous. Clay minerals are oxides of aluminium and silicon, with smaller amounts of metal ions substituted within the crystal. Although the terms *clays* and *clay soils* have been used in the literature to mean the same thing, we will use the term *clays* in this book, as the appendage of *soils* to *clays* is redundant, i.e. clays represent a certain class of soils much the same as *sands* and *silts* represent different classes of soils. We define *clays* to mean soils that contain clay-sized particles ranging from alumino-silicates (clay minerals), amorphous materials, organic matter, oxides, and hydrous oxides, to carbonates.

3.1.2   *Clay attributes required for barriers/buffers*

This chapter develops the basic information on the nature of clays, with a focus on those clays that exhibit the kinds of attributes needed to meet the requirements for clay buffers and barriers. From Chapter 2, we learn that in addition to good physical and chemical buffering capabilities, the clays used for clay barriers and buffers should be able to (a) naturally attenuate contaminants in their transport into and through the clay barriers; (b) impede the flow of water and leachates; (c) provide a stable support–containment system to the contained wastes or canisters; (d) remain robust, i.e. maintain their design properties and characteristics; and (e) withstand degradation with time or with exposure to external thermal, hydraulic, mechanical/physical/chemical forces and microbial activities. To illustrate one of the problems facing clay barriers and buffers, i.e. maintenance of the design properties of a clay used for clay barriers or buffers, we can examine the creep performance of a lacustrian clay under a constant load (Figure 3.1) when subjected to a leachate attack.

The clay minerals in the lacustrian clay shown in Figure 3.1, given in order of abundance, are illite, chlorite, kaolinite and vermiculite, and the constant axial load applied as the creep load was 4.8 kPa. The creep performance of the control sample, i.e. sample not subject to the leachate attack, is shown by the dashed line. A maximum strain of about 0.05 per cent is obtained at termination of the creep test (20×10^5 minutes). The results shown in Figure 3.1 indicate that when a companion sample was subjected to ingress of a leachate of 0.25 N $Na_2SiO_3 \cdot 9H_2O$ after 12.615×10^5 minutes, the creep strain increased in accord with the amount of leachate entering the sample. A creep strain of more than four times the control strain was obtained after

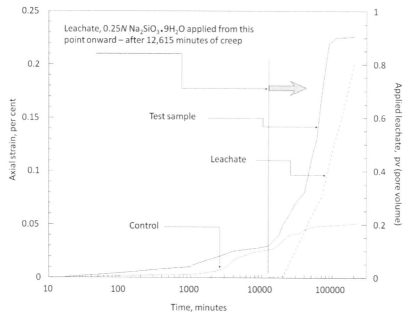

Figure 3.1 Effect of exposure of a natural lacustrian clay, under creep load (4.8 kPa), to a leachate (0.25 N $Na_2SiO_3 \cdot 9H_2O$) after 12,615 minutes of creep. Clay minerals in the clay were illite, chlorite, kaolinite and vermiculite, in decreasing order of abundance. Note the dramatic jump in axial creep strain as the amount of leachate entering the sample increases. The performance of the control sample of the same clay, which was not subject to the ingress of leachate, is shown by the dashed line.

permeation with 0.8 pore volumes of the leachate. The results shown in Figure 3.1 raise some pertinent questions such as:

- What are the processes within the sample that caused the leached sample to undergo a creep strain of more than four times the unleached sample?
- How are the effects of these processes affected by the type (composition) of clay and the chemistry of the leachate?
- What are the features of the clay fractions (constituents) that control the outcome of the processes involved?
- If the chemistry of the leachate were different, how differently would the sample respond?

In general, scrutiny of the features and attributes sought in the clays as candidates for engineered clay barriers relate to (a) how they react in the presence of water and solutes; (b) how they respond to heat and pressure; and (c) how they respond to acid–base and oxidation–reduction reactions, and microbial activities over the short term and especially over the long-term service life of the clays. The choice of the candidate clay for the engineered clay barrier or buffer will be determined according to the criteria and performance specifications provided by the stakeholder.

3.2   Clays and clay solids

We have stated previously that the term *clays* refers to the group of soils that are classed as clay sized in the particle size classification scheme, and that these include constituents that range from clay minerals to carbonates, for example:

- *clay minerals* – secondary minerals derived as the product of chemical weathering of primary minerals found in metamorphic and sedimentary rocks. The types of clay minerals obtained depend upon the composition of the solution, e.g. which ions are lost or retained in the leaching process and also subsequent to the leaching process;
- *organic matter* – humic substances and polysaccharides;
- *inorganic non-crystalline material (amorphous materials)* – hydrous oxides of iron, aluminium, and silicon, and allophones;
- *inorganic crystalline materials other than clay minerals* – oxides and hydrous oxides of iron, aluminium, and silicon, carbonates, sulphates, phosphates and sulphides.

Although the non-clay mineral constituents, by requirement of classification, are also clay sized in effective particle diameter, they do not have the same features and attributes as clay minerals. They are, however, important constituents that make up the composition of a clay. The terms *clay solids, clay particles* and *clay fractions* have been used to refer to these constituents – including the clay minerals. By and large, clay minerals are the dominant clay solids in clays, and common usage of the term *clays* implies soils with clay minerals as the major clay solids component.

3.3   Clay minerals

3.3.1   Basic structural units

Clay minerals are layer silicates (phyllosilicates). These consist of two basic crystal structural units: tetrahedral and octahedral units, as shown in Figure 3.2. The various layer silicates can be conveniently grouped into mineral structure groups based upon (a) the basic crystal structural units forming the elemental unit layer; (b) the stacking of the unit layers; and (c) the nature of the occupants in the interlayers.

The mineral structure groups of interest in so far as engineered clay buffers and barriers are concerned are the smectites, the kaolinite group, the

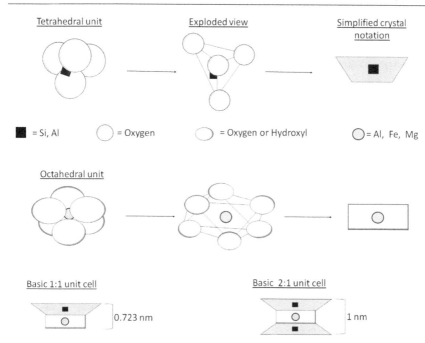

Figure 3.2   Basic crystal structural units in layer silicates.

hydrous micas and the chlorite group. In the kinds of clays used for buffers and barriers, both swelling and non-swelling clay minerals will be present. The smectites are the primary swelling-type clay minerals, whereas the other clay minerals (such as the kaolinites, illites (hydrous micas) and chlorites) are essentially the non-swelling type of minerals. All of these clay minerals consist of tetrahedral and octahedral layers stacked in 1:1 or 2:1 arrangements (see Figure 3.3).

In the sketches shown in Figure 3.2, the basic tetrahedral unit consists of a tetrahedron of oxygen atoms surrounding a silicon Si^{4+}; the 'exploded view' shows this arrangement more clearly. For convenience in representation, the simplified crystal notation is often used. In a similar fashion, the octahedral unit consists of an octahedron of oxygen atoms or hydroxyl groups 'encasing' a cation, which may be Al^{3+}, Fe^{3+} or Mg^{2+}, as illustrated in the 'exploded view' in the centre of the diagram. Again, for convenience in representation, the simplified crystal notation is most often used. The basic unit cells consist of stacked tetrahedral and octahedral units. The 1:1 layer stacking shown at the left gives us the basic unit cell typical of a kaolinite, and the 2:1 layer stacking shown at the right is the basic unit cell for smectites, illites and chlorites. These unit cells are joined in spatial articulation to form basic unit layers, and the stacking of these unit layers provides one with layer silicate particles, as shown in Figure 3.3.

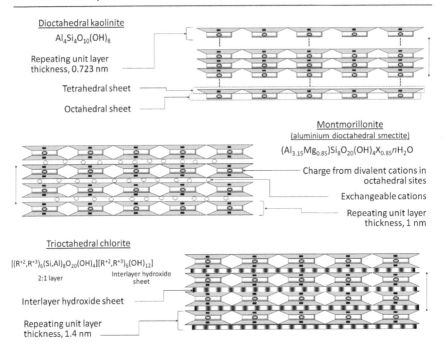

Figure 3.3   Basic repeating unit layers for 1:1 layer dioctahedral kaolinite, 2:1 layer aluminium dioctahedral smectite (montmorillonite) and 2:1 layer trioctahedral chlorite with their ideal structural formulae. Note that the repeating layers are stacked vertically and stretched spatially to form the respective mineral particles (structural formula information from Newman and Brown, 1987).

3.3.2   Non-swelling layer silicates – kaolinite, illite and chlorite

The most common non-expanding clay minerals are illite (hydrous mica, hydro-mica), chlorite and kaolinite. They are commonly described with the following characteristic compositions (Pusch, 1994; Gast, 1977):

- illite – $M^I_y Si_{(8-y)} Al_y O_{20}(OH)_4 (M^{III}, M^{II})_4$;
- chlorite – $(Si,Al)_4 O_{10}(Mg,Fe,Al)_6 O_{10}(OH)_8$;
- kaolinite – $Si_2 Al_2 O_5 (OH)_4$

where M^I = monovalent cations (K^+ in illite), M^{II} = Mg^{2+} and Fe^{2+}, and M^{III} = Al^{3+} and Fe^{3+}.

The different intraparticle bond strengths and different particle dimensions have direct impact on their specific surface areas (SSAs) and the hydration potentials.

Kaolinite

The clay mineral *kaolinite* belongs to the kandite group of clay minerals that are classified as dioctahedral 1:1 layer silicates (see top diagram in Figure 3.3). The major source of this kind of clay mineral is acid leaching of feldspars and micas in parent rocks. Typically, kaolinites are found in warm and moist climates. They are platey in morphology, with particle sizes dependent on the amount of disorder in the unit layers. On average, the typical particle will have equivalent diameters that are about 10 times the thickness of the particle. As shown in Figure 3.3, the typical structure for the 1:1 layer silicate consists of uncharged tetrahedral and octahedral sheets forming the basic unit layer – with Si ions in the tetrahedrons and two-thirds of the octahedral positions occupied by Al ions.

Illite

Illites belong to the mica mineral structural group. They are 2:1 layer silicates that are platey in shape with variable thicknesses. They are hydrous clay micas that do not ordinarily expand from the 1.0 nm basal spacing. Whereas their 2:1 layer structures are similar to that of aluminium dioctahedral smectites (see middle diagram in Figure 3.3), what distinguishes them from these smectites is the nature of the cations in the interlayers. Although the interlayer cations can consist of K, Na and Ca ions, it is the high and dominant concentration of interlayer potassium ions that characterizes the 2:1 layer silicate as an illite mineral. These interlayer K ions are not exchangeable.

When illites contain swelling layers, they are known as *interlayered illites*. The more common interlayered illites are illite–montmorillonite mixed-layer minerals. There also exist another kind of mixed-layer clay mineral: the interlayered chlorite. Interlayered chlorites, as the name implies, consist of chlorite–montmorillonite mixed-layer minerals. They are sometime known as swelling chlorites. In all of these interlayered clays, some degree of swelling can be expected.

Chlorite

The interlayer hydroxide sheet between the unit layers (see bottom part of Figure 3.3) can be a brucite or gibbsite layer. When Al ions fill two-thirds of the available position in the interlayer hydroxide sheet, this layer is known as a gibbsite layer with the chemical formula $Al_2(OH)_6$. With Mg ions in the octahedral layer instead of Al ions, the layer is known as a brucite layer, with chemical formula of $Mg_3(OH)_6$. These interlayer hydroxyl sheets do not have a plane of atoms that can be shared with the adjacent tetrahedral sheets.

3.3.3   Smectites

Smectites are swelling-type clays that are most often favoured for use as the clay component of engineered clay barriers. They are dioctahedral and trioctahedral aluminium silicate clay minerals containing magnesium and calcium. The dioctahedral minerals that are generally obtained from the transformation and weathering processes of volcanic material and igneous rocks are montmorillonite, beidellite and nontronite. Dioctahedral smectites consist of stacked layers, each of them consisting of two sheets of SiO_4 tetrahedrons confining a central octahedral layer of hydroxyls and Fe, Mg or Li ions. The terms *lamellar* and *layer* are often used interchangeably in the literature to mean the same thing. It is not uncommon to see notations as *layers* to mean *lamellae*, and *interlayer* space to mean *interlamellar* space. Referring to Figure 3.3, we note the following:

- *montmorillonite* – only Si in the tetrahedrons and Al in the octahedrons;
- *beidellite* – Si and Al in the tetrahedrons and Al in the octahedrons;
- *nontronite* – Si and Al in the tetrahedrons and Fe in the octahedrons.

The trioctahedral minerals are basically obtained or inherited from parent material, and are not commonly found as clay fractions. They are saponite, sauconite and hectorite:

- *hectorite* – Si in the tetrahedrons and Mg and Li in the octahedrons;
- *saponite* – Si and Al in the tetrahedrons and Mg in the octahedrons;
- *sauconite* – Zn in the octahedral sheet.

3.4   Surface properties of clay minerals

3.4.1   Reactive surfaces and surface functional groups

Reactive surfaces, as they relate to clay solids, refer to those clay solids' surfaces that, by virtue of their properties, are capable of reacting physically and chemically with solutes and other dissolved matter in the porewater. This is of particular significance in clay minerals. Chemically reactive groups (molecular units) associated with the surfaces of clay minerals, defined as *surface functional groups,* provide these surfaces with their *reactive* capabilities. It is not uncommon to see reference in the literature to these groups simply as *functional groups.* Clay solids with reactive surfaces include layer silicates (clay minerals), organic matter, hydrous oxides, carbonates and sulphates. For the discussion in this section, we will be concerned with the surface hydroxyls (OH group) that are the most common surface functional group in clay minerals with disrupted layers (e.g. broken crystallites) – often referred to as *hydroxylated surfaces.*

The source of charges that characterize the nature of the reactive surfaces of the clay minerals can be traced directly to the structure of the basic unit layers that constitute the layer lattice structures of these minerals. To illustrate this, we can look at basic unit layer of the group of clay minerals with the 1:1 structure, such as the kaolinite shown in Figure 3.3. The tetrahedral sheet at the top basal surface of the mineral particle is a siloxane surface, and the octahedral sheet at the bottom basal surface of the particle is a gibbsite surface. The siloxane surface is typical for minerals whose structures have bounding tetrahedral sheets, and is characterized by the basal plane of oxygen atoms bounding the tetrahedral silica sheet. Siloxane surfaces are reactive surfaces because of the structural arrangement of the silica tetrahedra and the nature of the substitutions in the layers. The regular structural arrangement of interlinked SiO_4 tetrahedra, with the silicon ions underlying the surface oxygen ions, produces cavities that are bounded by six oxygen ions in a ditrigonal formation. When no substitution of the silica in the tetrahedral layer and lower valence ions in the octahedral layers occurs, the surface is considered to be uncharged. When replacement of the ions in the tetrahedral and octahedral layers by lower valence ions occurs through isomorphous substitution, resultant charges on the siloxane surface are obtained. These provide the basis for a reactive mineral particle surface. These resultant charges are generally positive in nature. Table 3.1 provides a summary of the charge characteristics for many of these clay minerals. Note that in the fifth column of the table, the total number of electrostatic charges on the clay particles' surfaces divided by the total surface area of the particles involved provides us with the quantitative determination of the surface charge density – the inverse of which provides us with the reciprocal of charge density.

At the edges of the kaolinite particles, the broken edges have Al and Si centres which have hydroxyl terminals, i.e. they are terminated by hydroxyls. Clay minerals with the 1:1 structure, such as kaolinites (see top diagram in Figure 3.3) show siloxane and gibbsite surfaces on opposite basal surfaces of the particles, by nature of the 1:1 structural arrangement. The siloxane surface is defined by the basal plane of oxygen atoms that bound the tetrahedral silica sheet as shown in Figure 3.3. These basal planar surfaces are typical of minerals whose structures have bounding tetrahedral sheets. The 2:1 layer silicate minerals (such as illites, montmorillonites and vermiculites) have siloxane-type surfaces on both bounding surfaces. The more common types of silanol groups found within the structure of clay particles are: isolated (i.e. single silanols), geminal (silanediol) and vicinal groups. Silanol (SiOH) and siloxane (SiOSi) functional groups can exist together on the surfaces of the silica tetrahedra. The hydrophobicity of the siloxane surface is due to the presence of siloxane groups. Siloxanes tend to be unreactive because of the strong bonds established between the Si and O atoms and the partial ϖ interactions.

Table 3.1 Charge characteristics (SSA and CEC) for some clay minerals (adapted from Yong, 2001)

Soil fraction	Cation-exchange capacity (CEC) (mEq/100g)	Surface area (m²/g)	Range of charge (mEq/100g)	Reciprocal of charge density (nm²/charge)	Isomorphous substitution	Source of charges
Kaolinite	5–15	10–15	5–15	0.25	Dioctahedral; two-thirds of positions filled with Al	Surface silanol and edge silanol and aluminol groups (ionization of hydroxyls and broken bonds)
Clay micas and chlorite	10–40	70–90	20–40	0.50	Dioctahedral: Al for Si Trioctahedral or mixed Al for Mg	Silanol groups, plus isomorphous substitution and some broken bonds at edges
Illite	20–30	80–120	20–40	0.50	Usually octahedral substitution Al for Si	Isomorphous substitution, silanol groups and some edge contribution
Montmorillonite[a]	80–100	800	80–100	1.00	Dioctahedral; Mg for Al	Primarily from isomorphous substitution, with very little edge contribution
Vermiculite[b]	100–150	700	100–150	1.00	Usually trioctahedral substitution Al for Si	Primarily from isomorphous substitution, with very little edge contribution

Notes

Note that ratios of external–internal surface areas are highly approximate, as surface area measurements are operationally defined, i.e. they depend on the technique used to determine the measurement.

a Surface area includes both external and intralayer surfaces. Ratio of external particle surface area to internal (intralayer) surface area is approximately 5:80.

b Surface area includes both external and internal surfaces. Ratio of external to internal surface area is approximately 1:120.

When surface silanol groups dominate, the surface will be hydrophilic. Whilst surface silanol groups are weak acids, the acidity will be decreased if strong H-bonding is established between silanol groups and neighbouring siloxane groups. In silanol surfaces, the OH groups on the silica surface are the centres of adsorption of water molecules, and if internal silanol groups are present, hydrogen bonding (with water) could also be established between these internal groups in addition to the bonding established by the external silanol groups.

The surface of the edges of the particle of kaolinite contain both silanol and aluminol groups. The gibbsite sheet, which acts as the bounding surface, will also have aluminol groups. The Al^{3+} in the exposed edges of the octahedral sheets complex with both H^+ and OH^- in the coordinated OH groups, whereas the Si^{4+} will complex only with OH^-. Although association of the surface hydroxyls with a proton occurs below the point of zero charge (*zpc*) results in the development of a positively charged surface, the donation or loss of a proton by the surface hydroxyls above the *zpc* will result in a negatively charged surface. The aluminols can accept or donate protons, and according to Sposito (1984) adsorption of water onto these aluminol sites will produce Lewis acid sites. Although these surface aluminol groups will show some of the characteristics of the edge aluminol groups, they do not appear to affect the net negative charge distributed on the bounding surface (Greenland and Mott, 1985).

3.4.2 Net surface charges

The various functional groups at the basal and edge surfaces of the clay mineral particles and the effects of substitution in the lattices of the mineral particles are physically expressed as negative and positive charges, which are distributed on the surfaces of the particles. The charge density for any clay mineral particle is the sum of all the charges acting on the total surface of the particle, i.e. the sum of all the positive and negative charges. Strictly speaking, one should use the terms *net surface charge densities* in referring to the sum of all charges acting on the total surface. However, as the term *charge density* has been used in the literature to mean the net surface charge density, this term will be used in this book. Charge reversal due to changing pH values is a significant characteristic of kaolinites and hydrous oxides. In the case of kaolinites, charge reversal at the surfaces of the clay particles because of pH changes is the result of proton transfers at the surfaces.

We define *potential-determining ions* (*pdis*) as those cations and anions involved in surface coordination reactions. The total surface charge density of a soil particle σ_{ts} in the absence of *pdis* can be considered to consist of σ_s the permanent charge due to the structural characteristics of the clay particle (isomorphous substitution), and σ_h the resultant surface charge density due to hydroxylation and ionization (net proton surface charge density), i.e:

$$\sigma_{ts} = \sigma_s + \sigma_h \tag{3.1}$$

expressing the net proton surface charge density σ_h as:

$$\sigma_h = F(\Gamma_H - \Gamma_{OH}) \tag{3.2}$$

where F refers to the Faraday constant and Γ is the surface excess concentration, i.e. surface concentration in excess of the bulk concentration. These surface excess concentrations are the adsorption densities, and Γ_H and Γ_{OH} refer to adsorption densities of H^+ and OH^- ions and their complexes. When $|\Gamma_H| = |\Gamma_{OH}|$, the *point of zero net proton charge (pznpc)* is reached and the pH associated with this is designated as the pH_{pznpc}. From equations 3.1 and 3.3:

$$\sigma_{ts} = \sigma_s + F(\Gamma_H - \Gamma_{OH}) \tag{3.3}$$

The *pznpc* should not to be confused with the *point of zero charge (zero point charce, zpc)* or the *isoelectric point (iep)*. The pH_{zpc} represents the pH at which titration curve intersect differs from the pH_{iep}, which represents the pH at which the zeta potential ζ is zero. The zeta potential ζ refers to the electric potential developed at the solid–liquid interface as a result of movement of colloidal particles in one direction and counterions in the opposite direction. The distinction between pH_{zpc} and pH_{iep} is relevant, as slightly differing definitions exist in the literature for the *zpc* and *iep*. It would seem that these differences are related to methods of determination of these particular charge density relationships and the role of counterions in the inner and outer Helmholtz planes, i.e. the methods for determination of the influence of the charge densities and (influence) of the ions in the inner and outer Helmholtz planes are not the same. The zeta potential ζ is calculated from experimentally derived measurements made with a zetameter using the Helmholtz–Smoluchowski relationship. One could argue that the operationally defined nature of these unique pH points means that pH_{zpc} and pH_{iep} are operationally defined.

The *zpc* and *iep* can also be distinguished according to whether or not specific adsorption of cations or anions is considered. As specifically adsorbed ions are *pdis*, when H^+ and OH^- ions constitute the only potential-determining ions, the pH condition at which the adsorption densities of H^+ and OH^- ions and their complexes are equally balanced is characterized as the pH_{iep}. In terms of Γ, the surface excess concentration, this means that the pH_{iep} is obtained when $|\Gamma_H| = |\Gamma_{OH}|$. This distinguishes it from the situation where adsorbed ions from the porewater contribute to particle surface charges resulting in changes in the potential of the particle. Specifically adsorbed cations will decrease the *pznpc*, whereas specifically adsorbed anions will increase the *pznpc* – as shown by the net proton surface charge density relationship σ_h shown in Figure 3.4. The solid curve in the figure represents the proton balance condition in which H^+ and OH^- ions are the only *pdis*. Note that when the specifically adsorbed ions are cations (*wc*), a lower *pznpc* is obtained (*pznpc*$_{wc}$), and when the specifically adsorbed ions are anions (*wa*), a higher *pznpc* is obtained (*pznpc*$_{wa}$), as shown by the lower and upper dashed lines in the figure. Strictly speaking, as the *zpc* is really the point of zero net charge, pH_{pznc} should be used in place of pH_{zpc}. When $pH_{pznc} = pH_{iep}$, we obtain the *pristine point of zero charge (ppzc)* pH_{ppzc} (Bowden *et al.*, 1980).

3.4.3   Cation exchange and cation exchange capacity

Cation exchange occurs when positively charged ions in the porewater are attracted to the surfaces of the clay fractions because of the net negative charge imbalance of the charged reactive surfaces of the clay particles. This stoichiometric process responds to the need to satisfy electroneutrality in the system, i.e. replacing cations to satisfy the net negative charge imbalance of the charged reactive surfaces of the clay particles. In general, the number

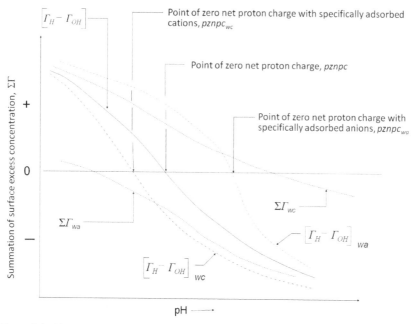

Figure 3.4   Net charge (σ_h) curves as determined by proton balance – with and without specifically adsorbed cations and anions. The solid curve represents the condition with only H^+ and OH^- ions as *pdis*. The subscripts 'wc' and 'wa' refer to with specifically adsorbed cations and with specifically adsorbed anions respectively.

of charged sites considered as exchange sites is determined by isomorphous substitution in the layer lattice structure of the clay minerals. *Exchangeable cations* are those cations associated with the charged sites on the surfaces of clay particles, and the quantity of exchangeable cations held by the clay is called the *cation-exchange capacity (CEC)* of the clay. The CEC is usually expressed as milliequivalents per 100 g of clay (mEq/100 g soil). Table 3.1 gives the range of CECs for the various clay minerals of interest in engineered clay barriers. Although exchangeable cations are generally associated with clay minerals, other clay fractions also contribute to the exchange capacity of a soil, as will be discussed in a later section.

The predominant proportion of exchangeable cations in clays are calcium and magnesium, followed by a smaller proportion of potassium and sodium. Replacement of exchangeable cations with the same positive charge and similar geometries as the replacing cations can be viewed according to the relationship $M_s/N_s = M_o/N_o = 1$, where M and N represent the cation species and the subscripts s and o represent the surface and the bulk solution. For clay fractions with pH-dependent net surface charges, the *CEC* of the clay is a function of the pH of the system. The clay fractions that are included in this list are kaolinites, natural organic matter, and the various oxides or amorphous materials. In kaolinites, for example, the values of CEC can vary by a factor of three between the CEC at pH 4 to pH 9 (Yong and Mulligan, 2004).

A common technique used for measurement of CEC in clays is to use ammonium acetate (NH_4OAc) as the saturation fluid. In theory, cation sorption should occur on all available sites, and one must determine that this occurs without creation of artefacts. Reactions between the saturating cation solution and clay fractions can produce erroneous results, for example the dissolution of $CaCO_3$ and gypsum in carbonate-rich clays when NH_4OAc is used as a saturation fluid. Because variations in measured CEC can occur owing to experimental conditions, the results obtained are sometimes referred to as operationally defined values.

Under a given set of conditions, different cations are not equally replaceable and do not have the same replacing power. The replacing power of some typical ions is shown as a lyotropic series, as follows (Yong, 2001): $Na^+ < Li^+ < K^+ < Rb^+ < Cs^+ < Mg^{2+} < Ca^{2+} < Ba^{2+} < Cu^{2+} < Al^{3+} < Fe^{3+} < Th^{4+}$.

The replacement positions are to a very large extent dependent on the size of the hydrated cation. Changes in the relative positions of the lyotropic series depend on the kind of clay and the ion being replaced. The number of exchangeable cations replaced depends on the concentration of ions in the replacing solution. In heterovalent exchange, the selective preference for monovalent and divalent cations is dependent on the magnitude of the electric potential in the region where the greatest amount of cations is located. When the outside concentration varies, the proportion of each exchangeable cation to the total CEC are determined by exchange–equilibrium equations. Perhaps the most commonly used relationship is the Gapon equation (Yong, 2001):

$$\frac{M_e^{+m}}{N_c^{+n}} = K \frac{\left[M_o^{+m}\right]^{\frac{1}{m}}}{\left[N_o^{+n}\right]^{\frac{1}{n}}} \tag{3.4}$$

where (a) the superscripts m and n refer to the valence of the cations; (b) the subscripts e and o refer to the exchangeable and bulk solution ions; and (c) the constant K is a function of specific cation adsorption and nature of the clay surface. K decreases in value as the surface density of charges increases. Na^+ versus Ca^{2+} represents a particularly important case of exchange competition. When the amount of exchangeable calcium on the clay mineral is decreased, its release becomes more difficult. On the other hand, when the degree of saturation with sodium ions is decreased they become easier to release. Potassium is an exception because its effective ionic diameter of 2.74 Å is about the same as the diameter of the cavity in the oxygen layer. This allows the potassium ion to just fit into one of these cavities – making it very difficult to replace.

For other cations it is the size of the hydrated ions, rather than the size of the non-hydrated ones, that controls their replaceability. Thus, it appears that for ions of equal valence, those that are least hydrated have the greatest energy of replacement and are the most difficult to displace. Li^+, although being a very small ion, is considered to be strongly hydrated and therefore to have a very large hydrated size. The low replacing power of Li^+ and its ready replaceability can be taken as a consequence of the large hydrated size, but there are in fact indications that Li^+ and Na^+ are only weakly hydrated in interlamellar positions.

3.4.4   Anion sorption and exchange

The negatively charged reactive surfaces of the layer silicates are not generally expected to attract anions – with the exception of broken bonds at the edges of the clay particles. When these are available, for example $Al\cdot(OH)H_2O$, anion attraction is somewhat similar to those associated with anion attraction to oxide surfaces. The 1:1 layer silicates, because of their larger proportion of edge–surface areas in comparison with the smectites (2:1 layer silicate), for example, have a much higher capability for attraction of anions. Anion sorption capacity for illites appear to be attributable to their hydrous mineral characteristic. Although almost negligible in the case of smectites, three types of anion exchange may occur in smectites and kaolinite:

- Replacement of OH ions of clay mineral surfaces. The extent of the exchange depends on the accessibility of the OH ions; those within the lattice are naturally not involved.
- Anions that fit the geometry of the clay lattice, such as phosphate and arsenate, may be adsorbed by fitting onto the edges of the silica tetrahedral sheets and growing as extensions of these sheets (Pusch and Karnland, 1988; Pusch, 1993). Other anions, such as sulphate and chloride, do not fit that of the silica tetrahedral sheets because of their geometry and do not become adsorbed.
- Local charge deficiencies may form anion exchange spots on basal plane surfaces.

The last mechanism is considered to contribute to the net anion exchange capacity of smectites. The other two may be important in kaolinite but are assumed to be relatively unimportant in smectite clays. The latter minerals commonly have an anion exchange capacity of 5–10 mEq/100 g but can be considerably higher for very fine-grained kaolinite and palygorskite.

3.4.5 Specific surface area (SSA)

Because of their generally planar shapes, clay mineral particles will exhibit larger surfaces in comparison with other clay-sized non-clay mineral particles that do not have the characteristic planar shape. In theory, if we have information on (a) the shapes and sizes of the individual clay particles and (b) their distribution, it is possible to obtain a numerical calculation of the surface area of a representative elementary volume (REV) of a clay. One could use knowledge of the unit cell of a mineral to determine its representative surface area, as has been undertaken by Greenland and Mott (1985). They used the a and b dimensions of a unit cell for a dioctahedral 2:1 mineral such as a smectite, and the Avogadro's number of these unit cells, to arrive at a calculated SSA of $757\,m^2/g$. It is important to distinguish between surfaces available for interaction with water or other fluids, i.e. exposed surfaces, and non-exposed surfaces. Figure 3.5 shows a highly simplified illustration of the available surfaces of a 2:1 layer silicate in various modes or arrangements. The uncleaved state of the mineral particle exhibits an apparent planar shape. The surface area for this particle is defined by the two basal planes and the sides of the particle. In the uncleaved state, when combined with other particles in a microstructural unit, the total available or exposed surface area becomes less than the sum of all the particles involved

Cleavage or separation of the dioctahedral 2:1 layer silicate mineral into n number of individual unit layers or particles with smaller numbers of unit layers will result in an increase in exposed particle surface areas (see top part of Figure 3.5). In swelling clays, under the right circumstances, apparent cleavage or separation of the layer silicate into individual unit layers can

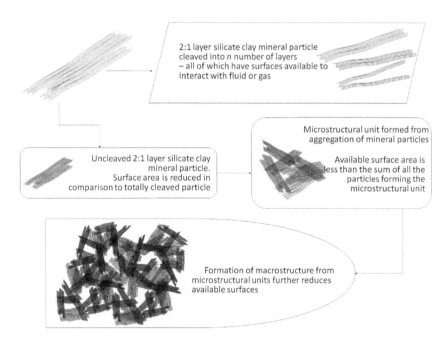

Figure 3.5 Surface areas of 2:1 layer silicate clay mineral particles in various configurations: (a) cleaved into *n* numbers of layers – shown in the top right-hand corner; (b) uncleaved as a single mineral particle – shown in the left centre; (c) aggregation of particles forming a coherent microstructural unit – shown at the right centre; and (d) aggregation of microstructural units forming a macrostructural unit – shown at the bottom.

happen, for example when double-layer swelling is allowed to occur. The footnotes in Table 3.1 highlight this phenomenon and a detailed discussion of this important property is provided later in this chapter. This is one of the characteristics of swelling clays that make them attractive for use as engineered clay barriers and buffers.

Laboratory measurements are generally used to determine the SSA of clays since theoretical calculations are not only tedious, but also unrealistic if the clays contain different clay and non-clay minerals. In the procedure that is commonly used, a gas or liquid is used as the adsorbate for the clay solids. One seeks to determine the amount of adsorbate that forms a monolayer coat on the surfaces of the clay solids (particles). Although the choice of adsorbate is an important factor in the determination of the SSA of a clay sample, one needs to be sure that all the individual clay particles' surfaces are available for interaction with the adsorbate. This means that the clay particles must be in a totally dispersed state. In essence, the availability of clay particles in a totally dispersed state and the choice of adsorbate are the two most important factors in any laboratory measurement of the SSA of a clay sample. This means that laboratory determinations of the SSA of clay and other types of clays will produce operationally defined measurements of SSA.

In earlier periods, nitrogen gas was commonly used as the adsorbate. With this technique, one notes that the number of molecules of nitrogen sorbed by the clay particles will be dependent on the partial pressure of the gas and also on the test temperature. As more than one layer of gas will be sorbed by the clay particles, determination of the amount (volume) of gas equivalent to a sorbed monomolecular layer of gas requires one to use the relationship developed by Brunauer *et al.* (1938). This relationship, known as the BET equation for multilayer sorption, is probably the best known of the relationships used. In more recent times, polar fluids have been used as the adsorbates. The techniques described by Mortland and Kemper (1965) for ethylene glycol and Carter *et al.* (1986) for ethylene glycol-monoethyl ether are commonly used. As all of the techniques require one to obtain a uniform monolayer coating by the adsorbate of all the clay particles, it is easy to see that the values of SSA determined are directly dependent on the procedure used and the analytical technique used to reduce the data obtained. This is why the determination of SSA is known as operationally defined.

3.5   Non-clay minerals

3.5.1 Carbonates

Carbonate minerals found in clays include calcite ($CaCO_3$), magnesite ($MgCO_3$), siderite ($FeCO_3$), dolomite [$CaMg(CO_3)_2$], trona (Na_2CO_3 HCO_3CH_2O), nahcolite ($NaHCO_3$) and soda ($Na_2CO_3C_{10}H_2O$). Of the preceding list, $CaCO_3$ is probably the most common carbonate mineral found in clays. Its influence on the pH of a clay makes it an important constituent in clays, as many of the chemical processes and reactions occurring in the clays are pH sensitive. They retain heavy metals well and exhibit relatively high solubility. Dissolution of calcite in the presence of CO_2 in the clay occurs according to the following relationship:

$$CaCO_3 + CO_2 + H_2O \leftrightarrow Ca^{2+} + 2HCO_3^{-}$$

Carbonates in clays function either as individual particles that connect with other particles, or as coatings on clay particles. In any event, they will alter the SSA of the clay because of aggregation of particles resulting from carbonate connections. In all likelihood, the SSA will be decreased if carbonates are present in the clay–water system in comparison with a carbonate-free clay. The results reported by Quigley *et al.* (1985) regarding the influence of carbonate content on the activity and SSA of a freshwater varved clay showed the aggregating effect of carbonate presence in the clay. Using a *relative activity* parameter (ratio of the plasticity index divided by the SSA) they showed that as the content of carbonate increased from zero to about 10 per cent by weight, the relative activity of the clay decreased dramatically – from just over 0.8 to about 0.37. However, when the carbonate content was increased after 10 per cent to about 25 per cent, the relative activity decrease was very small – from about 0.37 to about 0.34. The *relative activity* a useful index as it combines the influence of the carbonates on the activity of the clay (I_p) with the aggregating effect (by the carbonates) through the use of the SSA. The CEC may or may not be changed, as the CEC of carbonates is relatively small, and its contribution to the overall CEC of the soil may not be effectively felt.

3.5.2 Organic matter

The proportion of organic matter in clays is very small, ranging up to about 5 per cent by weight. Despite the small amounts, organic matter plays an important role in clay stability through formation and maintenance of beneficial clay structure (Hayes and Swift, 1985), improved water retention, and processes associated with contaminant attenuation. It is derived from vegetation and/or animal sources. In the clays of interest, we are concerned with organic matter that is classed as *transformed organics*. These are essentially *amorphous organic matter,* i.e. they show no morphological resemblance to the parent material and do not exhibit properties and characteristics of the parent material. They consist of humic acids, fulvic acids, humins and decayed material such as polysaccharides, lignins and polypeptides. The chemical and biologically mediated reactions involved in obtaining transformed organics (Flaig *et al.*, 1975) include demethylation, and oxidative–reductive and other electron transfer reactions catalysed by enzymes. As a result of the transformation processes, we can expect amorphous organic matter to be rich in aromatic polymers, which, in turn, will have an abundance of functional groups. Determination of amorphous organic matter is performed using alkali treatment procedures on organic matter extracted from the parent clay as shown in Figure 3.6. The standard procedure for determination of humic and fulvic acids, humins, and polysaccharides follows the protocols shown in the right-hand branch of the diagram shown.

According to Yong and Mourato (1988), classification of polysaccharides as a humic substance sometimes poses a dilemma. Polysaccharides are long molecules that are obtained as by-products of microbial metabolism, generally synthesized as a by-product from the breakdown of animal or vegetally derived organics. According to the basic definition of a humic or amorphous substance, polysaccharides do not fall in the class of non-identifiable degraded organics, as these polymeric carbohydrates have a definitive structure.

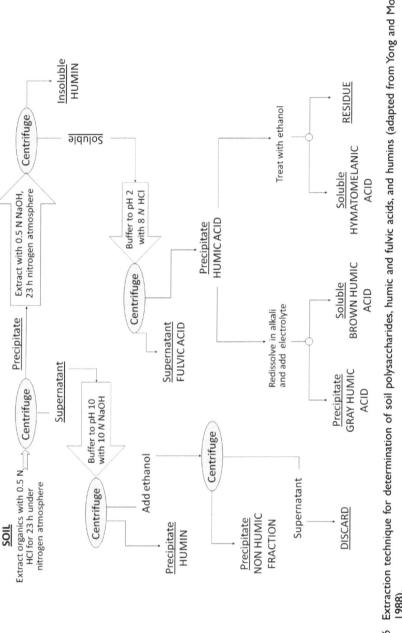

Figure 3.6 Extraction technique for determination of soil polysaccharides, humic and fulvic acids, and humins (adapted from Yong and Mourato, 1988).

3.5.3   Oxides, hydroxides and oxyhydroxides

The literature often uses the term *oxides* to include the group of oxides, hydroxides and oxyhydroxides. Oxides of aluminium are derived primarily from weathering of alumino-silicate minerals. These possess octahedral sheets containing OH^- ions with two-thirds of the positions occupied by Al^{3+} ions. One obtains gibbsite or bayerite, depending on how the OH^- ions in the octahedral sheets are positioned. Gibbsite is obtained when the OH^- ions in each of the octahedral sheets that are stacked on top of each other are directly opposite to each other (i.e. between the stacked sheets) and bonded by hydrogen bonds. Bayerite is obtained when the OH^- ions in each of the octahedral sheets are positioned in the space formed by the OH^- ions in the opposing stacked sheet.

Oxides of aluminium, iron and manganese are more common in natural clays, with iron oxides being the most common. Because of their generally positive surface charge characteristics, amorphous oxides form coatings around clay mineral particles with net negatively charged surfaces. The net result is a change in the charge characteristics of the clay particles, and also a change in the physical, chemical and interaction properties of the clay. It is important to distinguish between the mineral and amorphous forms of oxides, as distribution of the oxides in the clay will depend on whether it is in the mineral or amorphous form.

3.5.4   Surface properties of non-clay minerals

Organic matter

Organic matter exhibits a great variety of surface functional groups. They are organic molecular units that are part of the organic matter. The variety and range of proportions of each of the kinds of functional groups is by and large a function of differences in organic matter composition, i.e. source material, degradation, and extraction and testing procedures. The more common functional groups include hydroxyls, carboxyls, phenolic, alcoholic, ketones and amines. These consist of a combination of carbon and nitrogen with oxygen and/or hydrogen. They control most of the properties of organic molecules and their reactions with other materials in a clay–water system.

The type of charge and surface functional group of organic matter can significantly influence the surface properties of clays through aggregating and coating effects. Yong and Mourato (1990) reported on the interactions and kaolinite structure development with anionic carboxylic and non-ionic hydroxylic *non-humic organics (NHOs)* produced by soil bacteria, showing that:

- development of edge attraction between kaolinite protonated edge hydroxyls and negative functional groups of the NHO occurred, and that IR (infrared) studies showed that the higher-energy free hydroxyls on the edge–surface participated more in complexation than the less energetic ones;
- direct clay lattice interactions with NHO are most probably due to chemisorption, water bridging (H bonding) and ion–dipole interactions.

Oxides

Oxides contribute significantly to the surface properties of clays. The surfaces of oxides of iron, aluminium, manganese, titanium and silicon consist primarily of broken bonds. Typically, interaction of the oxide surfaces with water occurs between the broken (i.e. unsatisfied) bonds on the oxide surfaces and the hydroxyl groups of dissociated water molecules. The surface charges of the amorphous oxides coating the other clay solids are pH dependent. As a result, interactions between the coated clay particles and water will be conditioned by the pH of the medium. When the net surface charge of the amorphous oxide is positive, coating of the net negatively charged clay particle will result in a change in the surface charge of the coated particle. The surfaces of the hydrous oxides (e.g. iron and aluminium) show coordination to hydroxyl groups, which will protonate or deprotonate in accordance with the pH of the surrounding medium. Exposure of the Fe^{3+} and Al^{3+} on the surfaces provides development of Lewis acid sites when single coordination occurs between the Fe^{3+} and the associated H_2O. The model shown in Figure 3.7 is derived from analyses of performance of a sensitive marine clay (Yong *et al.* 1979). The structural organization of the various elements of the amorphous material model is similar to the Cloos *et al.* (1969) model for amorphous silico-aluminas.

The *coating* and *bridging* effect, when the opposite charges of clay particles and amorphous oxides interact, is also shown in Figure 3.7. The diagram shows the amorphous oxide material consisting of a core unit and an outer layer. The core of the amorphous oxide consists primarily of silicon in tetrahedral coordination with some isomorphic substitution of Si with Fe or Al, and is partially coated with Fe or Al in octahedral coordination. The outer layer of the amorphous oxide consists of Fe and Al. This outer layer is destroyed when the clay is exposed to acid leaching. The central core that remains is negatively charged, as shown in the schematic at the bottom of Figure 3.7.

The amphoteric nature of the surface of amorphous materials means that one needs to be aware of the possibility of charge reversal when pH values in the clay changes. Yong and Ohtsubo (1987) have shown that because of the pH-dependent charge of amorphous iron oxides, the sequence by which the amorphous material is exposed to clay mineral particles will determine

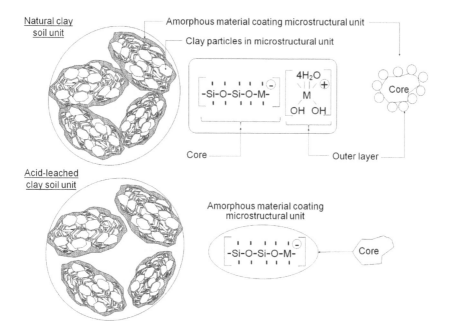

Figure 3.7 A natural sensitive marine clay with amorphous material coating the microstructural units of clay particles. The top of the diagram shows the natural clay structure and the bottom schematic shows the clay structure after acid leaching (adapted from Yong *et al.*, 1979).

whether effective coating of the clay mineral particles may or may not be obtained. Using kaolinite as the test specimen, they showed that when amorphous iron oxide was introduced to the clay particles at pH 3, attraction between the net negative surface charge of the kaolinite and the net positive surface charge of the amorphous material allowed for effective coating of the clay particles. However, when the amorphous iron oxide was introduced to the clay particles at pH 9, as the net surface charge of the kaolinite remained negative and that of the amorphous iron oxide was negative, little, if any, coating of the kaolinite particles was achieved. Because of the amphoteric nature of the surface of amorphous oxide material, reversal of the sign of the surface charges occurs as one progresses from below the pH_{iep} to pH levels above the pH_{iep}.

3.6   Clay macro- and microstructure

3.6.1   *Clay fabric and clay structure*

The influence of clay structure on the properties and performance of clays in general has been the subject of study by many researchers over the many years for different purposes. With respect to clay and geotechnical engineering, early studies on clay structure in geotechnical engineering, for example, provided us with descriptions of flocculent, honeycomb and 'cardhouse' structures (Terzaghi and Peck, 1948). Studies such as those reported by Lambe (1953, 1958), Pusch (1966) and Yong and Warkentin (1966) to a very large extent paid attention to the contributions made by the different clay minerals in combination with other clay fractions on the engineering properties and performance of these clays. Arising from these and later studies, it became abundantly evident that:

- clay structure plays a significant role in the establishment of clay properties and its behaviour;
- there are at least two levels of detail of clay structure (macrostructure and microstructure) that could be studied to provide one with the information needed to better understand clay properties and behaviour, depending on the type of clay and its application.

There is no universally accepted agreement on the concepts and definitions of the macrostructure and microstructure of clays. There is common agreement, however, on the concept and treatment of the structure of soils such as sands and maybe even silts. For gravity-controlled systems, theories of particle-packing (e.g. Deresiewicz, 1958) and the application of particle contact mechanics models allow us to develop detailed analyses for granular soil strength and stability. In the case of clays, conventional soil mechanics theories permit one to apply deterministic analytical continuum mechanics theories to determine stability and even permeability of the clays. Assumptions of elastoplastic, semiplastic, hypoplastic, hyperplastic and rigid plastic types of behaviour allow one to describe the load deformation and rheological behaviour of clays. For many clay-engineering applications, consideration of the clay as a continuum is appropriate, for example in (a) laboratory and field testing procedures that are designed to measure and/or determine the macroscopic performance of the clay; (b) laboratory and/or field data reduction techniques that are relying on data-reduction models developed for uniform and homogeneous media; and (c) analytical tools used for assessment of clay performance based on continuum mechanics.

Difficulties can arise when analysis, and especially prediction, of the response behaviour of clays to situations when particle interactions are not controlled by gravitational forces, and when the integrity of the total clay–water system is determined by actions and reactions at the molecular level. The two following points are significant and must be considered:

1   Clays are not homogeneous continua; they are heterogeneous and structured, requiring consideration of the structure of the clay – the case of sensitive marine clays is a good example.

2   The behaviour of clays responds to both gravitational and molecular forces, and hence requires analyses that include intermolecular forces in the overall analytical package; this is particularly true for clays with active clay minerals, i.e. clay minerals with active and reactive particle surfaces, such as smectites.

The study of *clay behaviour*, which began in earnest in the 1950s, showed the significance of the role of microstructure in the control of properties, behaviour and performance. In general, several types of particle structural units exist in a clay. These range from individual mineral particles to *microstructural units (msu)*. These microstructural units are known as flocs, clusters, aggregate groups or peds, depending on one's perception and background discipline. They consist of aggregations of particles formed into *msu* and are the backbone of the macrostructure of clays, as illustrated in the layer-silicate mineral particle system shown in Figure 3.8. As particle sizes are in the micron range, and as the SSAs of these particles and structural units are in the order of ten to hundreds of square metres per gram of clay, it is evident that intermolecular forces will dominate the interactions between these particles and microstructural units, as seen earlier in the sections 'Clay minerals' and 'Surface properties of clay minerals' regarding layer–lattice structural sheets identified as *stacked layers*. The microstructure of a clay is

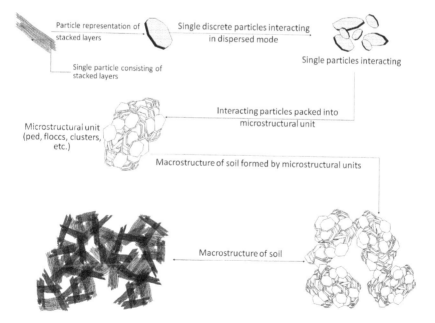

Figure 3.8 Hierarchic representation of soil structure beginning with interacting discrete particles and progressing from there to microstructural units and macrostructure of clay.

seen to be defined by the properties of individual microstructural units and the manner in which they are bonded and interact.

As we have seen from the preceding sections, the reactive nature of the surfaces of clay particles sets the stage for interparticle action that requires understanding of the microstructure of clays. A common appreciation of the microstructure of clays exists, which includes the geometry and properties of the particles and voids in the description or characterization of the microstructure of a clay. Some investigators also include the interparticle force fields in this concept, implying that they are determinants of the particle arrangement. To distinguish between the geometrical interpretation of clay structure and the one which includes the interparticle forces, the terms *clay fabric* and *clay structure* have been used, with the former referring to the geometrical concept and the latter including the interactions between particles. In other words, *clay fabric* refers to the geometrical arrangement of the particles constituting the *msu* and the distribution (arrangement) of the *msu* to form the macrostructure of the clay. Clay structure refers not only to these *msu*, but also includes the various interparticle forces that maintain the *msu* and their arrangements.

3.6.2   *Structure-forming factors and interparticle bonds*

The macrostructure of clay is a function of the assemblage of microstructural units, as illustrated in Figure 3.8. It is understood that the term *microstructures* means *msu*. It follows that the physical integrity of a clay is directly related to the nature and distribution of the microstructures, and the bonding and interaction forces not only between the clay particles within the microstructures, but also between microstructures themselves. There are two groups of bonds that exist between particles in microstructures and between microstructures themselves, both of which are responsible for the development of the properties and characteristics of the microstructures and the macrostructure itself. The first group of forces and bonds deal directly with particle–particle interaction. The second group of forces and bonds are those that are developed between particles with mediating forces and bonds from the solutes in water and the water molecules themselves. The discussion for the first group is given herein. We will discuss the second group of forces in greater detail when we deal with clay–water interactions later in this chapter, and also when we discuss water movement in partly saturated clays in Chapter 4, and contaminant–clay interactions in Chapter 5.

Short-range forces, such as those developed as a function of ion–dipole interaction, dipole–dipole interaction and dipole–particle site interaction, are of considerable importance in the arrangement of particles in a microstructure and more complete discussion of these forces (listed below) will be found in the next subsection.

- *Primary valence bonds between particles*. Except for cementation bonds, primary valence bonds are the strongest interparticle attraction forces; the activation energy for breakage normally exceeds 125.5 kJ/mol. They are dominant in heavily consolidated clays, in which the crystal lattices of adjacent particles are in contact. For smectites this state requires an effective pressure of about 200 MPa.

- *London–van der Waals forces*. These forces are of importance because they operate in the range from 0.2 nm to more than 10 nm, regardless of whether the particles are charged or uncharged. The bond strength is less than 4.184 kJ/mol.

- *Hydrogen bonds*. Hydrogen bonds are weak (4.184–12.55 kJ/mol) but their number can be large and the net attraction force will thus be high. In principle, the coupling between adjacent smectite particles for example could be through water-bridging bonds.

- *Bonds by sorbed cations*. These are coulombic bonds. The integrated attraction force between neighbouring but not contacting particles is significant. The approximate activation energy is intermediate between those of hydrogen bonds and primary valence bonds, i.e. of the order of 41.84–62.76 kJ/mol.

- *Dipole-type attraction between particles with different charge*. Depending on the equilibrium or near-equilibrium pH of the immediate microenvironment, the edges of clay particles can be positively or negatively charged.

- *Bonds by organic matter*. The attraction forces are primarily due to hydrogen bonds and purely physical coupling is obtained by embracing hyphae and flagellae. The bonds are flexible and can sustain large strain. However, their strength is very much dependent on the environment and can be short lived.

- *Cementation bonds*. Precipitated matter binding particles together can develop bonds with strengths that can approach that of primary valence bonds (> 125.5 kJ/mol). Their practical importance can be substantial, depending on the amount and nature of the precipitation, as shown, for example, by the difference between plastic clay and claystone.

3.6.3   Electrified interface and interactions

The nature of the distribution of the various ions in the porewater of a clay–water system is conditioned by the various reactions between these ions and the charged surfaces. These reactions are both electrostatic and chemical. Although some minor differences in details concerning the nature and distribution of the various ions adjacent to the reactive surfaces appear in the literature (e.g. Stern, 1924; Grahame, 1947; Kruyt, 1952; Sposito, 1984; Greenland and Mott, 1985; Ritchie and Sposito, 2002), there is general agreement on (a) the *altered* or *structured* water layers adjacent to the

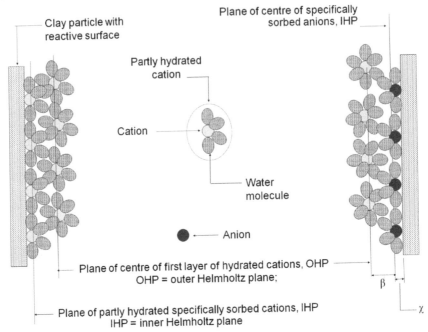

Figure 3.9 Specifically sorbed ions, IHP and OHP. Left-hand diagram shows partly hydrated specifically sorbed cations and fully hydrated cations. Right-hand drawing shows anions as specifically sorbed anions and hydrated cations. The specifically sorbed ions are potential-determining ions (*pdis*).

reactive surfaces and (b) the swarm or cloud of counterions forming a diffuse layer of ions. These interactions are portrayed in Figure 3.9.

The figure shows that the partly hydrated cations and anions in the inner Helmholtz plane are potential-determining ions (*pdis*) and are bonded to the reactive surfaces by ionic and covalent bonds. These *pdis* contribute directly to the charge and potential on the surfaces of reactive particles. The Stern layer, which includes the inner Helmholtz plane, *ihp*, and the outer Helmholtz plane, *ohp*, according to the Grahame (1947) model, is shown in Figure 3.10. The structured water adjacent to the surface of the particle is due to the specifically adsorbed ions at the interface.

According to Ritchie and Sposito (2002), the complexes formed between the surface functional groups and the ions at the electrified interface are inner sphere complexes – assuming direct contact between them without interruption from any water molecule. When a layer of water molecules interrupts contact between these, the complexes are classified as outer sphere complexes. The *ihp* and *ohp* shown in the figures identify the positions of the inner sphere and outer sphere complexes with corresponding distances of χ and β respectively. The thickness of the Stern layer is obtained as $\delta = \chi + \beta$. Electrostatic bonding mechanisms for the counterions beyond the Stern layer

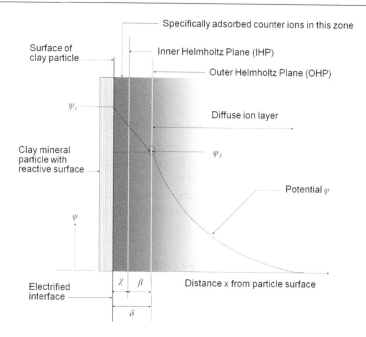

Figure 3.10 Generalized view of the electrified interface with aqueous (porewater) solution containing dissolved solutes – using the Grahame (1947) view of the Stern layer.

required to satisfy the net negative charge of the reactive particles result in the formation of a diffuse ion layer.

The surface potential ψ_s at the surface of the clay particle varies with electrolyte concentration and the nature of the charge of the clay particle, i.e. whether the mineral particle has a constant charge surface or a pH-dependent charge surface. The surface potential drops from ψ_s at the surface to a potential ψ_δ at the *ohp* – as shown in Figure 3.10. The potential ψ (electric potential) beyond the *ohp* can be described by the Gouy–Chapman diffuse double-layer model – as will be seen in the next section. Electrokinetically, ψ_δ is considered to be equal (or almost equal) to the zeta potential ζ.

The coulombic interaction energy E_{ihp} established between the ions of i and j can be calculated as (Yong, 2001):

$$E_{ihp} = \frac{z_i z_j e^2}{4 \pi \varepsilon r} \tag{3.5}$$

where z_i and z_j are the valencies of the i and j species of ion in the bulk solution, e is the electronic charge, ε is the dielectric constant, and r is the distance between the centre of the ith ion and jth ion.

Taking r as the distance between the centre of a dipole and the corresponding ion, the ion–dipole interaction E_{id} is given as follows:

Ion–dipole interaction E_{id} given as: $E_{id} = -\dfrac{\mu \, ze \cos \theta}{4\pi\varepsilon \, r^2}$

Dipole–dipole interaction E_{dd} is given as follows: $E_{dd} = -\dfrac{\mu_1 \mu_2}{4\pi\varepsilon \, r^3} D$

where μ is the dipole moment and D is a function of angles of the dipoles.

Where do all of these energy terms fit in? As will be seen in Figure 3.11 in the next section, these interaction energies need to be incorporated into calculations for pressures developed in swelling clays due to interparticle action.

3.6.4   London–van der Waals energy and total intermolecular pair potential

There is another important force between ions and between molecules which is called London–van der Waals force (potential or energy). This force (potential) is sometimes called *London dispersion forces* or simply *van der Waals forces*. We use the term *van der Waals force (potential)* in this book. This force, which operates between all molecules regardless of the electrostatic conditions of molecules, exceeds the dipole-dependent forces. The van der Waals interaction energy E_{vdw} is given by $E_{vdw} = -c_i c_j / r^6$, where r is the distance between molecules, and c_i and c_j are the London dispersion force constants of molecules i and j. These consist mainly of the ionization potential, dielectric constant and polarizability.

The repulsion force (potential) that operates between a pair of molecules can be determined or expressed in two different ways: (a) using an empirical form of hard sphere potential or power law potential and (b) using the exponential potential established on the basis of quantum mechanics. The power law potential E_{rp} is described by $E_{rp} = A/r^n$, where r is the distance between the molecules, A is coefficient and a commonly adopted value of $n = 12$. The exponential potential is given as follows: $E_{rp} = b\exp(-cr)$, where b and c are coefficients.

The total intermolecular pair potential E_{tip} used for uncharged colloidal particles is commonly described as follows:

$$E_{tip} = \frac{A}{r^{12}} - \frac{c_i c_j}{r^6} \qquad (3.6)$$

This type of the total intermolecular pair potential is called the

Lennard–Jones form or 12–6 potential. In a clay–water system, because we deal with charged ions and charged clay minerals, it is proper to use the following relationship as the total intermolecular pair potential:

$$E_{tip} = \frac{z_i z_j e^2}{4\pi\varepsilon r} - \frac{c_i c_j}{r^6} + b\exp(-cr) \tag{3.7}$$

This intermolecular pair potential will describe the mode of ion desorption on clay minerals with proper constants, as will be described in Chapter 5.

3.7   Clay–water interaction, water uptake and swelling

3.7.1   Swelling clays

Swelling clays are the most commonly used clays for engineered clay barriers and buffers. The unit layer structure of smectites and the various features that render some of them as swelling clays (e.g. montmorillonites, beidelites and nontronites) have been discussed in the earlier sections of this chapter. A significant feature in swelling clays can be seen in the water uptake characteristics of the 2:1 unit layer structure of montmorillonites saturated with different exchangeable cations. The basal spacing $d(001)$ of 0.95–1.0 nm for the anhydrous smectite will expand from 1.25 to 1.92 nm depending on the amount of water intake (hydration). Except for Li and Na as exchangeable cations in the interlayer, basal spacing expansion for montmorillonites containing other exchangeable cations appear to reach a maximum value of about 1.92 nm, which is about the thickness of four layers of water.

The size of a unit particle of montmorillonite varies according to the exchangeable cation in the interlayer. The number of unit layers in face–face orientation for montmorillonite ranges from a single-unit layer for Li and Na as exchangeable cations, up to between 6 and 16 for Ca as the exchangeable cation, depending upon the technique used for determination of unit-layer stacking (Sposito, 1984). An important characteristic of swelling clays is the manner in which water uptake is achieved. Depending on the nature of the exchangeable cations in the interlayer spaces and the initial water content of the swelling, hydration processes will dominate initial exposure to water or water vapour. Water uptake or sorption will be an interlayer phenomenon. If the exchangeable cations are Li or Na, continued water uptake after hydration will be due to double-layer forces. This phenomenon of uptake and swelling due to double-layer forces will be discussed in the next subsection when we deal with double-layer models. The hydration shell surrounding small monovalent cations consists of about six water molecules for dilute solutions (Sposito, 1984), reducing to about three water molecules for concentrated solutions. No secondary hydration or solvation shells are associated with added water intake. Water uptake beyond interlayer separation distances of about 1.2 nm occurs because of double-layer swelling forces, resulting in the formation of a solution containing dispersed single-unit layers. Basal spacings at 100 per cent humidity level for montmorillonite have been reported by Suquet *et al.* (1975) and by Quirk (1968) for lithium and sodium montmorillonites to be > 4.0 nm. Mooney *et al.* (1952) indicate that the 1.24–1.25 nm spacing corresponds to one monolayer of water in the interlayer region, and that basal spacings of 1.5, 1.9 and 2.2 nm correspond to two, three and four layers of water between each alumino-silicate layer. The volume change swelling in the interlayer between 1.0 and 2.2 nm should be identified as *crystalline swelling* only if there exists a definite hydration structure to the water. Table 3.2 shows the basal spacings in nanometres for montmorillonite–water complexes saturated with various cations at 25°C and equilibrated at different relative humidities (p/p_o).

When divalent cations constitute the interlayer cations, both primary and secondary hydration or solvation shells are obtainable – moving together as a solvation complex. There are from six to eight molecules in the primary shell, and about 15 water molecules in the secondary shell. In dilute suspensions, the homoionic forms of sodium montmorillonite, which are unit layer particles, are different from calcium montmorillonite particles, which consist of from six to eight unit layers stacked in face–face array (Sposito, 1984). This is in accord with the observations reported by Farmer (1978), who indicated that most lithium and sodium smectites swell in dilute solution or in water into gel-like state with average interlayer separation in proportion to $1/\sqrt{c}$, where c is the electrolyte concentration in the liquid phase. Initial water uptake by nearly anhydrous clays is strongly exothermic, with the water being firmly held in the coordination sphere of the cation and in

Table 3.2 Basal spacings in nanometres for montmorillonite–water complexes saturated with various cations at 25°C and equilibrated at different relative humidities (p/p_o)

Exchangeable cation	p/p_o			
	0	0.5	0.7	1
Li	0.95	1.24	d	M
Na	0.95	1.24	1.51	M
NH$_4$	1	d	d	1.5
K	1	1.24	d	1.5
Cs	1.2	1.28	1.28	1.38
Mg	0.95	1.43	d	1.92
Ca	0.95	1.5	1.5	1.89
Ba	0.98	1.26	1.62	1.89

d, diffuse reflections; M, macroscopic swelling.

contact with the surface oxygens. The apparent interlayer separation relationship with c and the fact that the interlayer space contains no discernible hydration states, allows us to separate crystalline swelling and double-layer swelling phenomena. This means to say that expansion beyond four layers of water is most likely due to osmotic forces, resulting in dilution of the ionic concentration in the interlamellae. One can conclude that the amount of water uptake and its distribution will be determined by (a) the nature of the clay mineral; (b) the factors associated with either interlayer or interparticle phenomena; and (c) the microstructure of the clay. Thus, for example, with Li and Na as the exchangeable ions, the basal spacings reach 1.24 nm at a relative humidity of 0.5, whereas with Mg and Ca as exchangeable ions, the basal spacings reach 1.43 and 1.5 nm, respectively, for the same relative humidity. Nuclear magnetic resonance ('magic angle spin-echo') studies show an obvious difference in the coordination of the silicons in sodium and calcium montmorillonite.

From a practical point of view, the different physical nature of the interlamellar or interlayer water is believed to be because the stacks of layers in calcium montmorillonite (i.e. the clay particles) are much stronger and less easily disrupted than in sodium montmorillonite. In both types, the interlayer water is believed to be more viscous than ordinary water and to be largely immobile under normal hydraulic gradients. The interlayer space offers large amounts of hydration sites due to the crystal lattice constitution. This determines the charge and coordination of adsorbed cations and water molecules. The hydration properties control the swelling potential, plasticity and rheological behaviour, and are hence of fundamental importance.

The coordination of interlamellar cations, the crystal lattice and the water molecules are strongly dependent on the size and charge of the cations, and on the charge distribution in the lattice, as discussed previously. When submerged in free water, the hydration of an initially dry montmorillonite crystallite proceeds until the maximum number of hydration layers is formed, provided that there is no geometrical restraint. In humid air, as shown in Table 3.2, the number of hydration layers depends on the relative humidity, as a balance exists between the water molecule concentrations in the moist air and on the basal surfaces of the crystallites as well as between the hydration stages of the basal surfaces and the interlayer space.

Spin-echo proton measurements allow for a distinction between inter- and extralayer water. These measurements suggest that the relaxation time T_2 is around 20–40 μs for protons in interlamellar water in montmorillonite compared with 2.3 s in free water (Pusch, 1993). This obvious difference indicates strong structuring and very limited mobility of interlayer water. The small difference in proton mobility of one and three hydration layers indicates that the interlayer water possesses approximately the same physical properties irrespective of the number of hydration layers.

3.7.2   Molecular dynamics simulation for the study of interlayer water

Another way to study interlayer phenomena and to better understand the nature of interlayer water is to use molecular dynamics simulations on a clay–water system. Molecular dynamics simulations allow one to gather information on molecular positions and velocities at the microscopic level, and in the case of a clay–water system this includes the clay mineral with its multiple interlayer water molecules. The simulation consists of solution of the Newtonian motion equation for the atoms in the system to obtain their time-dependent behaviour. By introducing proper relationships between macroscopic properties and the position and velocity of all the atoms in the system, we can deduce the various macroscopic properties of a clay–water system such as free energy, heat capacity and adsorption sites, as well as density and viscosity. When solving the Newtonian equation, it is necessary to have knowledge of the forces acting for all the atoms. These forces are assumed to be derived from the potential energies of interaction, thus making it very important to decide the proper form of interaction potential energies for all atoms. The interactions associated with certain water molecules in a clay–water system are assumed to come from all atoms around the water molecule, i.e. from atoms associated with the clay minerals and from the surrounding water molecules.

In general, the interaction energy acting between two atoms is called a *pair interaction potential energy*. The proper pair interaction potential energy in water systems proposed by Kawamura (1992) and Kumagai *et al.* (1994) has been applied to a clay–water system (Ichikawa *et al.*, 1999; Nakano and Kawamura, 2006). The total interaction energy for a certain atom in a clay–water system is given as the sum of the pair interaction potential energy that consists of the five components: (a) Coulomb, van der Waals; (b) the short-range repulsion energy; (c) the radial covalent bond energy between O–H, Si–O and Al–O; (d) the angular covalent bond for H–O–H bond of water molecule; and (e) the Si–O–Si bond of clay minerals.

The interaction potential energy for a pair of atoms is given as follows:

$$u_{ij}(r_{ij}) = \frac{z_i z_j e^2}{4\pi\varepsilon_0 r_{ij}} - \frac{c_i c_j}{r_{ij}^6} + f_0(b_i + b_j)\exp\left(\frac{a_i + a_j - r_{ij}}{b_i + b_j}\right) \tag{3.8}$$
$$+ D_{1ij}\exp(-\beta_{1ij}r_{ij}) + D_{2ij}\exp(-\beta_{2ij}r_{ij}) + D_{3ij}\exp[-\beta_{3ij}(r_{ij} - r_{3ij})^2]$$

where r_{ij} is the distance between two atoms, ε_0 is the dielectric constant of vacuum, z is ionic valence, e is the elementary electric charge, f_0 is a constant for unit adaptations between each term (6.9511×10^{-11} N), a, b and c are parameters defined for each atom, D_1, D_2 and D_3 are parameters establishing the magnitude of each potential, r_{3ij} is the threshold distance activating the dissociation of the OH bond, and β_1, β_2 and β_3 are parameters. The first term represents the Coulomb potential, the second is van der Waals potential, and the third is short-range repulsion potential. The fourth, fifth and sixth terms represent each component of the radial covalent bond potential. The fourth and fifth are repulsive and attractive terms respectively. The sixth term is introduced only to stabilize the covalent bond of OH.

The angular covalent bond potential energy for H–O–H bond of water molecule and Si–O–Si bond of clay minerals is given as follows:

$$u_{jik}(\theta_{jij}, r_{ij}, r_{ik}) = -f_k \left\{ \cos[2(\theta_{jij} - \theta_o)] - 1 \right\} \sqrt{k_1 k_2}$$ (3. 9)

$$k_l = \frac{1}{\exp[g_r(r_{ijl} - r_m)] + 1}$$ (3.10)

where θ_{jij} represents the angle of H_w–O_w–H_w and Si–O_c–Si, θ_0 is the adjustment parameter for determining the angles of H_w–O_w–H_w or Si–O_c–Si, k_1 and k_2 are introduced to provide the effective range of the three-body potential, subscript 1 and 2 imply each of two H_w or two Si atoms, r_{ijl} represents the interatomic distance between O–H or O–Si, and r_m is an adjustment parameter for determining the bond length of O–H or O–Si, g_r is an adjustment parameter for determining the magnitude of the effective range of the three-body potential and f_k is a parameter.

In applying potential energy to molecular dynamics simulations, the proper values of the parameters must be determined. These parameters have been considered to differ slightly with the components of atoms in minerals. They have to be decided for each minerals, using, for example, trial and error methods that consider the physical and chemical characteristics of minerals. The values required for caesium beidelite have been determined by Nakano and Kawamura (2006), as shown in Table 3.3.

Interlayer water

Interlayer water is assumed to have different physical properties from those of bulk water, especially at locations near the surface of clay minerals because clay minerals and water molecules interact strongly. The interaction energies involved include (a) Coulomb; (b) van der Waals; (c) short-range repulsion; and (d) covalent bond energy acting between atoms, i.e. O, H, Si, Al and metals of clay minerals as well as O and H of interlayer water. The results shown in Figure 3.11, obtained with the help of molecular dynamics simulations, describe the density, viscosity and self-diffusion coefficient of interlayer water at a distance from surface of clay minerals. At the surface of

Table 3.3 Interaction potential parameters

Atom	w (10^3 kg/mol)	z (e)	a (nm)	b (nm)	c [(kJ/mol)$^{0.5}$/nm^3]
O_w	16.00	−0.92	0.1728	0.01275	0.05606
H_w	1.01	0.46	0.0035	0.004400	0
O_c	16.00	−1.125278	0.1868	0.01510	0.05524
H_c	1.01	0.46	0.0074	0.00320	0
Si	28.09	2.10	0.0987	0.00830	0
Al	26.98	1.95	0.1089	0.00880	0
Na	22.99	1.00	0.1314	0.0115	0.01637
Ca	40.08	2.00	0.1494	0.0094	0.01228
Cs	132.90	1.00	0.1884	0.01300	0.04501

Atom–atom	D_1 (kJ/mol)	β_1 (nm^{-1})	D_2 (kJ/mol)	β_2 (nm^{-1})	D_3 (kJ/mol)	β_3 (nm^{-1})	r_3 (nm)
O_w–H_w	57394.9	74.0	−2189.3	31.3	34.74	128.0	0.1283
Si–O_c	205951.2	50.0	−13734.3	22.4	0	0	0
Al–O_c	151533.2	50.0	−8104.1	22.4	0	0	0
H_c–O_c	57394.9	74.0	−3277.6	31.3	34.74	128.0	0.1283

Atom–atom–atom	f_k (10^{-19} J)	θ_0 (degrees)	r_m (nm)	g_r (nm^{-1})
H_w–O_w–H_w	1.15	99.5	0.143	92.0
Si–O_c–Si	0.61	120.0	0.177	168

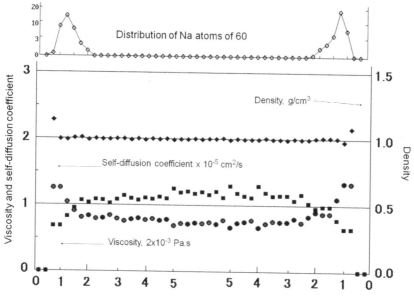

Figure 3.11   Density, viscosity, self-diffusion coefficient and distribution of Na in interlayer water of Na-adsorbed beidellite $[Na_{0.5}Al_2(OH)_2(Si_{3.5}Al_{0.5}O_{10})]$ that have been calculated using molecular dynamics simulations. Simulations have been performed for a clay–water system that includes a half layer of beidellite at both sides of the interlayer water, of about 10.7 nm in thickness, in a box of $51.9 \times 53.8 \times 116.8$ Å. The total number of atoms was 32,460, the number of H_2O molecules was 10,000 and the number of the adsorbed Na atoms required to maintain electric neutralization of the system was 60. The number of O, Si, Al and H constituting a clay mineral were 1440, 420, 300 and 240, respectively (offered by Professor Katsuyuki Kawamura, Tokyo Institute of Technology).

clay mineral, the density and the viscosity are larger. The self-diffusion coefficient is smaller than those located about 2.0 nm away. Beyond this distance, the coefficients have the same values as those of bulk water. No adsorbed Na atoms are found beyond the 2.0 nm distance. One concludes that the thickness of the electric diffuse double layer formed by the adsorbed metals will be about 2.0 nm. These results shown in Figure 3.11 have been calculated on the basis of the clay mineral beidellite for the case where the thickness of the interlayer is 10.7 nm. One deduces from the results shown that the flow of bulk water will occur in about 70 per cent of the interlayer space, with the rest being taken up by the diffuse double layers at both sides of the interlayer space. The thickness of diffuse double layer will depend on the clay minerals involved, as the chemical components in minerals differ with each clay.

3.7.3 Surface complexation – diffuse double layer and models

Diffuse double-layer models

The potential ψ shown in Figure 3.10 is a characteristic of the interaction energy of the diffuse ions in interparticle space. Evaluation of the complexes formed between the surface functional groups of the clay particles and the ions in the porewater can be performed using a variety of surface complexation models such as the single-, double-and triple-layer models (e.g. Kruyt, 1952; Bockriss and Reddy, 1970; Singh and Uehara, 1986; Yong, 2001). The Gouy–Chapman diffuse double layer model, which is perhaps the most familiar model, allows one to compute the potential ψ shown in Figure 3.10 as a function of the distance *d* from the charged particle surface. Because of chemical bonding processes and complexation in the δ region shown in the figure, simple electrostatic interaction calculations do not apply in this region. The following assumptions and conditions for determination of ψ apply:

- Interaction between the charged ions and solutes (cations and anions) in solution and the charged (clay) particle surfaces are Coulombic in nature. These interactions are given in terms of the potential ψ and are described by the Poisson relationship with respect to variation of ψ with distance *x* away from the particle surface (Figure 3.10).
- The ions in solution are considered to be point-like in nature, i.e. zero-volume condition.
- The density of the charges ρ due to the assumed point-like ions that contribute to the interactions (i.e. space charge density) can be described by the Boltzmann distribution.

With these conditions, one can obtain the Poisson–Boltzmann relationship for ψ as follows:

$$\frac{d^2\psi}{dx^2} = -\frac{4\pi}{\varepsilon}\sum_i n_i z_i e \exp\left(\frac{-z_i e\psi}{\kappa T}\right) \qquad (3.11)$$

where n_i and z_i are the concentration and valency of the *i*th species of ion in the bulk solution, and *e*, κ, ε and *T* represent the electronic charge, Boltzmann constant, dielectric constant and temperature, respectively. The relationship for ψ can be obtained (e.g. Kruyt, 1952; van Olphen, 1977; Yong, 2001) as follows:

$$\psi = -\frac{2\kappa T}{e} \ln \coth \left(\frac{d}{2} \sqrt{\frac{8\pi e^2 z_i^2 n_i}{\varepsilon \kappa T}} \right) \tag{3.12}$$

The relationship between the surface charge density σ_s and surface potential ψ_s is:

$$\sigma_s = \left(\frac{2n_i \varepsilon \kappa T}{\pi} \right)^{\frac{1}{2}} \sinh \frac{z_i e}{2\kappa T} \psi_s \tag{3.13}$$

The application of DDL models for computation of swelling pressures requires several conditions to be met if good agreement between computed and actual measured values is to be obtained. These are discussed in the section 'Swelling pressure measurements and calculations', below.

DLVO model

The DLVO (Derjaguin, Landau, Verwey and Overbeek) model, which is an interaction energy model, takes into account the nature of the charged (clay) particle surfaces, the chemical composition of the clay–water system, and the clay fabric (particle arrangement and particle separation distances) in its calculation of the interparticle or interaggregate forces. The calculations include van der Waals attraction and the DDL repulsion developed in the diffuse ion layer as the primary factors in the development of the energies of interaction developed between the particles. The particle interaction models reported by Flegmann et al. (1969) are used as the basis for calculation of the maximum energies of interparticle action for similar charged surfaces in face–face and edge–edge particle arrangement. In the case of dissimilar charged surfaces (edge–face), the relationship given by Hogg et al. (1966) is used.

Assuming (a) constant surface potential surfaces and (b) potential determining ion influence on surface potentials, the energy of repulsion between interacting parallel-faced particles, E_r^{ff}, is obtained as follows:

$$E_r^{ff} = \frac{4n_i \kappa T (yz_i)^2 \exp \left(\frac{-D_H x}{2} \right)}{D_H \left[1 + \exp(-D_H x) \right]} \tag{3.14}$$

where y is a dimensionless potential, $y = -(\varepsilon\psi/\kappa T)$ and D_H is the Debye–Hueckle reciprocal length. The energy of attraction between interacting parallel-faced particles, E_a^{ff}, utilizes the London–van der Waals attraction energy for two similar flat plates, and is obtained as follows:

$$E_a^{ff} = \frac{A}{12\pi x^2} \tag{3.15}$$

where A is the Hamaker constant. To calculate the face–edge repulsion and attraction energies, two interacting spheres are adopted as the model – with one sphere having a very large radius a_f with potential ψ_f in comparison with the other with a much small radius a_e with its corresponding potential ψ_e. By using a sphere with a very large radius a_f in comparison with the other interacting sphere, it is assumed that this can take the place of a flat particle. The energies of face–edge repulsion E_r^{fe} and attraction E_a^{fe} are given as follows:

$$E_r^{fe} = \frac{\varepsilon a_f a_e \left(\psi_f^2 + \psi_e^2\right)}{4(a_f + a_e)} \left[\frac{2\psi_f \psi_e}{\psi_f^2 + \psi_e^2} \ln \frac{1+\exp(-D_H x)}{1-\exp(-D_H x)} \right.$$

$$\left. + \ln\left[1-\exp(-2D_H x)\right] \right. \tag{3.16}$$

$$E_a^{fe} = \frac{-A}{12} \left(\frac{r_a}{x_a^2 + x_a r_a + x_a} + \frac{r_a}{x_a^2 + x_a r_a + r_a} + 2\ln\frac{x_a^2 + x_a r_a + x_a}{x_a^2 + x_a r_a + x_a + r_a} \right) \tag{3.17}$$

where $x_a = x/(2a_f)$ and $r_a = a_e/a_f$.

The calculation of the edge–edge repulsion and attraction energies assumes small interacting spheres with identical radii of a_e. The energies of repulsion E_r^{ee} and attraction E_a^{ee} are obtained as follows:

$$E_r^{ee} = \frac{\varepsilon a_e \psi_e^2}{2} \ln\left[1+\exp(-D_H x)\right] \tag{3.18}$$

$$E_a^{ee} = -\frac{A}{12} \left(\frac{1}{x_a^2 + 2x_a} + \frac{1}{x_a^2 + 2x_a + 1} + 2\ln\frac{x_a^2 + 2x_a}{x_a^2 + 2x_a + 1} \right) \tag{3.19}$$

3.7.4   Swelling pressure measurements and calculations

There have been considerable studies in previous years focused on determination of the level of agreement between diffuse double layer DDL model calculations and laboratory-measured swelling pressures of swelling soils. Agreement between measured and calculated values was reached when laboratory procedures/tests conformed to DDL model conditions for calculations. Some of the conditions necessary to obtain good agreement include:

• *laboratory experiments* – homoionic montmorillonite free from impurities; porewater solution that consists of monovalent cations; homogeneous clay–water system; clay particles must be in face–face arrangement;
• *assumptions used in theoretical calculations* – clay particle are plate-like in shape; dimensions of particles are such that the lateral dimensions far exceed their thicknesses; the surface charge of the particles is of constant sign and negative; the surface charge is uniformly distributed; ideal behaviour of ions; attractive forces between particles can be neglected; the van't Hoff pressure is taken as the swelling pressure.

Figure 3.12 shows the results from a series of very high pressure consolidation tests on a sodium montmorillonite saturated with a 10^{-3} mol/L NaCl reported by Alammawi (1988). Included in the figure are calculated pressures using a parallel-particle configuration as the calculation model. Assuming the high pressure consolidation pressures to be equivalent to swelling pressures, calculations using the DLVO theory (see equation 3.14 for face–face particle configuration) and a modified Gouy–Chapman model were made to compare with the experimentally measured consolidation pressure values. The modified Gouy–Chapman model consisted of incorporation of the energies E_{thp} and E_{ohp} to the Gouy–Chapman calculation of the energies of interaction in the region outside the Stern layer. Also included in the figure are the results reported by Bolt (1956), shown as 'Bolt' in the figure. These are from the experimental results for sodium montmorillonite at the same salt concentration. The 'van Olphen' results are interpreted from calculations reported by van Olphen (1977) for pressures required to remove the first four water layers next to a typical montmorillonite unit.

At higher ion concentrations in the porewater (e.g. concentrations $\geq 10^{-2}$ mol/L NaCl), studies have shown that even when meeting all of the preceding conditions, laboratory-measured swelling pressures can be higher than calculated values from the DDL models, indicating that the Boltzmann and van't Hoff relations become less valid with increasing electrolyte concentrations. At the higher concentrations, ionic activities need to be used instead of ionic concentrations in the calculations. Efforts made to incorporate ionic activity in the DDL calculations have encountered difficulties in

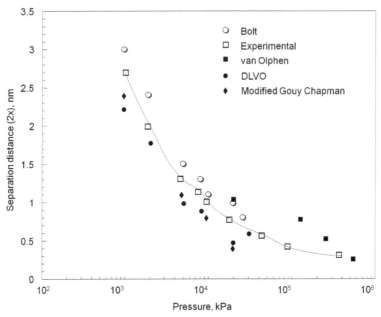

Figure 3.12   Comparison of measured swelling pressures at various interparticle distance separations, measured as 2×, with DDL and DLVO models for 10⁻³mol/L NaCl montmorillonite.

determination of the activity coefficients for the ions between adjacent clay particles. Bolt (1955) has developed approximate expressions that take into account the influence of dielectric saturation, polarization of ions, and ionic interactions in the structure of the electric double layer.

In 'real-life' and natural clay conditions, face–face particle arrangements are not easily attained on a general scale, and most certainly not on a homogeneous scale. Microstructural units consisting of packets of particles are the more consistent type of structure found in engineered clay barriers. In addition, the clays used in clay barriers do not consist of pure montmorillonites. Laboratory tests using target clays destined for application in the field are conducted to obtain information on swelling and swelling pressures generated when the clays are volume-change restricted. Empirical relationships for swelling pressure in relation to water content are generally obtained – for subsequent use in predictions of clay barrier performance during water uptake. Computations of swelling pressures generated in a the laboratory tests with the target clays, using any of the surface complexation models, are undertaken as 'scoping' calculations and also to compare with the empirical swelling pressure relations. It is obvious that the extent to which the test procedures and the target clays used meet, or closely meet, the conditions previously stated will determine the kind of comparisons obtained between computed and measured values.

3.8   Clay–water characteristics

Interactions between clay particles and water are characterized in terms of energy relationships. In soil physics, these are called *soil–water characteristics*. In the context of clays used for engineered clay barriers, the term *clay–water* is used in this book. Care should be taken to distinguish this from the term *clay water* (without the dash), which is quite often used in the literature to mean the water in the pore spaces of a clay. To avoid confusion, we use the term *porewater* to mean the water in the clay pores, i.e. the water within the macro- and micropores of the clay. There are several components of energy that contribute to the total energy relationship in a clay which is defined as the clay–water potential. The energy relationship provides one with a useful means to assess the capability of a clay to retain water. It is a measure of the water-holding capability of the clay under the various conditions defined by its water content and the nature of the clay.

3.8.1   *Components of clay water potential*

To explain what one means when one discusses *clay–water potential*, we introduce a simple capillary experiment using a glass column containing clean sand. If the bottom of the column is open and placed in a shallow pan of water, we will observe water rising in the sand-filled glass column to a height that we will call h – as shown in the left-hand diagram of Figure 3.13. We can have a similar glass column containing a dry compact clay also placed in a shallow pan of water. In this column, we will see that the height of water rise (water uptake) H in the column will far exceed that of the sand column in the equilibrium state. The height of capillary rise in the sand column h is given as follows: $h = 2\sigma\cos\alpha/r\gamma_w$, where the surface tension of the water is given by σ, r is the effective radius of the average pore size in the sand column, α is the contact angle established between the water and the surfaces of the sand particles and γ_w is the density of water.

The height of the capillary rise in the sand column in Figure 3.13 is determined by the factors detailed in the relationship for h. In geotechnical and soil engineering, it is not uncommon to ascribe the height of capillary rise h to *capillary suction* – meaning that capillary forces are responsible for suction of the water up the sand column. One could define a capillary potential ψ_c that represents a measure of the energy by which water is held by the sand particles by 'capillary forces'. Buckingham (1907) defined it as the potential due to capillary forces at the air–water interfaces in the sand pores holding water in the sand.

The requirements of analyses of the clay–water system as a whole are better satisfied if the height of water uptake in the clay column is determined in terms of the work required to move water up to the height H. One defines the total work required to move water into (and out of) a clay as

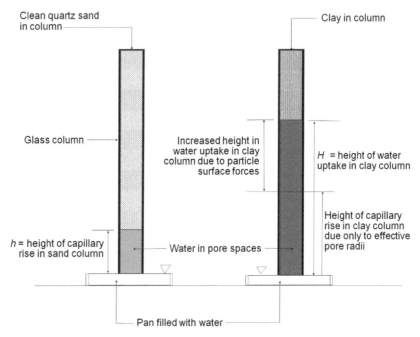

Figure 3.13 Capillary rise experiment with quartz sand column shown on the left, and water uptake experiment with inorganic clay shown on the right.

the *clay–water potential* ψ. The clay–water potential ψ describes the water-holding capability of clay, i.e. it describes the energy by which water is held to (or attracted to) the clay fractions in the clay. It is particularly useful in providing a simple picture of the kinds of internal forces that will contribute to water movement and retention in clays.

Potentials are defined with a reference base in mind. This base is normally considered to be a reference pool of free water under one atmospheric pressure, and at the same elevation and temperature of the clay. As a clay–water system is a three-phase system, there are various component potentials within the total potential ψ. Consider the clay column on the right-hand side of Figure 3.13, which shows the water rising up to a height of *H*. This 'water rising' phenomenon is called water uptake – to indicate that the water is taken into the clay by generally described *internal forces*. The internal forces are essentially those that have been described in the previous sections as originating from the interaction of the reactive surfaces of the clay particles with water in combination with the microstructure of the clay. We can imagine the total potential ψ responsible for raising water to the height *H* as being a combination of the microstructure and particle interaction contributions. The right-hand diagram shows an imaginary capillary rise due to effective pore sizes defined by the microstructure, and the additional height of water uptake due to surface forces (of the clay particles).

The components of the total clay–water potential ψ are described as follows, bearing in mind that the reference base is a reference pool of free water at the same elevation, etc.:

- ψ = total potential = total work required to move a unit quantity of water from the reference pool to the point under consideration in the clay. It is a negative quantity.
- ψ_m = matric potential = property of the clay matrix and pertains to sorption forces between clay fractions and clay water (porewater). This is often mistakenly assumed to be the capillary potential. For granular materials, this assumption may be quite valid. However, in the case of clays, and especially in the case of swelling clays, complications surrounding analysis of interlayer swelling and microstructural effects and influences do not permit easy resolution in terms of capillary 'forces'. According to Sposito (1981), the matric potential ψ_m includes the effects of dissolved components of the clay–water system on the chemical potential μ_w.
- ψ_g = gravitational potential = $-\gamma_\omega g h$, where γ_ω is the density of water, h is the height of the water in the clay above the free water surface, and g is the gravitational constant. If the point in the clay under consideration is below the surface, h is a negative quantity, and hence the relationship becomes positive.
- ψ_π = osmotic potential for swelling clay, and in the case of non-swelling clay, ψ_π is taken to be the solute potential ψ_s, which is $= nRTc$, where n is the number of molecules per mole of salt, c is the concentration of the salt, R is the universal gas constant and T is the absolute temperature. It is not unusual for the literature to report on the use of ψ_s as the osmotic potential ψ_π.
- ψ_p = pressure potential due to externally applied pressure and transmitted through the fluid phase of the clay–water system.
- ψ_a = pneumatic (air) pressure potential arising from pressures in the air phase.

3.8.2   Measurements of clay–water potentials

The three more common types of systems used to obtain a measure of the clay–water potential include (a) tensiometers; (b) pressure plates or pressure-membrane systems; and (c) thermocouple psychometry. The tensiometer basically measures the 'water tension' in a clay and is only useful for measurements of clay–water tensions, in which pressure differences are less than one atmosphere. As water and solutes can pass through the membrane which separates the tensiometer clay sample from the water, measurements obtained with the tensiometer include the effects of the dissolved solutes. At equilibrium, the condition for the clay water (porewater) will be given as:

$$\mu_w(\text{pore water}) = \mu_w(\text{tensiometer}) = \mu_w^o + \frac{1}{\rho_w^o}(P_t - P^o) \tag{3.20}$$

where μ_w is the chemical potential of the porewater, the superscript o refers to the standard state for the corresponding parameters, and P_t is the pressure in the tensiometer. If τ_w represents the porewater tension that is measured by the tensiometer at gauge pressure in pascals or atmospheres, the following is obtained (Sposito, 1981):

$$\tau_w = P^o - P_t = -\rho_w^o \left[\psi_p(P,\theta) + \psi_m(P^o,\theta) \right] \tag{3.21}$$

The pressure membrane technique uses air pressure to drive the porewater from the sample contained in the pressure cell. For each applied pressure, when equilibrium is reached, the water remaining in the sample is considered to be held within the clay by internal forces (i.e. forces originating within the clay) that are greater than those applied by the air pressure introduced into the cell. Figure 3.14 shows some typical suction–water retention characteristic relationships between three kinds of soils, generally obtained with pressure membrane and tensiometer-type techniques. Water retention or water-holding capacity is viewed from the soil particle frame of reference in terms of suction. This is the opposite of the frame of reference that views water-holding capacity in terms of the work required to move water into or out of the sample. In that sense, although both measurements should give equal results in terms of magnitude of effort required, suction measurements are expressed positively, whereas potential measurements have the opposite (negative) sign.

The thermodynamic analysis of the processes associated with the pressure membrane procedure shows that if P_w represents the applied pressure, and if the sample is fully saturated and initial pressure in the pressure membrane apparatus is zero, i.e. $P_w(\text{initial}) = 0$, then P_w provides a direct measure of the matric and solute potentials, ψ_m and ψ_s respectively. As in the case of tensiometer measurements, as the effect of dissolved solutes is included in the measurements obtained, one might want to keep this in mind and to be sure to distinguish measurements that do or do not include the effect of dissolved solutes. There are at least two different concepts of the matric potential ψ_m. These revolve around whether the matric potential does or does not include the effects of solutes. So long as one is careful in differentiating between the various effects, either concept is acceptable.

The thermocouple psychrometer measures the clay–water potential by determining the relative humidity of the immediate microenvironment surrounding the psychrometer. The psychrometer probe essentially consists of a

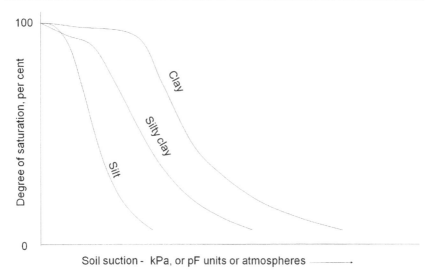

Figure 3.14 Typical soil suction curves showing differences in water-holding capacities of different types of soil.

small ceramic bulb within which the thermocouple end or juncture is embedded. Cooling of the juncture is obtained by passing an electrical current through it (Peltier effect). Cooling of the juncture below the dew point will result in condensation at the juncture. The condensed water will evaporate when the electrical current is removed or discontinued. There is an inverse relationship between the rate of evaporation of the condensed water and the vapour pressure in the psychrometer bulb. Evaporation of the condensed water at the juncture will result in a drop in the temperature, the magnitude of which will depend on the relative humidity and temperature of the immediate volume surrounding the psychrometer. The drier the surroundings, the faster is the evaporation rate and hence the greater is the temperature drop. The drop in temperature is measured as the voltage output of the thermocouple. The relationship between ψ and the relative humidity is given as:

$$\psi = \frac{RT}{V_m} \ln \frac{p}{p^\circ} \tag{3.22}$$

where R is the universal gas constant, T represents the absolute temperature, V_m is the molal volume of water, p is the vapour pressure of the air in the soil voids, p° is the vapour pressure of saturated air at the same temperature, and the ratio of p/p° is the relative humidity.

3.8.3   Macropores, micropores, potentials and swelling clays

Measurements using tensiometers, pressure membranes, psychrometers, etc., are conducted as 'bulk' (macrostructural type) measurements. Because of the technique or the size of the measuring tool, what is measured is the equilibrium status of the porewater in the macropores (pore spaces separating microstructural units). The equilibrium states of interlayer water and the micropore porewater (water in the pore spaces separating clay particles in the microstructural units) are not measured or determined. The equilibrium state of the macropore porewater is defined by the energy states in the micropores and in the interlayers. At that time, the osmotic potential ψ_{π} in the microstructural units will be balanced by the ψ_m in the macropores. To explain this, we need to consider the different pore sizes in a representative elementary volume (macrofabric). At least three kinds of pore spaces make up a total macrofabric: (1) pore spaces obtained as interlayer separation distances, i.e. interlayer spaces, (2) micropores representing pore spaces between particles in the microstructural units; and (3) macropores as shown in Figure 3.15.

It is often not easy to differentiate between interlayer pores (spaces) and micropores, especially when the unit layers of individual swelling clay particles are fully 'exploded' by DDL forces. Consequently, it is more

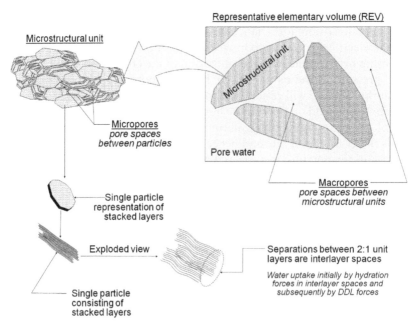

Figure 3.15   Macropores, micropores and interlayer spaces in a representative elementary volume of a swelling clay.

convenient to group interlayer pores with micropores under the general name of *micropores*. The study conducted by Muurinen (2006), using different concentrations of NaCl to saturate a sodium bentonite, provides a very useful technique for differentiating between macropore porosity and micropore porosity. By analysing the Cl concentration in the test samples using the Donnan model, and further assuming two Donnan membranes in the system instead of the classical single Donnan membrane, Muurinen was able to obtain proper accord between predicted and measured values. Figure 3.16 shows the distribution of both the macropore and micropore porosities that combine to make up the total porosity of the bentonite samples tested – at various densities. It is interesting to note that as the density increases, the proportion of micropore porosity increases in relation to the macropore porosity.

Of the various components of the clay–water potential ψ that are responsible for the water-holding capacity of clays, the matric ψ_m and osmotic ψ_π potentials can be considered to be the ones that are most responsible for this property. In the absence of externally applied gradients and under isothermal conditions, these two components are most often considered as being responsible for partly saturated moisture movement in clays – especially in swelling clays. Determination of the role of the various potentials in water movement in partly saturated swelling clays is complicated by a combination of swelling pressure and volume expansion. In the absence of temperature

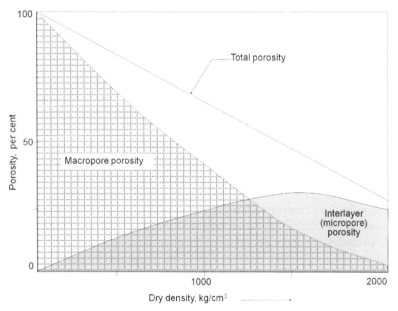

Figure 3.16   Proportioning of total porosity of a bentonite into macropore and micropore porosities (adapted from information reported by Muurinen, 2006).

and pressure gradients, volume expansion upon water uptake will result if the clay is not constrained – dependent on water inlet source distribution, clay microstructure and chemistry of the uptake water. Assuming no restrictions on source water availability and volume expansion, volume change due to swelling terminates upon dissipation of those swelling forces responsible for the volume expansion, i.e. when the clay–water potential ψ is no longer effective in provoking water uptake.

Volume expansion will be restricted if external pressures or geometrical constraints are applied to restrict volume change, and also if water availability is terminated. When such happens, full dissipation of those forces responsible for swelling (volume change) does not occur. Water uptake in the interlayer spaces of the swelling clay due to hydration forces will provide for a different form of water structure. This volume expansion is defined as crystalline swelling due to ψ_m. For the swelling pressures at low water contents for the test sample shown in Figure 3.12, the results presented as 'van Olphen' in the figures are derived from calculations interpreted from van Olphen (1977). He, along with many other researchers on swelling clays, maintains that the dominant mechanism responsible for interlayer separation (twice the distance shown in the figures) during water uptake at interlayer spacings of up to about 1 nm, is the result of the actions due to the adsorption energy of water at the clay particles' surfaces. These assertions, taken in conjunction with computations from the energy relationships shown in section 3.6, provide one with the capability to distinguish between (a) crystalline swelling, i.e. swelling due to sorption of the first two to three water layers between the unit layers of the 2:1 dioctahedral series of alumino-silicate clays (interlayer or interlamellar uptake of water), and (b) interparticle and intercluster (microstrustural unit) water uptake.

Water uptake leading to crystalline swelling is consistent with hydration water uptake. The calculations reported by Alammawi (1988) for a sodium montmorillonite with 10^{-3} mol/L NaCl shown in Figure 3.17 indicate that dilution of the ions occurs rapidly after the first two layers of water. The significant reduction in the concentration of ions beginning with the third water layer suggests that crystalline swelling is confined to the first two water layers, and that double-layer swelling occurs from the third water layer onward. This is consistent with the interlayer separation distances reported by Suquet et al. (1975) and Quirk (1968) for lithium and sodium montmorillonite. One can advance the argument that crystalline swelling results from hydration/sorption forces, and that these are associated with the matric potential ψ_m. Interlayer or interlamellar swelling beyond this point (of crystalline swelling) is due to DDL forces and these can be described by the osmotic potential ψ_π. The presence of air–water interfaces is not a necessary requirement for water uptake by forces associated with the matric potential ψ_m. This means to say that the interlayer spaces become saturated during hydration.

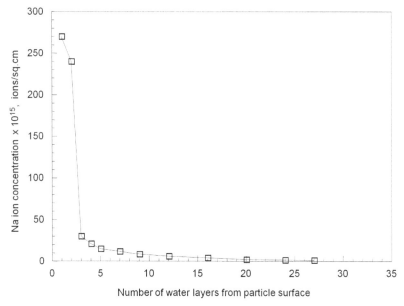

Figure 3.17   Distribution of Na ions at distances away form the surface of a montmorillonite particle (given in terms of number of water layers) in interaction with 10^{-3} mol/L NaCl (data from Alammawi, 1988).

3.8.4   Swelling and compression

In engineered clay buffer and barrier systems such as those used in HLW repositories and in HSW landfills, swelling of the buffer/barrier mass consisting of swelling clay materials upon water entry and uptake will produce compressive forces throughout the wetted portion. These will be transmitted to the unwetted portion of buffer mass at the wetting front, resulting in progressive compression of the unwetted portion of the buffer mass ahead of the wetting front. The compressive forces resulting from the constrained swelling of the buffer material will be directly related to at least two factors: (a) the amount of actual volume change occurring in the wetted portion and (b) the rigidity and degree of uniformity of the unwetted portion. The two factors are completely intertwined, i.e. they are in actuality a single complex phenomenon of continuous adjustment of compressive and expansive performances. If one assumes complete rigidity of the unwetted portion, there will be no expansion or volume change in the wetted portion. The consequences of this assumption are:

*   total swelling pressures, i.e. compressive forces, will be developed in accord with the initial particle spacing of the material in place; complete rigidity means at least one of two things – the macrostructure is composed of particles and microstructural units that have unyielding interparticle

bonds, or there is no microstructural influence or control on material behaviour and all particles act individually and are dispersed uniformly;
• total swelling pressures in the wetted portion are uniformly distributed, and transmitted without loss of pressure through the rigid unwetted portion to the other end of the column specimen.

In short, the result of this assumption is that the compressive forces resulting from the constrained swelling will be constant and essentially not dependent on the amount of water uptake, as shown in Figure 3.18. With the rigid block and/or no-volume-change assumptions, the magnitudes of the double-ended arrows shown in the figure are all equal and constant so long as saturation of the wetted portion is obtained. Furthermore, the amount of advance of the wetting front does not impact on the pressures and forces generated, as these are constant. One can conclude that the rigid block assumption is an extreme assumption.

As the physical structure of clays in general, and especially swelling clays, consist of an agglomeration of microstructures, one can assume that compressibility of the unwetted portion of the clay column (top of Figure 3.18) ahead of the wetting front would occur for the following reasons:

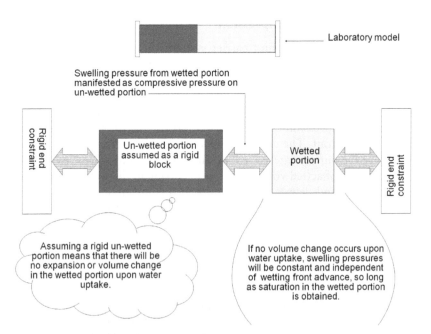

Figure 3.18 Assumption of properties and behaviour of unwetted portion as a rigid block in the face of wetting of the buffer mass.

- Swelling will occur upon water uptake, leading to a tendency for volume increase in the wetted portion.
- Volume increase restraint in the wetted portion by the unwetted portion means that swelling pressures will be generated in the wetted portion.
- The swelling pressures become the compressive pressure acting on the unwetted portion of the sample.
- Microscopic non-uniformity in microstructures and their distribution exists in the buffer mass. These are responsible for the wide distribution in the sizes and proportions of macro- and micropores.
- The heterogeneity, i.e. non-uniformity in porosity and density, of the sample will ensure that compression is not uniform in the unwetted portion of the sample because the compressible nature of the material is controlled by the microstructural non-uniformity of the specimen. In a clay buffer mass, this kind of heterogeneity would result in unequal longitudinal compression in the unwetted portion of the buffer mass. As this will vary as the wetting front advances, it follows that the rheological performance of the material in both the wetted and unwetted portion of the buffer mass will be a transient phenomenon.

3.9   Hydraulic conductance and permeability

The flow of fluid in a saturated soil mass has been treated in groundwater flow and soil mechanics literature as saturated seepage flow. In this section, we are concerned with at least two significant factors in the determination and analysis of the transmission properties of clays, and, in particular, swelling clays, aside from the impact of partitioning on interparticle fluid transport. These are fundamental considerations in the design and specification of engineered clay barriers and buffers.

3.9.1   *Water permeability and unsaturated hydraulic conductivity*

Water permeability

Water permeability is normally defined for flow of water in saturated clay and is simply called *permeability* in this book. It is not uncommon to find the term *permeability* used in association with water movement in unsaturated clays. Permeability is considered to be an impedance factor to flow of water. Therefore, permeability changes with the microstructural features of clays such as density of clays, size and shape of particles, states of aggregate and type of clay minerals. In other words, permeability is seen to be a function of the volumetric pore density, pore size, pore shape, tortuosity and characteristics of water such as viscosity and density of water.

The conventional procedure for determining the permeability of clays in the water-saturated state is to conduct an experiment in which the rate of fluid flow (flux v) through a laterally confined sample is measured in relation to the hydraulic gradient. From the measurements, the Darcy (1856) model for data reduction is used to obtain the coefficient k – 'un coefficient dependant de la perméabilité de la couche de sable'. The applicable relationship obtained with this model states that $v = ki$, where i is the hydraulic gradient. This is, in essence, a flux relationship in which the rate of flow v is related to the hydraulic gradient applied i through a compliance k, the hydraulic conductivity coefficient, otherwise known as the Darcy permeability coefficient. It is interesting to note that although not much has changed since the time of Henry Darcy with respect to the $v = ki$ Darcy permeability relationship (Darcy model), considerable recognition and appreciation of the restraints, conditions, test requirements, and limits of applicability of the Darcy model now exist, not the least of which is the fact that the model only provides the relevant data allowing one to calculate the coefficient k – 'the coefficient dependent on the permeability of the layer of sand'. This means to say that one does not measure the (Darcy) permeability coefficient k. It is a computed parameter. Figure 3.19 shows the basic elements of the Darcy experiment

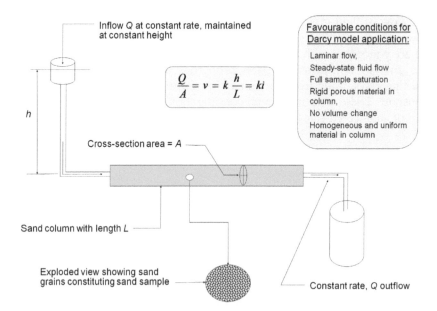

Inflow Q at constant rate, maintained at constant height

$$\frac{Q}{A} = v = k\,\frac{h}{L} = ki$$

Favourable conditions for Darcy model application:

Laminar flow,
Steady-state fluid flow
Full sample saturation
Rigid porous material in column,
No volume change
Homogeneous and uniform material in column

Cross-section area = A

h

Sand column with length L

Exploded view showing sand grains constituting sand sample

Constant rate, Q outflow

Figure 3.19 The basic Darcy flow model system for determination of 'un coefficient dependant de la perméabilité de la couche de sable', $k = v/i$, where v is the velocity and i is the hydraulic gradient. The top right-hand description provides a description of some of the essential conditions for favourable application of the Darcy model.

and the conditions thought favourable for application of this Darcy model for data reduction to obtain the coefficient of permeability k.

As tests are conducted on a bulk sample, and as k is computed (not measured) from recordings of flow rate under a specific gradient, no attention is given to the nature of the material or its constitution. Ideally, if one has a rigid porous material that is homogeneous and uniform, the Darcy model for data reduction should provide one with the straight line shown in Figure 3.20, assuming that the hydraulic flow conditions meet the ones shown in top right-hand corner of Figure 3.19. It has been shown for example that in non-swelling clays, linearity may not be obtained immediately between velocity v and the hydraulic gradient i as shown in Figure 3.20. Linearity occurs only after one reaches a critical gradient i_c – dependent on the type and density of clay tested. The non-linearity between zero and the critical gradient is due to the impedance offered by the interlayer DDL forces. As shown in the figure, the greater the swelling potential of the clay is, the shallower the slope will be. For situations where the clay has a high swelling potential or high swelling pressure upon wetting, the standard procedures need to be modified or even changed drastically. Because of the high swelling pressures that can develop, rigid confinement is required if the cross-sectional area presented to the permeating fluid is to be preserved.

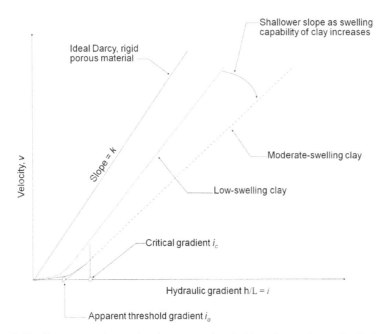

Figure 3.20 Illustration of critical and apparent threshold gradients obtained in hydraulic conductivity tests on low to moderate-swelling clays. Note that the ideal Darcy condition applies to rigid porous materials and other soil materials that fit the conditions shown in Figure 3.19.

The question arises as to whether initial water uptake in response to the microstructure and DDL forces (internal gradients) is accommodated in the Darcy model. To answer this question, calculations or measurements are needed to establish the magnitude of the internal gradients – to be incorporated with the externally applied positive gradients for proper calculation of the effective Darcy permeability coefficient. Evidence concerning the lack of proportionality between flow velocities and hydraulic gradients in recent historical experimental studies of flow in saturated clays reported for example by Philip (1957), Lutz and Kemper (1959), Hansbo (1960) and Yong and Warkentin (1966) shows that at low water contents, linear proportionality between flow velocity and hydraulic gradient is not established until some critical gradient is reached, as demonstrated in Figure 3.20.

Permeameters are generally utilized in laboratory experiments that are structured to determine the permeability of clays using either constant head conditions – as shown in the Darcy-type experiment in Figure 3.19 – or falling head conditions. Even although the procedures for test implementation are well documented in the standards of most countries [e.g. ASTM (American Society for Testing of Materials), EC standards], there is still some considerable debate on the sizes of permeameters and nature of sample confinement when it comes to testing the hydraulic conductivity of clays. Instead of permeameters, oedometers have been used as a popular piece of test equipment for determination of the permeability of swelling and non-swelling clays. Although these are not the classical oedometers used in soil engineering in consolidation testing of clays, they are similar in principle. The advantage of this equipment is its robust confinement against the swelling pressure developed during fluid permeation of the test sample. Such a system is shown in Figure 3.21, together with a double-membrane triaxial-type permeameter. There are particular reasons for choosing one or the other of these types of permeameters, as will be discussed in the next section. Although there are no particular standards for the dimensions of samples used for testing in the two kinds of cells shown in the figure, it is necessary to ensure that (a) the cross-sectional area of the test sample must be sufficiently large that one has a representative portion of the sample, and (b) the height of sample should be sufficiently high so as to permit proper hydraulic transport of the permeating fluid. Sample sizes used in the oedometer- and triaxial-type permeameters have been 30 cm and 40 cm in diameter, respectively, with heights that are between 1 and 1.5 times greater than the diameters of the samples.

Swelling clays present some unique challenges when it comes to determining their inherent properties when these clays are exposed to water. As swelling clays will swell upon exposure to water, and hence will expand three dimensionally, one needs to define the specific state of the clay for which the properties are to be determined. If volume change is to be controlled or denied, swelling pressures will be generated against the constraining devices, thereby invoking boundary stresses.

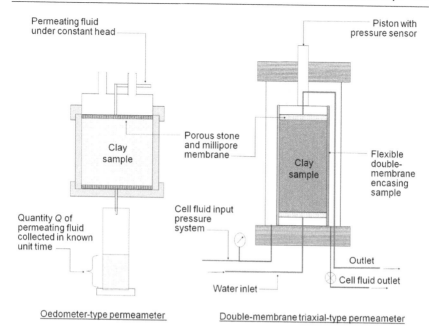

Oedometer-type permeameter | Double-membrane triaxial-type permeameter

Figure 3.21   Oedometer type (left) and double-membrane triaxial type (right) of permeameter for determination of permeability of swelling clays. For high-swelling clays, sample sizes should be at least 30–40 cm in diameter (D), and heights (H) of samples should be greater than or equal to I D. For other swelling clays, diameters of the samples should be greater than 40 cm and H/D ratios should be greater than I.

Oedometer-type permeameter

The robust oedometer-type permeameter shown in Figure 3.21 is designed to provide radial rigidity, i.e. no radial expansion of the confined sample. The piston pressure cell system at the top of the cell allows one to record swelling pressures generated axially in the permeability tests. The Darcy model is used to determine the permeability coefficient, as shown in Figure 3.19. In the oedometer set-up shown in Figure 3.21, the permeating fluid is applied under a specified constant head to the test sample, and the fluid quantity Q accumulated within a unit time period is collected at the outlet end in a calibrated flask. As the cross-sectional area A of the sample is known, the permeating velocity $v = Q/A$ for the unit time period can be related to the test-specified gradient i via the compliance k as given in the Darcy model: $v = Q/A = ki$. With this model for data reduction, the compliance k is generally known as the *Darcy coefficient of permeability*. The literature commonly omits the 'Darcy' designation and refers to k as the coefficient of permeability.

A necessary limiting condition for constant head permeation in permeability testing of clays, and particularly in swelling clays is the magnitude of the hydraulic gradient applied to the test sample. The top right-hand box in Figure 3.19 gives the ideal conditions required for the permeating fluid – steady-state laminar flow. There should be no local disturbance within the test sample due to the fluid pressure since this will cause local piping and erosion and hence subsequent local volume changes. Defining the *excess hydraulic gradient* (EHG) to be the gradient difference above that required to equalize the swelling pressure, experience has shown that excess hydraulic gradients above 60 or so, applied to dense swelling clays, would begin to cause local channelling and piping. For less compact swelling clays, the upper limit of the EHG will be lower. To illustrate the concept of the EHG, we take as an example a hydraulic gradient of 30 required to counter the swelling pressure developed in a saturated swelling clay sample contained in the oedometer shown in Figure 3.21. Assuming that one desires to apply an EHG of 50 to the test sample, this means that a total hydraulic gradient of 80 would be required to implement the test plan. This also means that if one is to conduct the test so that the EHG does not exceed 60, the total hydraulic gradient applied to the sample must not exceed 90.

Triaxial-type permeameter

The triaxial-type permeameter has been used for permeability tests in sand–bentonite mixtures and also for medium-swelling clays. The primary difference in sample performance between this and the oedometer-type device is the ability of the test sample to expand three-dimensionally upon uptake of water. The question of whether one wants to have three-dimensional confinement or lateral confinement can be answered by controlling the cell pressure that exercises triaxial confinement. Monitoring of the confining fluid volume and control of the vertical piston permits one to exercise a no-volume-change condition on the sample being permeated. In addition to the control of volume change, the triaxial-type permeameter allows one to apply a back pressure in the test sample to counter the swelling pressure. The test protocol for this device is similar to that of the oedometer-type device.

Unsaturated hydraulic conductivity

The unsaturated hydraulic conductivity for partly saturated clays is also determined in terms of a coefficient k that relates fluid (water) flux with its hydraulic gradient, according to the Darcy model. In the physical sense, this means that unsaturated hydraulic conductivity is determined in terms of the resistance to water flow through the partly saturated clay, and is influenced by the volumetric water content, microstructural features of the clay such as density, size and shape of particles, pore size, shape and distribution, tortuosity, and viscosity and density of the fluid involved. Impedance to flow is generated with a portion of the void space that is occupied with water. As mathematical modelling of the structural characteristics and water content dependency is not easily accomplished, it is more expedient to determine unsaturated hydraulic conductivity through experimentation.

3.9.2 Gas permeability

Gas flow in partly saturated clays occurs under the imposition of pressure gradients. As the rate of flow is considered to obey Darcy's law, the gas permeability is defined as a coefficient that relates the flow rates with their respective pressure gradients. Gas flow is accompanied by changes in volume, as gas is a compressible fluid. Gas permeability can be experimentally measured with oedometer- and triaxial-type permeameters (see section 3.9.1), where measurements of gas flow are made for samples subject to pressure gradients generated by pressure imposed on opposites sides of the sample.

There are two types of expression that can be used to calculate the mean gas permeability over a certain thickness of samples. One is derived for steady state flow and the other is derived for unsteady state flow, on the assumption that both the Darcy model $v = ki$ and the state equation for an ideal gas are applicable for gas flow in the samples, i.e. $pV = nRT$, where p is the pressure, V is the volume, n is the number of moles, R is the gas constant and T is the temperature.

For steady-state experiments, the permeability at a certain pressure p in the samples is obtained by directly combining Darcy's law with the equation of state for gas (Horseman et al., 1999; Tanai and Yamamoto, 2003) as follows:

$$k = \frac{2\mu qpL}{P_1^2 - P_2^2} = \frac{2\mu QLp}{A\left(P_1^2 - P_2^2\right)}$$

(3.23)

where P_1 and P_2 are the external pressures applied at the opposite sides of the sample, p is the pressure at a certain point in the sample, q is the gas flux, Q is the outflow of gas, L is the length of samples, A is the cross-sectional area of samples, and μ is the viscosity of the gas. Applying the mean pressure of samples, for example:

$$p = \bar{p} = \frac{1}{2}\left(P_1 + P_2\right)$$

to p, we obtain

$$k = \frac{\mu \bar{Q} L}{A\left(P_1 - P_2\right)} = \frac{\mu \bar{q} L}{P_1 - P_2} \tag{3.24}$$

where the overbars associated with Q and q are the mean values.

In the unsteady-state experiments, gas chambers are set at the opposite sides of samples. Pressures p_{vo} and p_{so} ($p_{vo} > p_{so}$) are applied to both chambers at the beginning. The pressure in the chamber that applies gas to the samples is decreased gradually as gas migrates into the samples, with pressure maintained constant in the chamber that captures gas outflow (Yoshimi and Osterberg, 1963). The unsteady gas flow in the samples can be determined, and the relationship for gas in the chamber that applies gas to the samples is obtained using the law of mass conservation as:

$$\frac{d(\rho V)}{dt} = -\rho q A$$

where ρ is the density of the gas in the chamber, V is the volume of the chamber, q is the flux defined by Darcy's law and A is the cross-sectional area of test samples. Applying both the state equation for gas and Darcy's law to this relationship, integrating the resultant relationship and rearranging, we obtain the gas permeability at a mean pressure:

$$p = \frac{p_{Vt} + p_{SO}}{2}$$

in the samples as follows,

$$\bar{k} = \frac{VL\mu}{Ap_{SO} t} \ln \frac{1}{C} \left| \frac{p_{Vt} + p_{SO}}{2\left(p_{Vt} - p_{SO}\right)} \right| \tag{3.25}$$

and

$$C = \left| \frac{p_{VO} + p_{SO}}{2\left(p_{VO} - p_{SO}\right)} \right| \tag{3.26}$$

where $\bar{k}$ is the mean gas permeability with time t that is defined as:

$$\bar{k} = \frac{1}{t} \int_0^t k \, dt \,,$$

μ is the viscosity of the gas, p_{Vt} is the pressure of accumulated gas in the chamber used to apply gas to the sample at a certain time t, p_{VO} is the pressure of accumulated gas in the chamber at $t = 0$, p_{SO} is the pressure in the chamber capturing gas outflow, L is the length of the sample, V is the volume of the chamber, and A is the cross sectional area of the sample.

One should note that the gas permeability given by equation 3.25 includes the effects of changes in pore spaces, solid density and water content with time resulting from expansion and shrinkage of samples during permeation of gas. As gas flow in clay buffers and barriers show similar behaviour to those observed in the experiments designed to determine equation 3.25, the values calculated from this equation can be usefully applied for analyses of gas flow in clay buffers and barriers.

3.9.3 General principles of permeability

As the microstructural features differ considerably between swelling and non-swelling clays, principally owing to the interlayer phenomena associated with swelling clays, it follows that microstructure influence on fluid trans-mission characteristics between them would differ somewhat. The Darcy model for determination of the Darcy coefficient k, which is more suited for testing of non-swelling (no-volume change) clays, does not directly consider the impact of the properties of the permeant on permeability, nor does it take into account the structure of the clay (micro- and macrostructure). One can study the influence of clay properties such as pore sizes, shape and distribution, and tortuosity, and permeant viscosity, on clay permeability determination using the Poiseuille relationship for viscous flow – in an adapted form, as shown by Yong and Mulligan (2004) to determine the link between clay microstructure and permeability.

The flow of viscous fluids in pores is generally described by the Navier–Stokes relation for incompressible fluids in slow steady-state flow. Application of this to the Darcy model gives us:

$$-\frac{\partial p}{\partial x_i} + \eta \nabla^2 V_i + F_i = 0 \tag{3.27}$$

where p is the pressure, V is the velocity, η is the fluid viscosity, F is the external forces, and i indicates directions of x, y and z.

The Poiseuille relation is obtained by solving the Navier–Stokes relation analytically for flow q through a narrow tube of radius r, and for flow through two parallel plates separated by a space of width h as follows:

$$q = -\frac{\pi r^4}{8\eta} \frac{dp}{dx} \text{ (for narrow tube);}$$

$$q = -\frac{b^3}{12\eta}\frac{dp}{dx} \quad \text{(for parallel plates)} \tag{3.28}$$

The Poiseuille relationships show that fluid flow in clays depends on three components: driving forces, characteristics of flow spaces, and viscosity of the fluid involved. In Darcian flow in clays, the driving force is considered to be the potential gradient $\Delta\psi/\Delta l$, and the characteristics of flow spaces are represented by the porosity n of the clay, the shape factor C_s, wetted surface area S_w, and tortuosity T. $\Delta\psi$ is the potential difference between the ends of a tube of length Δl. Yong and Mulligan (2003, 2004) have used a modification of the combined form of the Poiseuille and Kozeny–Carman (K–C) models to determine the effective permeability of clays as follows:

$$v = k^* i = \frac{C_s n^3 \gamma}{\eta T^2 S_w^2}\frac{\Delta\psi}{\Delta l} \tag{3.29}$$

where:

- k^* = the effective permeability coefficient, which takes into account both permeant and clay microstructure properties, is:

$$\frac{C_s n^3 \gamma}{\eta T^2 S_w^2}$$

- C_s = shape factor, and has values ranging from 0.33 for a strip cross-sectional face to 0.56 for a square face: this factor accounts for the fact that the cross-sectional face of any of the pore spaces in the clay mass may be highly irregular and allows one to choose a *typical value* for a representative pore cross-section area; Yong and Warkentin (1975) have suggested that a value of 0.4 for C_s may be used as a standard value – with a possible error of less than 25 per cent in the calculations for an applicable value of k^*;
- i = hydraulic gradient, the ratio of the potential difference $\Delta\psi$ between the entry and exit points of the permeant, and the direct path length Δl of the clay mass being tested;
- T = tortuosity, the ratio of effective flow path Δl_e to thickness of test sample Δl, and which is quite often taken to be $\approx \sqrt{2}$;
- γ and η = density and viscosity of the permeating fluid, respectively;
- n = porosity of the unit clay mass;
- S_w = wetted surface area per unit volume of clay particles in the microstructural units.

Equation 3.29 differs slightly from the standard K–C relationship in that the wetted surface area consideration in the K–C model assumes that $S_w = S(1-n)$ and that the radius r of the Poiseuille tube is $r = n/S(1-n)$. This gives us the relationship for k from the K–C model as:

$$k = \frac{C_s \gamma n^3}{\eta T^2 S^2 (1-n)^2}$$

(3.30)

where S = specific surface area of clay.

With this measure of surface area, we assume that all particle surfaces are in contact with the permeating fluid. Experience with the use of the classical K–C relationship has been good when the pore sizes are relatively uniform, i.e. when the particle sizes are relatively uniform. However, experience has also shown that when pore sizes vary significantly and especially when we have a high proportion of fine fractions in the clay, the relationship does not hold well. This should be intuitively obvious because the SSA of the clay will be proportionately higher because of the presence of the fines.

The adaptation introduced by Yong and Mulligan (2003) considers the surface area of the particles in terms of only the wetted surfaces, i.e. S_w. This allows one to have compositional differences and clay microstructure differences that can impact severely on the distribution of pore sizes and availability of clay particle surfaces for direct interaction with the permeant. We see from equation 3.24 that C_s, T and S are clay property parameters that are dependent on clay composition and clay microstructure. These can be expressed as a parameter $\beta = C_s/(TS)^2$. Along the same lines, the density and viscosity of the permeating fluid, γ and η respectively, are properties of the permeant and can be described by a parameter $\mu = \gamma/\eta$. Using these parameters, the relationship for k^* can be expressed as follows: $k^* = \mu \beta n^3$. Assuming that the properties of a leachate permeant are not too far distant from that of water at about 20°C, we can compute μ directly. Further assuming a tortuosity T value of $\sqrt{2}$, and $C_s = 0.4$, the graphical relationship shown in Figure 3.22 can be obtained. The graph shows the relationship between the clay permeability expressed as an effective permeability coefficient k^* and the amount of surface area wetted in fluid flow through the clay, all of which are determined in relation to the total porosity of the clay. The wetted surface area is the surface area of the pore channels in the clay through which fluid flow occurs. From the information shown in the figure, and converting the SSA units to cm²/cm³, it is seen that the wetted surface areas S_w are at least one order of magnitude less than the SSA of the various soils tested by Yong and Mulligan (2003). This tells us that the microstructure of the clays is influential in the development of flow channels.

In considering the characteristics of the permeant, it is important to note that the viscosity and density of fluids in clay pores differ amongst various

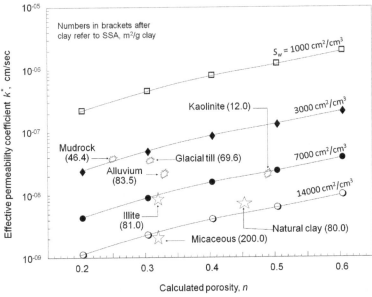

Figure 3.22   Variation of effective permeability coefficient k^* with calculated porosity – in relation to wetted surface area S_w. Numbers in parentheses after the names of clays refer to the SSA given in terms of square metres per gram of clay (m²/g clay).

fluids. With that in mind, it is appropriate to represent the Darcy relationship as $v = k^+(\gamma/\eta)$, where k^+ denotes the intrinsic permeability that is dependent on the characteristics of clay pore structure and the properties of the permeant expressed in terms of the ratio of its density to its viscosity (i.e. γ/η). The viscosity and density of fluids in a pore space changes with distance from the surface of a solid. It is not constant over the entire pore space occupied with fluids – as has been discussed in section 3.7.2. The influence of the profiles of viscosity and density of fluids in a pore space on permeability can be estimated by solving the Navier–Stokes relationship that describes the motion of fluids under the condition that viscosity and density are dependent on the position of the fluid. Such a solution has reported by Ichikawa *et al.* (1999), who used a perturbation theory (homogenization analysis) in their analysis of fluids in simple pore spaces formed between parallel planes in an infinitely small volume of clay.

The macromechanistic performance with respect to hydraulic conductivity in compact swelling clays is closely similar to that of the non-swelling clays. However, because of the considerably higher proportioning of micropores to macropores, not only will fluid flow be extremely slower, but also the effect of restructuring of the microstructural units will be greater with the swelling clays. Consequently, the changes in wetted surface areas would also be considerable – if and when restructuring of the microstructures occurs. Figure 3.23 illustrates the differences in proportions of micropore porosity and macropore porosity, taking note of the fact that the kaolinite used as an example of a non-swelling clay has platelets as particles, and that these do not have interlayers that 'explode' upon exposure to water.

3.10 Clay rheology

3.10.1 General

Clay rheology refers to the time deformation characteristics of clays under external loading conditions. This includes the constitutive performance of the clay and its yield failure characteristics. As normal practice calls for clay barriers and clay buffers to be initially emplaced in the dry, unsaturated or partly saturated state, and as the clays will eventually reach full saturation, one needs to be concerned with the rheological behaviour of both partly saturated and saturated clays.

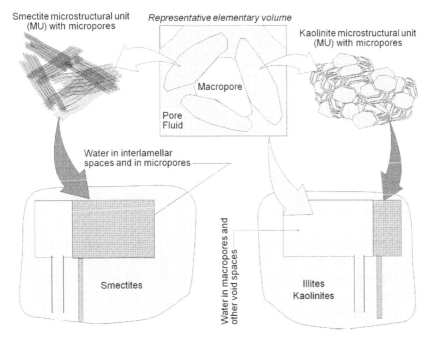

Figure 3.23 The proportions of water held in interlamellar spaces, micropores and macropores. The two bottom diagrams show the corresponding proportions of water from the sources identified in the representative elementary volume. Water held in the micropores (which includes the interlamellar spaces) is significantly less mobile than water held in the macropores and other void spaces.

From lessons in classical soil mechanics, we have learnt that the stress–strain–time behaviour of clays is different when the porewater is (a) not permitted to leave the clay during external load application or (b) allowed to be taken up or expelled from the clay – depending on the applied pressure and boundary conditions. The common designations for tests for these two conditions are 'undrained' and 'drained' test conditions. In addition, we have learnt that the rheological performance of clays can be significantly different between clays that are completely water saturated in comparison with partly saturated clays, i.e. clays that have a degree of water saturation that is below 100 per cent. Only the properties at complete water saturation will be treated here.

The most important rheological properties of completely water-saturated buffer materials are the following:

- stress–strain behaviour at compression (consolidation);
- stress–strain behaviour at expansion (swelling);
- stress–strain behaviour at shearing;
- shear strength behaviour;
- creep properties (change of strain with time at constant stress).

In common soil mechanics and geotechnical literature, the following properties are determined:

- compressibility
- shear strain
- shear strength
- swelling
- swelling pressure.

Following the Terzaghi–Rendulic model, one accepts the thesis that the rheological behaviour of non-swelling soils is controlled by effective stresses. With respect to swelling clays, however, it is not unreasonable to question whether the classical effective stress concept is viable. In short, one could ask whether the effective stress concept, which has been shown to be a very useful and valid concept when applied to commonly encountered non-swelling soils, is also valid for smectite-rich buffer materials as discussed in this chapter.

3.10.2 The effective stress concept

The effective stress concept $\sigma' = \sigma - u$, where σ' is the effective stress, σ is the total stress and u is the porewater pressure, has been the basis for many stress-related expressions used in soil mechanics. From classical soil mechanics, the total stress is given as:

$$\sigma = \frac{N}{A} = p\frac{A_s}{A} + u\frac{A - A_s}{A} \qquad (3.31)$$

where N is the total force on one side of a cubical element of clay, A is the total area of the side of the cubical element where N is applied, A_s is the total particle contact area in a cross-section through the element, p is the contact pressure between particles, and u is the porewater pressure.

The contact pressure p is very high owing to the very small contact area A_s. pA_s is thus the total force over the area A that is transmitted through the cross-section by the interacting particles. This force, divided by the total area, i.e. $p = A_s/A$ is termed the effective stress $\sigma_c{'}$ in the sample (subscript c stands for contact). The remaining horizontal area in the cross-section, $A–A_s$, is thus made up of water (see Figure 3.24). Designating $A_s/A = b$, equation 3.31 can be written as $\sigma = \sigma_c{'} + u(1-b)$. The effective stress theory is valid if A_s is very small in comparison to the total area, and thus $b \approx 0$. This relationship applies to most ordinary soils.

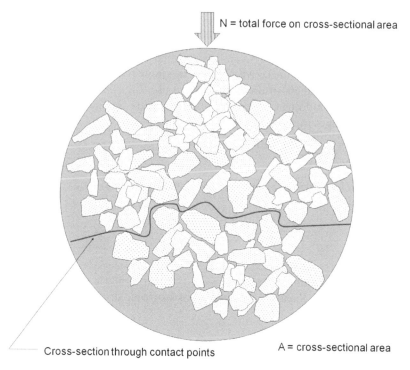

N = total force on cross-sectional area

Cross-section through contact points

A = cross-sectional area

Figure 3.24   Imagined section of microstructural network.

3.10.3 Relevance of the effective stress concept for smectitic clays

In smectitic clays the question is not whether the contact area A_s is small. This is because it is generally accepted that there is no direct particle–particle contact in such clays. The question that needs to be addressed is 'what is the state of the water within and between the lamellar stacks, and how are the stresses transferred?'. If the water between the stacks has a viscosity that is Newtonian and distributes stresses in the same way as free water (Poisson's ratio $v = 0.5$) it is probable that the effective stress theory is valid and $A_s/A = b = 0$. This would mean that compressive and shear stresses at the particle contacts are directly transferred between the crystal lattices. However, the water hull surrounding the stacks of smectite lamellae and the interlamellar water structure are believed to be responsible for stress transfer. If such is the case, this means that b is substantially higher than unity. For montmorillonite with a density at water saturation of 2000 kg/m³ the b value may be about 0.2, as the percentage of interlamellar water is about 80 per cent (Pusch and Yong, 2006). This means that the effective stress concept has a different meaning for smectitic clay, in contrast with that of ordinary soils. At present, however, there is no practical proven way of determining the value of b.

3.10.4 Basic stress–strain relationships

Rheological behaviour of clays can be described with partly elastic recoverable and partly plastic unrecoverable stress–strain relations (Pusch, 2002). Two basic types of stress/strain moduli are in use: the compression modulus K and the shear modulus G. The first one represents changes in volume and the second changes in shape.

Compression modulus K

Compression is expressed by use of K as defined by the following relationship:

$$K = \frac{d\sigma'}{d\varepsilon_v} = \frac{d\sigma'}{d\varepsilon_1 + d\varepsilon_2 + d\varepsilon_3} \tag{3.32}$$

where σ' represents the effective stress, ε_v is the volumetric strain and ε_1, ε_2, ε_3 are linear strains. K can be determined by isotropic drained triaxial tests.

Shear modulus G

Simple shearing gives us the shear modulus G as $G = d\tau/d\gamma$, where τ and γ are shear stress and shear strain respectively. G can be determined by conducting undrained shear tests. For 'linearly' elastic materials one can derive relationships between these basic moduli and Poisson's ratio v as well as other derived moduli (Pusch, 2002):

$$E = \frac{3G}{1 + G/3K} \tag{3.33}$$

$$v = \frac{1 - 2G/3K}{2 + 2G/3K} \tag{3.34}$$

The modulus of elasticity E can be determined by uniaxial compression tests, and the G modulus can be evaluated from E and K. E can also be derived from undrained triaxial tests and is expressed in terms of the tangent modulus (Figure 3.25).

$$E = \frac{d\left(\sigma_1' - \sigma_3'\right)}{d\varepsilon} = \frac{d\left(\sigma_1 - \sigma_3\right)}{d\varepsilon} \tag{3.35}$$

where σ_1 and σ_3 are the major and minor principal stresses respectively. In triaxial testing, σ_3 is the cell pressure.

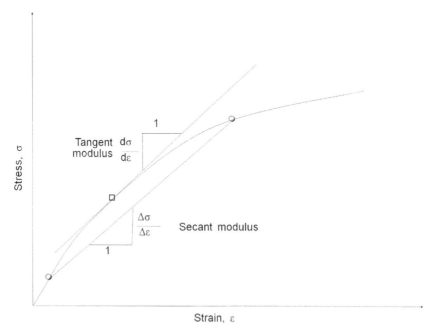

Figure 3.25   Secant and tangent moduli.

We can obtain an estimate of E using the empirical relationship $E = n\tau_{fu}$, where n is a coefficient ranging from 150 to 400 and τ_{fu} is the undrained shear strength. A value of 250 is generally assumed as a mean value for ordinary clay materials.

Oedometer modulus M

Oedometer compression testing (Figure 3.26) gives us the oedometer modulus M as $M = d\sigma'/d\varepsilon$.

The oedometer modulus M can be obtained indirectly, from a knowledge of K and G – the compression and shear moduli, respectively. This requires testing for determination of the stress–strain performance of the clay and obviously, determination of the compression and shear moduli. The relationship that is used is given as $M = K + (4G/3)$.

The results obtained for oedometer and shear strain curves are usually not straight lines. M and G can generally be expressed in terms of the following general expression where D represents any deformation modulus, as shown in Figure 3.27 (Pusch, 2002).

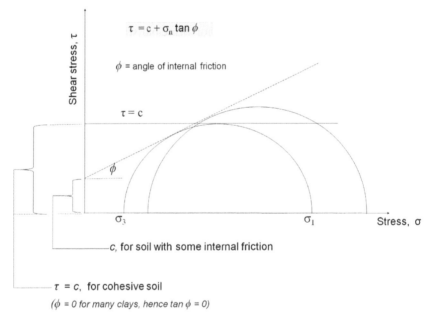

Figure 3.26 Uniaxial confined compression (oedometer testing) showing application of Mohr's circle for determination of the shear stress τ for a soil with some internal friction, and a cohesive (frictionless) soil. σ_1 and σ_3 are major and minor principal stresses respectively, and σ_n is the normal stress. c is the cohesion and φ is the angle of internal friction of the soil. Increasing sizes of Mohr's circle reflect increasing applied stress and corresponding increasing confining stress in oedometer testing.

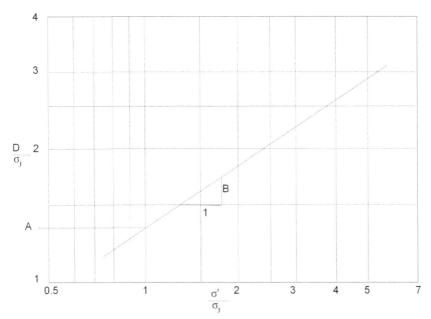

Figure 3.27 Evaluation of D from oedometer stress–strain curve (adapted from Janbu, 1967).

$$D = A\sigma_i \left(\frac{\sigma'}{\sigma_i}\right)^{B} \tag{3.36}$$

where σ_i is a reference stress (often taken as 100 kPa).

3.10.5 Oedometer – consolidation and compressibility

In its simplest form, the compressibility of soils is expressed in terms of the compression index C_c or the index ε_2, which can be evaluated from oedometer tests (see Figure 3.28). The following expression for C_c for a linearly scaled void ratio and log-scaled pressure is obtained (Pusch, 2002):

$$C_c = \frac{\Delta e}{\Delta\left(\log\sigma'\right)} \tag{3.37}$$

Doubling the load gives us:

$$C_c = \frac{\varepsilon_2\left(1+e_0\right)}{\log 2} \tag{3.38}$$

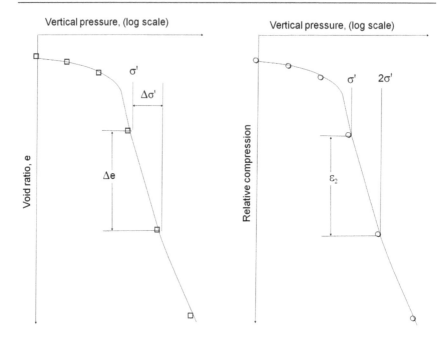

Figure 3.28 Determination of compression index C_c (left), and index ε_2 (right).

where ε_o = void ratio before load increase ($\Delta\sigma'$ or σ' in Figure 3.28).

The fact that the compression curve is not a straight line has led to more general expressions. For M, expressed as tangent modulus $M = d\sigma'/d\varepsilon$, one has:

$$M = m\sigma_j \left(\frac{\sigma'}{\sigma_j} \right)^{(1-\beta)} \tag{3.39}$$

where M is the oedometer modulus, m is the modulus number, β represents the stress exponent, σ' is the effective compressive stress and σ_j is the reference stress, which is commonly taken to be 100 kPa. In Germany a similar expression, $E_o = v \times \sigma^w$, is often used, with v referring to $\sigma' = 100$ kPa (Pusch, 2002). In this instance, E_o is called the modulus of elasticity, and the stress exponent w is a material constant that varies with the type of soil tested.

Integration of the expression $d\varepsilon = d\sigma'/M$ yields the compression ε strain as:

$$\varepsilon = \frac{1}{m\beta} \left(\frac{\sigma'}{\sigma_j} \right)^{\beta} + C \qquad \text{for } \beta \neq 0 \tag{3.40}$$

$$\varepsilon = \frac{1}{m} \ln \left(\frac{\sigma'}{\sigma_i} \right) + C \qquad \text{for } \beta = 0 \tag{3.41}$$

These expressions are used for calculating the settlement of foundations on soils such as the 16,000-T concrete silo hosting low-level radioactive waste at the Forsmark repository site in Sweden. m and β are evaluated from oedometer tests in the manner shown in Figure 3.29.

Time-dependent compression

Compression of water saturated clays is always delayed because the porewater overpressure at loading has to be expelled, yielding 'primary consolidation'. Creep caused by mutual displacement of particles forming the clay skeleton takes place parallel to this process and after dissipation of the porewater overpressure (i.e. excess porewater pressure). It is termed 'secondary consolidation'. The dissipation is a diffusion-type process and is expressed using the consolidation coefficient, c_v, which is evaluated from oedometer tests.

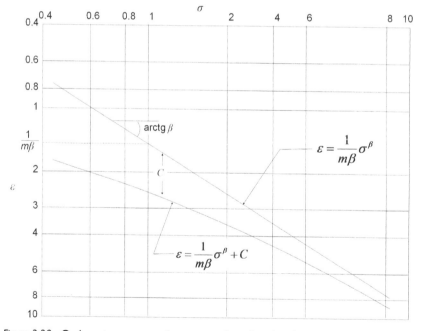

Figure 3.29 Oedometer compression curve plotted in log–log diagram. The primary curve is the lower one, which is transformed to the upper by finding a value $\Delta = C$ for which a straight line is obtained. Once this has been made, $1/m\beta$ is directly obtained for $\sigma' = 1$ bar $= 100$ kPa. ε is evaluated from the inclination of the line, which then yields the m-value (Janbu, 1967).

The time dependence of the consolidation of oedometer-tested samples is obtained from the two parameters c_v and T_v, which are related according to equation 3.42. For each load increment at stepwise loading, T_v is evaluated as: T_v = compression of time (t)/total compression after completion of primary consolidation.

T_v is related to c_v and the sample thickness d by the following relationship:

$$c_v = \frac{T_v d^2}{t} \qquad (3.42)$$

where c_v is the consolidation coefficient, t is the time after onset of consolidation, and d is half the thickness of the soil sample, or the thickness of the soil layer if one uses double-sided drainage.

Compression of clay-rich soils continues after the dissipation of the load-generated porewater overpressure. This gives us the phenomenon of 'secondary consolidation', as represented by α_s in Figure 3.30.

Secondary consolidation can be expressed using the coefficients α_s and C_α in the following fashion (using logarithmic time scales):

$$\alpha_s = \frac{d\varepsilon}{d(\log t)} \quad \text{and} \quad C_\alpha = \frac{de}{d(\log t)} \qquad (3.43)$$

where ε is the compressive strain, and e is the void ratio.

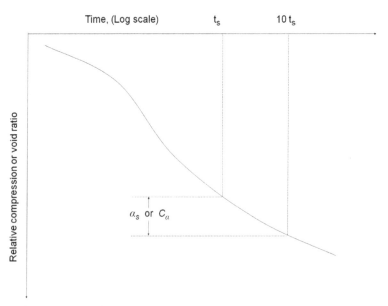

Figure 3.30 Depiction of the secondary consolidation phenomenon.

Consolidation-type performance of swelling clays

As swelling clays are the preferred types of clays used for clay barriers, one needs to be aware of the impact of interlayer phenomena on both compression and yield performance of the clay under external loading. Figure 3.31 shows a comparison of the consolidation characteristics of a laboratory-prepared bulk sodium montmorillonite against the measured swelling behaviour of the same montmorillonite with parallel-particle orientation, prepared using the techniques described by Warkentin and Schofield (1962).

The postulated microstructures at the beginning and end stages of consolidation and swelling are shown in Figure 3.31. For the bulk sample, the microstructure depicted is one that includes a random arrangement of particles, in a fashion that provides some pseudostructural strength to the sample. The oriented particle arrangement shown at the end stage of the swelling pressure test shows expansion of the layers forming the layer–lattice structure of the clay mineral. At the beginning of the swelling pressure test, interlayer expansion is at a minimum, expanding more as swelling against the prescribed restraining pressures is allowed. The influence of the chemistry of the environment with respect to the consolidation performance is seen in the next figure (Figure 3.32), where the consolidation characteristics

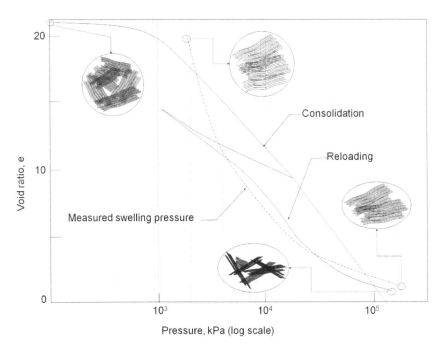

Figure 3.31 Comparison of consolidation and swelling pressure for sodium montmorillonite. Measured swelling pressure is from laboratory-prepared oriented particles at a salt concentration of 10^{-4} mol/L NaCl.

of a slurry-prepared preconsolidated natural bentonite sample is shown in relation to its pH environment. The minerals in the sample of bentonite consisted of montmorillonite, kaolinite, calcite and quartz, listed in decreasing order of abundance. Also included in the composition of the bentonite was a significant proportion of carbonates and a small proportion of organic matter. The sharp drop in consolidation characteristics in the sample as one goes from pH 8.1 to 6.8 is noteworthy.

Shear strain

Shear strain can be determined from direct shear tests and from triaxial tests using an apparatus similar to the one shown in the right-hand side of Figure 3.21. For triaxial tests the stress conditions in the sample are defined by the effective average normal stress $p' = (\sigma_1 + 2\sigma_3')/2$ and the deviator stress, $q = \sigma_1' - \sigma_3'$, which are changed by altering the major principal total stress σ_1 – keeping the cell pressure σ_3 (minor principal total stress) constant. The major principal stress is successively increased until failure is reached (see Figure

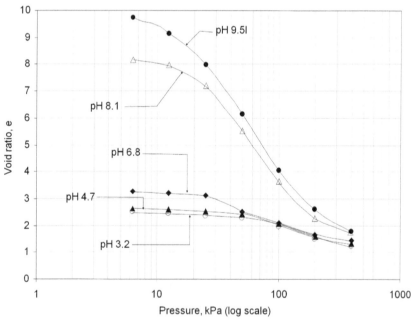

Figure 3.32 pH effect on consolidation of slurry samples of a bentonite lightly pre-consolidated to 25 kPa prior to consolidation testing. The bentonite contained 8 per cent carbonates and 1.4 per cent organic matter. Mineral composition of bentonite consisted of montmorillonite, kaolinite, calcite and quartz, in decreasing order of abundance (adapted from data presented by Ouhadi et al., 2006).

3.26). The axial strain $\Delta l/l$ and the volumetric strain $\Delta V/V$ are measures of the shear strain. Typical results are shown in Figure 3.33.

3.10.6 Creep behaviour

Secondary consolidation, which primarily results from creep strain on the microstructural scale, is particularly important for non-expandable clays with high water contents and for smectite clays. Smectite-rich buffers and backfills deform considerably with time when exposed to a deviator stress, even at low water contents, as will be discussed in Chapter 6. In determination of creep performance, standard incremental load compression–time tests are conducted. Increments of load are introduced at the end of each creep. Figure 3.34 shows the typical creep testing curves obtained under constant load creep tests. The characteristic creep curve can be broken into three distinct performance characteristics as shown in the figure: instantaneous elastic deformation or strain e_e, retarded deformation or strain e_r and a constant rate deformation or strain e_f.

A creep model that is applicable to most ordinary clays has been proposed by Singh and Mitchell (1968) has the following form:

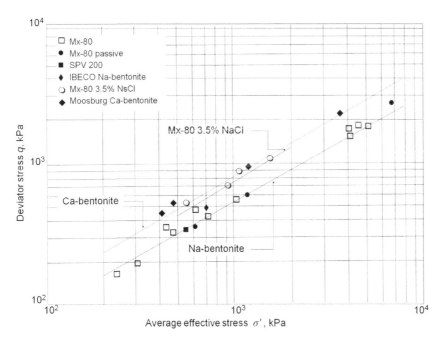

Figure 3.33 Relation between deviator stress q and effective stress σ′ for a number of clay soils including the smectite-rich MX-80 clay. The materials were isotropically consolidated prior to shear application.

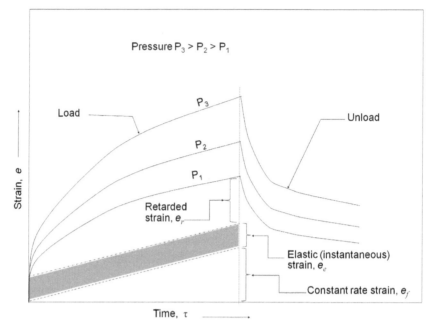

Figure 3.34 Load and unload deformation (strain) relationships in creep testing of clays, showing the various categories of strain (deformation) developed under an applied constant load.

$$\dot{\gamma} = \frac{d\gamma}{dt} = Ae^{\alpha\tau} \cdot t^{-n} \qquad (3.44)$$

where A, α and n are constants, τ is the shear stress, $\dot{\gamma}$ is the rate of shear strain and t is time. This expression can be normalized and related to a reference time t_r as follows:

$$\dot{\gamma}_0 = Ae^{\alpha\tau_0} \cdot t_r^{-n} \qquad (3.45)$$

If the shear stress is expressed as the degree of mobilization of the peak shear strength (τ/τ_f) one obtains (Singh and Mitchell, 1968; Börgesson *et al.*, 1988):

$$\dot{\gamma} = \dot{\gamma}_0 \cdot e^{\alpha\tau/\tau_f} \cdot e^{-\alpha\tau_0/\tau_f} \cdot \left[\frac{t}{t_r}\right]^{-n} \qquad (3.46)$$

where τ_f is the shear strength. Equation 3.45 can be expressed as follows:

$$\log\dot{\gamma} = \log\dot{\gamma}_0 - \alpha\frac{\tau_0}{\tau_f} + \alpha\frac{\tau}{\tau_f} - n\left(\log t - \log t_r\right) \qquad (3.47)$$

A graphical log–log plot of the strain rate $\dot{\gamma}$ and the time after onset of creep t gives a straight line as shown in Figure 3.35. To normalize data from creep tests, t_r and τ_0/τ_f should be given standard values such as $t_r = 10^4$s and $\tau_0/\tau_f = 0.5$.

Microstructure and creep analysis

In terms of the microstructure of clays, we consider the overall (macro) deformation, or strain e, of a clay to be composed of the collection or aggregation of strains of individual and groups of microstructural units. The term *group of microstructural units* used in the discussion refers to groups that have one or more microstructural units. Considering the macrostructure of clay to be composed of groups of microstructural units, and designating the strain of a grouping of such as microstrain ε, one can make the case that ε is similar in characteristics to the macrostrain e. Accordingly, one obtains:

$$\varepsilon = \varepsilon_e + \varepsilon_r + \varepsilon_f = a_i \xi_i + \alpha f_i(t)\xi_i + \beta b_i \xi_i t \tag{3.48}$$

where ε_e, ε_r and ε_p represent the elastic, retarded and constant-rate strains of the microstructural units respectively; a_i is the compliance of the elastic strain, $f_i(t)$ represents the creep compliance for the ith microstructural unit;

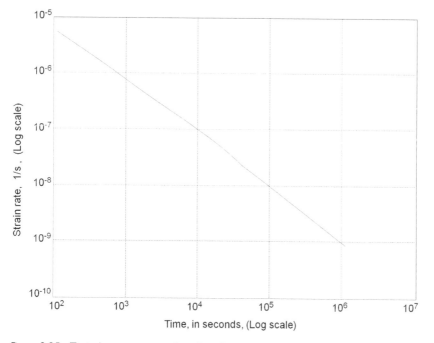

Figure 3.35   Typical creep curve of sodium bentonite, showing almost perfect log-time behaviour ($n = 1$ in equation 3.46).

b_i is the viscous coefficient responsible for continuous flow; ξ_i is the micro-stress acting on a basic grouping of microstructural units.

- $\alpha = 0$, $\beta = 1$ for $\xi_i <$ yield point of a basic grouping of microstructural units;
- $\alpha = 1$, $\beta = 0$ for $\xi_i >$ yield point of a basic grouping of microstructural units.

The macro or overall strain e will be given as:

$$e = \sum a_i \xi_i + \sum_1^m f_i(t)\xi_i + \sum_1^n \beta b_i \xi_i t \qquad (3.49)$$

where m represents the number of basic groups of microstructural units whose yield strengths have not been exceeded, and n is the number of basic groups of microstructural units whose interaction strengths have been exceeded. If we consider the number of groups of microstructural units to be semi-infinite, i.e. $n + m \rightarrow \infty$, it follows that if ξ_i exceeds the basic strength of the ith group, continuous flow of the clay under test will occur. In that instance, we can represent the previous equation (equation 3.49) as a continuous function of the following form:

$$e = A\sigma + \int_{-\infty}^t f(t - \tau)\sigma\, d\tau + Bt\sigma \qquad (3.50)$$

where A is an instantaneous compliance, B is a flow parameter, σ represents the total stress, t is the current time and τ is a time variable.

The probability of occurrence P_i of a group of microstructural units at a particular state is a function of its energy state E_i, i.e. $P_i = f(E_i)$. This is another way of saying that the energy state of a group of microstructural units defines its state of being. If each group of microstructural units with volume v constitutes an elemental system in a canonical ensemble of volume Ω, and if there are M elemental systems in the ensemble, the probability of occurrence of one system v in Ω in a state i is proportional to the number of states accessible to Ω. Accordingly:

$$P_i = c\omega(E^{(o)} - E_i) \qquad (3.51)$$

where c is a proportional constant independent of i and $\omega(E^{(o)} - E_i)$ represents the number of energy states accessible to the systems remaining in Ω. From a normalization procedure, where the summation includes all possible states of v, $\Sigma P_i = 1$, and with the use of a partition function,

$$\frac{1}{C} = \sum e^{-\lambda E_i}$$

it can be shown that:

$$P_i = \frac{e^{-\lambda E_i}}{\sum e^{-\lambda E_i}} \tag{3.52}$$

and that the mathematical expectation of the total deformation strain $<\varepsilon(t)>$ can be obtained as follows:

$$<\varepsilon(t)> = A\sigma + \int_0^t \frac{P_i(\tau)}{g_i} f(t-\tau)\sigma\, d\tau + B\sigma t \tag{3.53}$$

where g_i represents the number of groups of microstructural units at the same energy level and, as

$$\left[\frac{\partial \ln \omega}{\partial E'} \right]_0 = \lambda$$

is a constant at a fixed energy level $E' = E^{(0)}$, one obtains

$$\lambda = \frac{1}{KT}$$

where K represents a positive constant with dimensions of energy, T is the temperature and E' is the energy of $M-1$ elementary systems remaining in Ω for the situation where any one system has energy E_i.

3.10.7 General material model

Although a number of problems in classical soil mechanics and also in repository design and performance assessment can be treated and solved by applying simple, individual analytical models that account for compression, shearing etc., coupled processes including expansion and thermal processes require the use of general material models. This section will summarize the basic features of a general model for smectitic clays that would be useful for application to complex situations.

Properties modelled

The components and the material model can be grouped as follows (Börgesson *et al.*, 1988):

- Drucker–Prager plasticity behaviour;
- porous elasticitic behaviour;

- pore and particle properties behaviour;
- thermal and thermomechanical behaviour;
- initial conditions.

Drucker–Prager plasticity, extended version

The simple Mohr–Coulomb failure criterion is not sufficiently accurate for many purposes and an extended version of the Drucker–Prager plasticity model has been developed, in which the influence of the intermediate principal stress can be taken into account and that dilation can also be simulated (Börgesson *et al.*, 1988). The parameters used in the model are illustrated in Figure 3.36. The figure shows that the stress dependence is caused by the 'angle of internal friction' β in the $\sigma_j - p$ plane and that the parameter d illustrates the 'cohesion' (see Figure 3.26 for the Mohr circle representation). Introducing plastic flow $d\varepsilon^{pl}$, the flow direction is perpendicular to the yield surface angle, which means that $\psi = \beta$. In order to decrease the resulting dilatation it is necessary to assume non-associated flow by making $\psi < \beta$. When the stress path enters the plastic region ds^{pl}, the yield surface is moved upwards until it reaches the failure surface.

The model also includes a constant K, which controls the dependence of the yield surface on the intermediate principal stress. The yield surface is defined such that K is the ratio of the yield stress in triaxial tension to the yield stress in triaxial compression. The parameters needed for the extended Drucker–Prager model are thus β, d, K, ψ and the yield function f. Typical values for bentonite at high densities for the parameters are $\beta = 20°$, $d = 100\,\text{kPa}$, $\psi = 2°$ and $K = 0.9$. The yield function, i.e. the relationship between

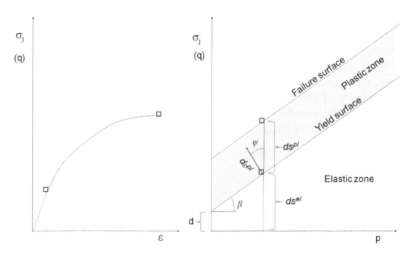

Figure 3.36   Basis for extended Drucker–Prager plasticity model (adapted from Börgesson *et al.*, 1988).

Table 3.4 Yield function data (Börgesson et al., 1988)

σ_j (kPa)	ε_j
113	0
138	0.005
163	0.02
188	0.04
213	0.1

σ_j and ε_p is given in the form of discrete data in Table 3.4. σ_j is the von Mises stress and ε_j is the plastic strain for a stress path that corresponds to uniaxial compression, i.e. when there is no confining pressure. Linear interpolation can be made using these data.

Porous elasticity

The stress–void ratio relation can be expressed by use of a 'porous elasticity' model 'elasticity', in which the volumetric behaviour is defined according to the following relationship (Börgesson *et al.*, 1988):

$$\frac{\kappa}{1+e_0}\ln\left(p_0\,/\,p\right)=j^{el}-1 \tag{3.54}$$

where e_0 is the initial void ratio, p_0 is the initial average stress, κ represents the inclination of the e-log p relation, and j^{el} is the elastic volume ratio. The key parameter κ is the logarithmic bulk modulus for which a typical value is 0.21.

Pore and particle properties

The water density ρ_w and the bulk modulus B_w of the porewater as well as the bulk modulus of the solid particles can be taken as: $\rho_w = 1000\,\text{kg/m}^3$, $B_w = 2.1 \times 10^6\,\text{kPa}$, and $B_s = 2.1 \times 10^8\,\text{kPa}$.

The flow of water through the clay is modelled by applying Darcy's law, with the hydraulic conductivity k being a function of the void ratio e (interpolation between the values). The relationships shown in Table 3.5 are valid (Börgesson *et al.*, 1988).

Table 3.5 Relationship between hydraulic conductivity k and void ratio (e)

e	k (m/s)
0.45	1.0×10^{-14}
0.70	6.0×10^{-14}
1.00	3.0×10^{-13}

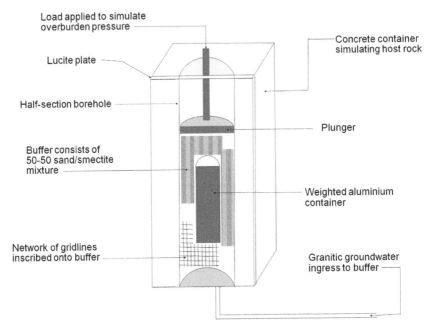

Figure 3.37   Deformation creep experiment using a 1:8 half-section model. The network of gridline inscribed onto the buffer permits one to record the distortion of the network of gridlines during the deformation creep test, to determine the movement of representative points in the buffer under load. The half-cell aluminium container, which measures 76.4 mm in diameter and 280 mm in height, is weighted to simulate canister loading (adapted from Yong *et al.*, 1985).

3.10.8 Complex stress field

The stresses imposed on clay barriers and buffers are not always simple compressive or simple triaxial stresses. Rheological (deformation–time and stress–strain) relationships obtained from laboratory tests, such as creep or triaxial tests, may not suitably represent the stress deformation conditions in these barriers and buffers. To demonstrate the preceding, we can use as an example a laboratory test using the set-up shown in Figure 3.37. This test set-up is a 1:8 scale of a repository system housing a canister containing HLW.

The half-repository set-up shown in Figure 3.37 contains a weighted aluminium container that is weighted with weights to simulate loading from a HLW canister. The amount of weights can be varied to study different load conditions – in addition to the added pressure placed on the container from the surcharge (overburden) weight. The network of gridlines inscribed onto the bentonite clay–sand buffer allowed one to observe grid-point deformations in relation to time of load application. This permitted one to develop

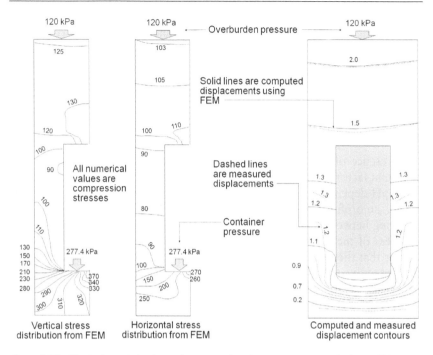

Vertical stress distribution from FEM

Horizontal stress distribution from FEM

Computed and measured displacement contours

Figure 3.38 Vertical and horizontal stress distributions computed from FEM, and comparison of computed and measured displacements – at end of primary creep period (18 days) – prior to ingress of granitic groundwater. The buffer material consisted of 50 per cent smectite and 50 per cent sand, with a dry density of 1.57 mg/m³ and a moisture content of 23.5 per cent. Swelling pressure of the mixed buffer material was measured at 47 kPa, and free swell was measured at 10.7 per cent.

or determine the kind of rheological relationship to be used in determining the stress field developed in the confining clay-sand buffer system. Figure 3.38 shows the displacement contours (right-hand figure) after 18 days of primary creep, together with horizontal and vertical stress fields determined from finite element analysis – using two types of stress–strain relationships.

Observations of the deformation–time relationships for the various nodes of the inscribed grid network showed a creep-type performance for the buffer region above the container. The deformation of the buffer directly beneath the container demonstrated a bearing capacity type of behaviour, suggesting thereby that a stress–strain relationship derived from consolidated drained triaxial tests would be appropriate. However, the deformation–time relationships obtained for the buffer surrounding the container proved to be complex, i.e. there were no immediately recognizable patterns of behaviour. The results shown in Figure 3.38 using finite element analysis used two types of constitutive relationships: (a) a creep-type rheological relationship for the buffer above the container and (b) stress–strain relationships from triaxial tests for the buffer surrounding the container and also below the container. The discrepancies between calculated and measured deformations in the buffer surrounding the container testify to the effect of the complex stress field in this region and the need for a better specification of the rheological behaviour of this buffer region.

3.11   Concluding remarks

The presence of various kinds of clay fractions and their interactions are important factors in the evaluation of the surface properties of clays. Because of the pH dependency of the surface properties of such fractions as organic matter, amorphous materials, and even some clay minerals, interactions occurring between various clay fractions will change the characteristic SSA and CEC of the clays. We see from the discussion in the early part of this chapter that the surface properties of the mineral particles are more or less dominated by the hydroxyl surface functional groups, whereas organic matter possess a greater variety of surface functional groups.

The energy characteristics defined by clay–water potentials provide us with information on how strongly water is held to the clay particles. As one might expect, the matric potential (component) ψ_m, which is a property of the clay particles' surfaces is by far the most important clay–water potential component. Computations of mid-plane potentials based upon interactions of counterions and clay particle reactive surfaces in DDL-type models are, by and large, in accord with the osmotic potential ψ_π. Much work remains to fully reconcile the matric component ψ_m with the potentials at the Stern layer or at the inner Helmholtz plane. Water-holding capacities and clay–water potentials, together with DDL-type models, provide us with the basic elements of interactions between water and clay particles.

The importance of role of microstructure in the control of the hydraulic conductivity and other transmission characteristics of clay, and in the development of clay integrity and rheological properties, cannot be overstated. We need to emphasize that the microstructural units and groups of microstructural units are never static – in the sense that restructuring of the microstructural units and groups of such units will always occur in response to internal and external provocative gradients. The particular characteristics of interlayers and interactions in interlayers and between stacked layers contribute greatly to the nature of swelling clays, and, in doing so, make it all the more important for one to obtain a better understanding of the nature of the clay and the importance of macro- and microstructures.

Chapter 4

Clay–water reactions and partly saturated water transport

4.1 Reactions in porewater

4.1.1 Chemical reactions and water uptake

Clay–water interactions and reactions involve, for example, acid–base reactions, oxidation–reduction reactions, and hydrolysis. The interactions and reactions between the clay minerals and water establish the internal energy state of a clay–water system, and affect not only the physical and mechanical properties of clays, but also their hydraulic properties and transport processes in clays. In engineered clay buffers and barriers, the initial water content of the clay is expected to be very low, as engineered clay buffers and barriers are emplaced in a semi-dry to an air-dried state. Interactions between water and clay minerals, and chemical reactions in the porewater and with the clay minerals, will occur first when water is taken into the clay. Various mechanisms and processes are involved in water entry and movement in anhydrous and partly saturated clays. To distinguish 'what is happening' at the various stages of water movement in clays, we need to review our perception and understanding of the various terms used to describe water movement, interactions and reactions in clay–water systems.

Chemical reactions in the porewater occur between the various chemicals and solutes in the porewater, regardless of their origin. Some of these may originate as chemicals or solutes tenuously attached to the clay minerals and some of these originate in the porewater itself. Chemical reactions also occur between the chemicals or solutes in the porewater and the chemically reactive surface functional groups associated with the clay minerals. *Interactions* between clay minerals and water, and between contiguous clay minerals, are most likely to be physico-chemical in nature. Interparticle action generally involves short- and long-range forces, previously described in Chapter 3.

We consider a natural clay–water system devoid of *ex situ* originating contaminants as the basis for our discussion in this section. The examination and discussion of contaminant–clay interactions are the subjects of concern in Chapter 5. Naturally occurring salts, such as those included in groups I and II in the periodic table, constitute the predominant solutes in the porewater of clays. Also included in the porewater are traces of some transition metals and heavy metals, many of which originate from lateritic clays and chemical weathering products of metalliferous rocks. At equilibrium, the chemistry of the porewater is a product of the reactions involving the chemicals in the porewater, the dissolved solutes, the chemically reactive surface functional groups of the clay minerals and the chemical reactions occurring in the porewater – including the biologically mediated chemical processes. At the same time when the reactions occur in the porewater, we will have diffusive flow of chemicals, dissolved solutes and decomposed colloids of clay minerals in the porewater.

Entry of water into the clay buffer/barrier and its subsequent movement in the clay mass involves a number of processes. These can be divided into three convenient categories based broadly on the initial conditions of the clay mass as follows: (a) initial uptake of water by semi-dry clays; (b) unsaturated flow, i.e. water movement in partly saturated clays; and (c) fully saturated flow, i.e. water movement in fully saturated clays. The term *partly saturated clays* is used in preference to *unsaturated clays* to recognize the fact that the range of water contents in partly saturated clays can vary from relatively dry to relatively wet, i.e. anywhere from just greater than the anhydrous state to near full water saturation of the clay. As water movement in a particular clay mass can be due to the action of internal and/or external forces and factors, we have used the term *water uptake* in Chapter 3 in reference to water entry into clays, and especially into swelling clays when the water is taken into the clay by internal forces. Strictly speaking, the term *water uptake* refers to the water drawn into a dry or partly saturated clay by internal forces resulting in wetting of the clay – and in the case of swelling clays, we have seen that these internal forces are associated with the matric and osmotic potentials, ψ_m and ψ_π, respectively. In water uptake and unsaturated flow in clays, vapour flow and liquid water flow coexist, with vapour flow being the more significant component in the water uptake phase. Regardless of whether it is liquid and/or vapour movement that results in water movement in clays, the portion of the clay mass in contact with the incoming water will swell. This will reduce its bulk density and increase its volume. This swelling and expansion will continue until full saturation is achieved. By this progressive swelling behaviour, an initially semi-dry smectite buffer encasing a HLW canister will become more homogeneous, through the gradual filling of the macro voids between microstructural units by the expanding smectite.

We cannot however specify hard and fast demarcation boundaries separating the groups simply because of (a) the absence of specific demarcation boundaries separating *semi-dry* from *partly saturated* states; (b) the wide range of degrees or levels of saturation that qualify under the *partly saturated* designation; (c) the heterogeneity of the clay mass that constitutes the clay buffer and/or barrier; and (d) the impact of local microstructural control on all the aspects of water entry and its subsequent movement.

4.1.2 Concept of acids, bases and pH

The concept of acids and bases first proposed by Arrhenius in 1883, and refined in 1887, has been shown to be useful in application to water. In this concept, an *acid* is defined as a substance that dissociates itself to produce H^+ ions, and a *base* is defined as a substance that dissociates itself to produce OH^- ions. The studies of the 'Arrhenius theory of electrolytic dissociation' by Sörenson in 1909 led to the development of the pH scale as a means to identify the degree of acidity. With this scale, one determines the pH of a solution to be the negative logarithm to the base ten of the molar hydrogen ion concentration. There are, however, some limitations in application of the Arrhenius concept to solvents and to the specific requirement for the presence of OH^- ions in a base.

The broader concept of acids and bases proposed by Brønsted (1923), leading to the Brønsted–Lowry definition of acids and bases, perhaps provides a much simpler set of definitions as follows: an *acid* is a substance that has the tendency to lose a proton (H^+), and, conversely, a *base* is a substance that has a tendency to accept a proton. Simply put, an acid is a *protogenic* substance, i.e. it is a *proton donor*. In the same vein, a base is a *protophilic* substance, i.e. it is a *proton acceptor*. One could say that a base has the ability to remove a proton from an acid. There are substances that can both donate and accept protons, i.e. they are both protogenic and protophilic. Water and alcohols are such substances. In such instances, these substances are called *amphiprotic* substances.

The Lewis (1923) concept of acids and bases avoids the requirement of a solvent or aqueous solution in defining acids and bases. This is important in the case of clays, and especially in the context of clay buffers and barriers. In the Lewis concept, an *acid* is a substance that is capable of accepting a pair of electrons for bonding, and a *base* is a substance that is capable of donating a pair of electrons. This means that *Lewis acids* are *electron-pair acceptors*, whereas *Lewis bases* are *electron-pair donors*. This also means that substances with unshared pair of electrons are bases, and substances lacking an octet are acids. With this set of definitions, metal ions M^{n+} are Lewis acids. Theoretically, all cations are Lewis acids. Water is a Lewis base and also a Brønsted base. Whereas Lewis bases are also Brønsted bases, Lewis acids are not necessarily Brønsted acids, as Lewis acids include substances that are not proton donors. The Lewis concept of acids and bases allows one to treat metal ligand bonding as an acid–base reaction.

Pearson (1963) classified a whole range of atoms, ions and molecules into three groups as hard, borderline and soft Lewis acids and Lewis bases. This classification is known by the acronym HSAB (hard, soft, acid, base) classification. With this classification, Pearson noted that hard acids prefer to bind with hard bases, and soft acids prefer to bind with soft bases.

- *Lewis hard acids* are generally small in size, with high positive charge, high electronegativity and low polarizability, and do not have unshared pairs of electrons in their valence shells. The acids include aluminium chloride, arsenic (III) ion, chloride ion, iron (III) ion, magnesium ion, manganese (II) ion, uranium (IV) ion and zirconium ion.
- *Lewis borderline acids* range from antimony (III) ion, copper (II) ion and iron (II) ion to sulphur dioxide and zinc ion.
- *Lewis soft acids* are generally large in size with low positive charge and low electronegativity – in contrast with Lewis hard acids. They have unshared pairs of electrons in their valence shells. The acids range from borane, cadmium ion, hydroxyl cation and iodine, to thallium (III) ion and 1,3,5-trinitrobenzene.
- *Lewis hard bases* usually have high electronegativity and low polarizability and are difficult to oxidize. The bases range from acetate ion, ammonia, carbonate ion and chloride ion, to hydroxide ion, nitrate ion and water.
- *Lewis borderline bases* include aniline, bromide ion, nitrogen and sulphide ion.
- *Lewis soft bases* usually have low electronegativity and high polarizability and are easy to oxidize – in contrast to the hard bases. The bases range from benzene, ethylene, hydride ion, iodide ion to trimethylphosphite.

4.1.3   *Chemical buffering capacity of clay*

The preceding concepts of acids, bases and pH are useful to express the chemical state in porewater of clays. Bearing in mind that porewater is in contact with the charged surfaces of clay minerals, increases of H^+ or OH^- in clay porewater cannot easily contribute to (a) the decrease of pH because of ion exchanges with the adsorbed cations and (b) the increases of pH because of dissociation of H^+. These phenomena in clay porewater are called the *chemical buffering action* of clay and this is expressed in terms of its chemical buffering capacity. In essence, the chemical buffering capacity of a clay identifies the capability of the clay to accept inputs of acids or bases without significant changes in its pH status. The *clay buffer capacity* can be defined as the number of moles of H^+ or OH^- that must be added to the clay to lower or raise the pH of the clay by one pH unit. To determine the buffer capacity, one titrates a clay suspension with a strong acid or base, and subsequently analyses the slope of the titration curve. Figure 4.1 shows the titration curves for two clays (a montmorillonite and an illite) and a blank solution that is devoid of any clay solids. We can define *clay buffering capability* in terms of the changes in the amount of hydrogen ions H^+ or hydroxyl ions OH^- added to the system as follows:

$$\beta = \frac{dOH^-}{dpH} = \frac{dH^+}{dpH}$$

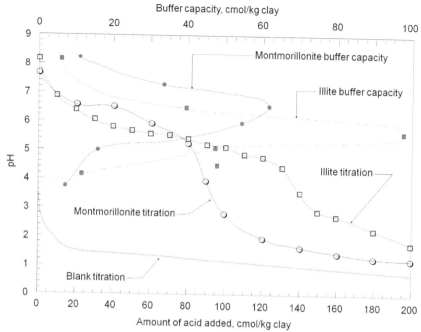

Figure 4.1 Titration and buffer capacity curves for an illite and a montmorillonite. For comparison, the titration curve for a blank (pure solution, without any clay minerals) is also shown. Note that the buffer capacities (curves), with top horizontal axis values, are given with respect to the pH ordinate axis.

A plot of the negative inverse slope of a titration curve against the relevant pH will define the buffer capability, as shown in Figure 4.1. The pH range over which the buffer capacity of each clay exists together with the magnitude of the buffer capacity are evident from the figure.

4.1.4 Acid–base reactions

It is important to consider the various chemical reactions in porewater to be acid–base reactions. Acid–base reactions are defined as the proton transfer reactions between a proton donor (acid) and a proton acceptor (base). Proton loss is called *protolysis* and the proton transfer reaction is a *protolytic reaction*. Water is both a protophilic and a protogenic solvent, in that it can act either as a base or as an acid. Technically speaking, water is an amphiprotic substance. It can undergo self-ionization, thereby producing the conjugate base OH^- and conjugate acid H_3O^+. That is $2H_2O$ (solvent) $\Leftrightarrow H_3O^+$ (acid) $+ OH^-$ (base).

This self-ionization is called autoprotolysis, the reverse of which is neutralization. Hydrolysis is also a neutralization process, and, technically speaking, classifies as an acid–base reaction in as much as it (hydrolysis) refers to the reaction of H^+ and OH^- ions of water with the dissolved solutes and other constituents present in the porewater.

For example, the hydrolysis of iron in porewater is shown as:

$$Fe^0 \Leftrightarrow Fe^{2+} + 2e^-$$

$$2H_2O + 2e^- \Leftrightarrow 2OH^- + 2H^+$$

Then:

$$Fe^0 + 2H_2O \Leftrightarrow Fe^{2+} + 2OH^- + 2H^+$$

$$\text{or } Fe^{2+} + 2H_2O \Leftrightarrow Fe(OH)_2 + 2H^+$$

Hydrolysis reactions of metal ions in the porewater are influenced by (a) pH of the active system; (b) type, concentration and oxidation state of the metal cations, and the redox environment; and (c) temperature. Favourable circumstances for hydrolysis reactions include high temperatures, low organic contents, low pH environment and low redox potentials.

4.1.5   Oxidation–reduction (redox) reactions

Oxidation–reduction (redox) reactions involve the transfer of electrons. As with the acid–base reactions, oxidation–reduction reactions are important processes that occur in clays that are used as buffers and barriers in multi-barrier systems for the containment of HLW and HSW. Redox reactions can be abiotic and/or biotic. Biotic redox reactions are of greater significance than abiotic redox reactions. Microorganisms play a significant role in catalysing redox reactions. Their utilization of redox reactions as a means of extraction of the energy required for growth serves as a catalyst for reactions involving molecular oxygen and organic matter (and organic chemicals) in the clays. The activity of the electron e^- in the chemical sytem is a dominant influence. Generally speaking, the transfer of electrons in a redox reaction is accompanied by proton transfer. In the case of inorganic solutes in the porewater, redox reactions result in the decrease or increase of the oxidation state of an atom – a matter of some significance for those ions that have multiple oxidation states.

The *redox potential Eh* is a measure of electron activity in the porewater. It allows one to determine the potential for oxidation–reduction reactions in a clay–water system, and is determined as follows:

$$Eh = 0.0591pE = E^0 + \left(\frac{RT}{nF}\right)\ln\frac{a_{i,ox}}{a_{i,red}} \tag{4.1}$$

where the expression pE is a mathematical term that represents the negative logarithm of the electron activity e^-, E^0 is the standard reference electrode potential, R is the gas constant, T represents the absolute temperature, F is the Faraday constant, n is the number of electrons, a refers to the activity of the ith species, and the subscripts ox and red refer to *oxidized* and *reduced* respectively.

The maximum amount of acid or base that can be added to a clay–water system without any measurable change in the Eh or pE of the system will establish the redox capacity of the clay. In a sense, this concept is similar to the buffering capacity concept of clays discussed in the previous section. Many chemical reactions occurring in a clay–water system are dependent on temperature, concentration of solutes in the porewater, ligands in the porewater, and on Eh and also pH – with the last two being important because proton transfer is neutralized by electrons. The Nernst equation, which is similar to equation 4.1, demonstrates this influence:

$$pE = 16.92 Eh = E^0 + \left(\frac{RT}{nF} \right) \ln \frac{[A]^a [H_2O]^w}{[B]^b [H^+]^h} \tag{4.2}$$

where the superscripts a, b, w and h refer to the number of moles of reactant, product, water and hydrogen ions respectively. The pE–pH diagram for an iron–water system for a maximum soluble iron concentration of 10^{-5} mol/L is shown in Figure 4.2.

4.1.6 Physical reactions and hydration

The charged surfaces of clay minerals attract water molecule, resulting in the formation of hydrated clay minerals. Ions in porewater are also hydrated (with water molecules) and are bound to hydrated clay minerals by such forces as Coulomb, van der Waals, short-range repulsion forces and covalent bonds, as described in section 3.7.2. For air-dried clays, the hydration state of ions in porewater is considered to be different from that which would occur in bulk solution. In general, the hydration state of ions or the association state of ions with clay mineral surfaces can be demonstrated by observing the status of oxygen around the ions in porewater – using neutron diffraction, the X-ray diffraction method, and the extended X-ray absorption fine structure spectroscopy method (see Nakano *et al.*, 2004).

Table 4.1 shows the coordination number of water molecules around Cs^+, Ba^{2+} and Sr^{2+} in porewater and their bond distances as determined by extended *X-ray adsorption fine structure* (EXAFS) analyses using the two-shell fitting technique on oxygen around metal ions in porewater. In addition, the number of the clay mineral oxygens associated with the metals and the bond distances between the metals and the clay mineral oxygen are

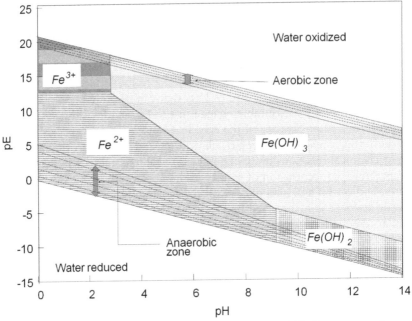

Figure 4.2 pE–pH chart for Fe and water with maximum soluble Fe concentrations of 10^{-5} mol/L. The zone sandwiched between the aerobic and anaerobic zones is the transition zone (adapted from Yong, 2001).

shown for air-dried samples in comparison with that for the paste samples. The bond distance between the metals and the clay mineral oxygen expresses the distance between metal and clay mineral surface. In other words, Table 4.1 shows the hydration state of Cs^+, Ba^{2+} and Sr^{2+} in porewater and the mode that the ion is attracted to the hydrated clay mineral surface. The hydration numbers of Cs^+, Ba^{2+} and Sr^{2+} of 6, 8 and 8 in bulk solution, respectively, have been measured using *neutron diffraction* or *X-ray diffraction* methods (Albrigh, 1972; Ohtomo and Arakawa, 1979). In comparison with these results, the hydration number of metals in porewater is smaller for air-dried clays, whereas the hydration number for the paste samples is nearly equal to that in bulk solution. The number of clay mineral oxygens associated with the metals is about three to four, except for that of Cs ion in the air-dried state. Cs ions specifically associate with six mineral oxygens, indicating their strong bonding to mineral surfaces in the air-dried state. The evidence suggests that metals in the porewater hover over clay mineral surfaces because of the hydration of clay minerals. This conclusion arises from the observation that the distances between metal and mineral oxygen is considered to be slightly larger than the bond distances of metal–oxygen in the hydrated metals in porewater. Accordingly, the metals are assumed to be

Table 4.1 Hydration of ions and their association with clay minerals

Sample	pH	Cs		Ba		Sr	
		Air dried	Paste	Air dried	Paste	Air dried	Paste
aHydration of metal ion in porewater							
Number of water molecules coordinated	4.5	4.5	5.6	5.1	6.9	5.6	6.8
	10	4.5	6.1	5.7	6.8	5.4	7.2
Distance between metal and water molecule (Å)	4.5	3.17	3.19	2.88	2.87	2.58	2.59
	10	3.18	3.18	2.85	2.87	2.58	2.61
bAssociation of metal ion with clay mineral							
Number of mineral oxygens associated with metal	4.5	5.9	3.9	3.7	3.7	2.7	2.4
	10	6.3	4.2	4.3	4.1	2.5	2.5
Distance between metal and oxygen on mineral (Å)	4.5	3.55	3.62	3.09	3.11	2.77	2.80
	10	3.56	3.61	3.07	3.11	2.76	2.83

The data have been obtained from the two-shell fitting technique of EXAFS spectra.
a Denotes the first shell of oxygen around the metal ion.
b Denotes the second shell of oxygen around the metal ion.

easy to partition in porewater when water will migrate to the air-dried clays from the surroundings.

4.2   Water uptake and partly saturated flow

4.2.1   Water uptake in anhydrous and partly saturated clays

For swelling clays such as the 2:1 dioctahedral series of alumino-silicate clays (montmorillonites and nontronites for example) water uptake by internal forces associated with the matric potential ψ_m can exceed 2 nm in thickness at the surface of the clay minerals. Uptake of water beyond that used to form the hydration layers for certain monovalent montmorillonites (e.g. sodium and lithium montmorillonites) is due to the mechanisms represented in the diffuse double-layer (DDL) models, determined by the strength of the electric field of the ions in the porewater and the repulsion forces between two sheets of clay layers (Figure 4.3).

For clays that are initially dry (anhydrous clays), the primary mechanism involved in initial wetting of clay mineral surfaces will be due to hydration forces such as the covalent bond, Coulomb, van der Waals, and the

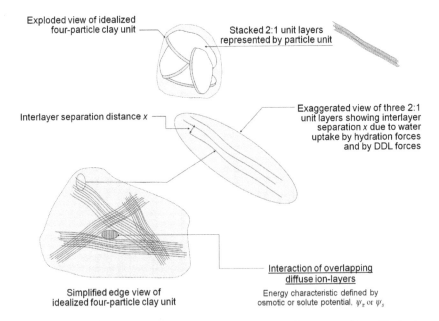

Figure 4.3   Exploded view of idealized four-particle clay unit, illustrating effect of hydration water layers and interaction of overlapping diffuse ion layers on restructuring of the original microstructure. The clay in the four-particle unit is a 2:1 layer silicate, e.g. sodium montmorillonite.

short-range repulsion force between water molecules and atoms constituting the clay minerals. The characteristics of water uptake beyond the hydration layers are different between non-swelling and swelling clays. For non-swelling clays, the thickness of the hydration water adsorbed onto the surfaces of the clay minerals will not exceed 1 nm. This water, which is often called *crystalline water,* possesses a structure that is unlike that of ordinary bulk water. In swelling clays, the interaction of the overlapping diffuse ion layers will characterize the swelling phenomenon, leading to clay volume expansion that is called *swelling.* The interlayer or interlamellar expansion due to crystalline water uptake (hydration) is a function of the layer charge, interlayer cations, properties of adsorbed liquid and particle size. The energy characteristics associated with this have been discussed in section 3.7. By and large, water uptake beyond hydration due to double-layer forces will increase interlayer separation space in proportion to $1/\sqrt{s}$, where s is the electrolyte concentration in the liquid phase. It is important to note that the characteristics of water uptake and uptake of porewater containing contaminants by swelling clays differ somewhat from those of non-swelling clays. This is owing to the energy requirements associated with the movement of water in the interlayers, as distinct from the movement of water between particles. Analyses and predictions of water and solute movement based on specification of solute and matric potentials ψ_s and ψ_m need to recognize how these measurements are made, and, in particular, the role of these potentials in water uptake and movement (water transfer).

4.2.2 *Water movement and water uptake in partly saturated clays*

For partly saturated clays, water uptake and movement are important study subjects because of the different types of mechanisms involved in water transport, dependent on the degree of saturation in the partly saturated clay. The term *water transfer* is used to indicate that the total amount of water moved from one location to another in partly saturated clays because the transfer can be due to either liquid transfer (water movement) or vapour transfer, or both. The term *water movement* is used to indicate movement of water or porewater from one location in the clay to another. Discounting advective flow, water movement in the liquid phase in clays can be ascribed to forces associated with the gradients established by the matric and solute potentials, ψ_m and ψ_s respectively. The simple osmometer experiment shown in Figure 4.4 demonstrates the contributions made to water holding capacity (or potential) by the two potential gradients.

The osmometer shown in the figure consists of two compartments. The right-hand compartment contains de-ionized water with access to (a) a partly saturated (moist) clay containing a specified concentration of solutes in the left-hand compartment (top device) and (b) a solution containing the same

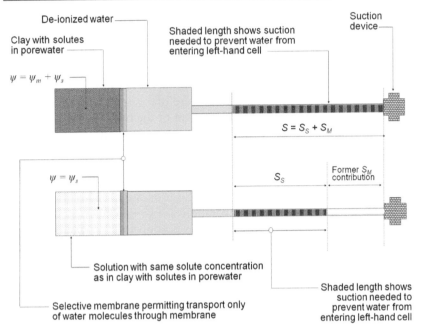

Figure 4.4 Osmometer-type cells showing development of suction required to counter flow of water into the left-hand side chambers because of the total potential ψ in the clay soil (top), and because of the solute potential ψ_s in the cell with the solution containing the same concentration of solutes (bottom). For a partly saturated clay (moist clay) $\psi = \psi_m + \psi_s$. However, for a fully saturated clay, ψ will be equal to ψ_s. S, total suction; S_s, solute suction; S_M, matric suction.

solutes as in the clay and at the same solute concentration in the porewater (bottom). The semipermeable membrane separating the right-hand and left-hand compartments is permeable to water but impermeable to solutes. Hence, only water can be transported through the membrane. The water in the right-hand compartment, which is connected to a suction measuring device, is devoid of solutes. The length of the shaded horizontal column represents the suction required to prevent water from entering the right hand compartment to the left-hand compartments in response to the potential gradients established by the clay or the solutes in the solution.

In the bottom osmometer, the solute concentration gradient set up by the solute potential ψ_s serves as the mechanism for movement of water from the right-hand compartment. The suction required to prevent this movement can be identified as the solute suction S_s. When the left-hand compartment contains a moist clay, as in the top osmometer, with the same kind and concentration of solutes, the required suction to prevent movement of water into the moist clay compartment is much higher, owing to the addition of the matric potential ψ_m in the moist clay. This simple 'thought experiment' illustrates the fact that even without external pressures and temperature gradients, liquid water moves in partly saturated and saturated clays in response to internal gradients developed as a result of the differences in potentials between adjacent points.

When internal gradients are greater than the external ones, water movement in partly saturated clays under low external gradients occurs along film boundaries in pore spaces that are not completely filled with water. This phenomenon is called *film boundary transport*. The activation energy required for movement of ions in boundary layer associated with wetted clay particles is given as:

$$E_i = \frac{ze^2}{\varepsilon} \sum_{i=1}^{i=n} \frac{1}{r_i} \tag{4.3}$$

where E_i is the electrical energy of ions, z is the valence, e is the electronic charge, ε is the dielectric constant, n is the number of charge sites on the particle and r_i is the distance between the ion and the charge site i. It is useful to note that the processes associated with assimilation of contaminants and degree of bonding between clay minerals and contaminants are to some degree affected by the properties of this layer.

4.2.3 Wetting front and water movement in non-swelling clays

The rate of water movement in partly saturated clays is generally determined in terms of changes in the volumetric water content at a point by mass conservation law. The equation of continuity, which states that the flow of water into or out of a unit volume of clay is equal to the rate of change of the volumetric water content, for a no-volume-change condition, is given as:

$$\frac{\partial \theta}{\partial t} = -\frac{\partial v}{\partial x} \tag{4.4}$$

where v is the macroscopic velocity (flux), x is the spatial coordinate and t is time.

For water movement in water uptake (including the film flow) in the absence of external gradients, and assuming that the gradient of the total potential ψ is responsible for movement of water in the partly saturated clay, the conventional procedure is to use the compliance relationship expressed in terms of the total water potential ψ as:

$$v = -K(\theta) \frac{\partial \psi}{\partial x} \tag{4.5}$$

where $K(\theta)$ represents the water entry or infiltration coefficient. This coefficient corresponds to the unsaturated hydraulic conductivity and is dependent on the volumetric water content θ of the clay. One obtains thereby:

$$\frac{\partial \theta}{\partial t} = \frac{\partial}{\partial x}\left(K(\theta)\frac{\partial \psi}{\partial x} \right)$$
$$= \frac{\partial}{\partial x}\left(K(\theta)\frac{\partial \psi}{\partial \theta}\frac{\partial \theta}{\partial x} \right) \tag{4.6}$$

If ψ is a single-valued function of θ, equation 4.6 can be written as:

$$\frac{\partial \theta}{\partial t} = \frac{\partial}{\partial x}\left(D(\theta)\frac{\partial \theta}{\partial x} \right) \tag{4.7}$$

where:

$$D(\theta) = K(\theta)\left(\frac{\partial \psi}{\partial \theta} \right)$$

is the water diffusivity coefficient. This is the common relationship cited for water movement in partly saturated clays.

The initial and boundary conditions are: $\theta = \theta_i = \text{const.}$ at $x > 0$, $t = 0$, and $\theta = \theta_{sat}$ at $x = 0$, $t > 0$, where θ_i is the initial water content and θ_{sat} is the saturated water content.

It is not uncommon to associate the compliance relationship shown in equation 4.5 with the Darcy flow relationship described in section 3.9. This is acceptable as long as the Darcy conditions are met (see Figures 3.19 and 3.20). Not unlike the ideal conditions required for proper application of the Darcy model, for unsaturated flow or infiltration, the key elements that need to be in place are (a) homogeneous and uniform sample; (b) no volume change upon infiltration or water uptake; and (c) absence of advective forces in water uptake. When these conditions prevail, time-wetting studies show that a linear relationship exists between the wetting front distance x from the source and the square root of time ($\sqrt{t}$) required for the wet front to reach x (as will be discussed later).

Determination of water entry and the characteristics of water movement into clays is facilitated by using a system that allows one to observe and record the rate of advance of the wetting front, i.e. the rate at which the wetting front advances into the clay. The advantage of the system shown in Figure 4.5 lies in its ability to impose both positive and negative hydraulic heads to the sample. Note that the tops of the tubes are sealed, and that

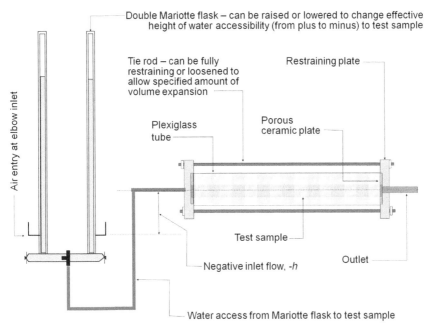

Double Mariotte flask – can be raised or lowered to change effective height of water accessibility (from plus to minus) to test sample

Tie rod – can be fully restraining or loosened to allow specified amount of volume expansion

Restraining plate

Plexiglass tube

Porous ceramic plate

Air entry at elbow inlet

Test sample

Negative inlet flow, -*h*

Outlet

Water access from Mariotte flask to test sample

Figure 4.5   Water infiltration and/or uptake experimental set-up. The double Mariotte flask arrangement allows for changes in the effective height of water accessibility to the test sample. A negative hydraulic head (–h) for water entry to test sample is shown. The amount of water uptake into sample is noted by reduced levels of fluid in the Mariotte tubes, and advance of the wetting front can be observed in the sample contained in the Plexiglas tube.

exposure to atmosphere is only at the elbow junction at the bottom end of the tubes. By this means, a constant head can be maintained at the level of the elbow atmospheric outlets. The negative hydraulic head situation shown in the illustration depicted in Figure 4.3 is best used to test the water uptake capability of the test sample, i.e. the capability of the internal forces associated with the matric ψ_m and osmotic ψ_π potentials to take water into the clay. In mechanical terms, one equates these to suction forces drawing water into the clay.

The information obtained from the experimental set-up shown can be graphed three-dimensionally to show wet-front advance in relation to distance from water source and volumetric water content at the point of interest. These kinds of graphical information are shown in the left-hand portion of Figure 4.6. We stress again that these are typical profiles for tests conducted on samples that (a) do not undergo volume change and (b) do not have any significant restructuring of the microstructure of the sample during the wetting process. The right-hand portion of the figure shows how one can interpret or deduce the characteristics of the water diffusivity coefficient

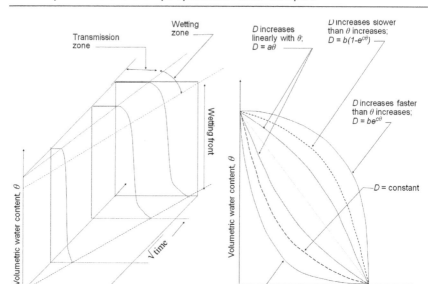

Figure 4.6 Characteristics of a wetting front profile in relation to the square root of time – shown in left-hand side of figure. Right-hand side of drawing shows the isolated view of the wetting front with various characteristic shapes – the nature of which tells one about the variation of the water diffusion coefficient D, in relation to the volumetric water content of the soil being infiltrated.

$D(\theta)$ by examining the shape of the wetting-front profile (including the wetting zone). One observes that in the wetting zone, $D(\theta)$ increases more or less rapidly than the corresponding increase in θ as water movement occurs. For the wetting-front profile itself, $D(\theta)$ is at best constant but will more probably decrease as θ increases. The kinds of changes in $D(\theta)$ gives one a clue as to the mechanisms involved in the movement of water in the partly saturated clay.

4.2.4   Vapour transport and water movement

Movement of water vapour in clay will add or subtract water from any one location. In the presence of high temperature gradients in clays, as, for example, in the containment of HLW canisters in deep underground repositories, vapour transfer is likely to be greater than liquid transfer in the relatively dry clay buffer. The reverse will be true in relatively wet but partly saturated clays. Vapour movement is by convective flow of the air in the clay and/or by diffusion of water molecules in the direction of decreasing vapour pressure. Vapour pressure gradients can develop not only because of temperature differences, but also because of salt concentration differences and water content differences – testifying to the fact that vapour transfer in partly saturated soils can also occur under isothermal conditions.

Diffusion of water vapour is generally modelled as a Fickian diffusion process. Under isothermal conditions, this is given as:

$$q_v = -D_v \frac{\partial c_{\theta v}}{\partial x}$$

where q_v is the vapour flux, D_v is the vapour diffusion coefficient, x is the spatial coordinate and $c_{\theta v}$ is the concentration of vapour in the gaseous phase. The state of vapour is considered to conform to the equation of state of the ideal gas. Therefore, the concentration of vapour is generally given as a function of pressure and temperature. Consequently, the Fickian diffusion equation is expressed using various alternative kinds of pressure. Under non-isothermal condition, a temperature factor is added to the equation (Nakano and Miyazaki, 1979).

Under isothermal conditions, in the absence of convective flow currents and assuming that linear superposition could be applied to water transfer, we can combine this Fickian process with the relationship given in equation 4.7 to obtain the combined water transfer (water movement and vapour transfer) relationship for partly saturated clays as follows:

$$\frac{\partial \theta}{\partial t} = \frac{\partial}{\partial x}\left((D_{\theta v} + D_{\theta l}) \frac{\partial \theta}{\partial x} \right)$$ (4.8)

where $D_{\theta v} = D_v(\partial c_{\theta v}/\partial \theta)$ and $D_{\theta l}$ is equal to $D(\theta)$ in equation 4.7 and the total water diffusivity coefficient D_θ is defined by $D_\theta = D_{\theta v} + D_{\theta l}$. As the water content at the surface will change with time (as shown in the next sections), the initial and boundary conditions are given as follows: $\theta = \theta_i = $ const. at $x > 0$, $t = 0$, and $\theta = f(t)$ at $x = 0$, $t > 0$.

The total water diffusivity coefficient D_θ has been calculated for two Kunigel bentonite samples: (a) a 100 per cent Kunigel bentonite sample and (b) a Kunigel bentonite sample that is mixed with 30 per cent sand. The calculated values are shown in Figure 4.7, using the results of changes of volumetric water content with time in unsteady infiltration experiments (Chijimatsu et al., 2000). The total water diffusivity coefficient becomes larger at both low water contents close to air-dried condition and high water contents near the fully saturated condition. This is on account of the larger $D_{\theta v}$ of predominant vapour flow in the unsaturated region, and the larger $D_{\theta l}$ of predominant liquid flow of water in the saturated region, as the total water diffusivity coefficient is expressed as the sum of $D_{\theta v}$ and $D_{\theta l}$. In the pure Kunigel sample, the region of predominant vapour flow lies in range from air-dried state to about a volumetric water content of 0.1. However, in the bentonite–30 per

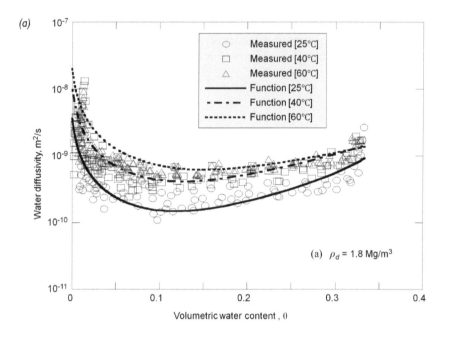

(a) ρ_d = 1.8 Mg/m^3

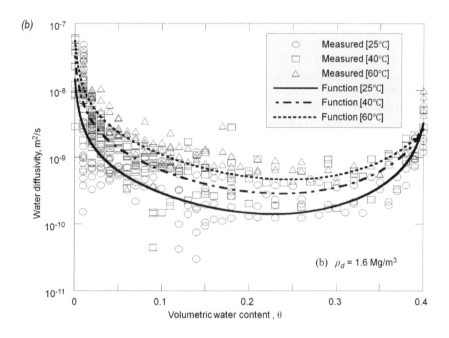

(b) ρ_d = 1.6 Mg/m^3

cent sand mixture, the region of vapour flow expands to a volumetric water content of about 0.25 because of the formation of macropores with sand inclusion.

Under thermal gradients, evaluation of water vapour movement becomes more complex because of pressures developed as a result of vaporization processes and also because of heat–mass coupling transfer processes. Analyses of water vapour movement as a distinct process separate from and independent of water (mass) transfer, as a result of thermal gradients, require considerable effort in tracking changes in water content at every instance of time and at every single discrete location as iterative procedures in updating the status of water vapour – at those same instances of time and discrete locations.

The general relationship for the isothermal vapour diffusivity $D_{\theta v}$ has been given as (see section 9.5):

$$D_{\theta v} = \frac{\alpha \omega D_{atm} \gamma g \rho_v}{\rho_L RT} \frac{\partial \psi}{\partial \theta}$$

where ψ is the clay–water potential, α the tortuosity, ω the volumetric air content, D_{atm} the molecular diffusion coefficient, γ the mass flow factor, g the gravitational acceleration, ρ_v the density of vapour, ρ_L the density of liquid and R is the gas constant.

The dependence of the clay water potential ψ on temperature T has been previously given by Philip and de Vries (1957) as:

$$\frac{\partial \psi}{\partial T} = \frac{\psi}{\sigma} \frac{\partial \sigma}{\partial T}$$

In general, the molecular diffusion coefficient at temperature T is given as follows:

$$D_{atm} = D_0 \frac{p_0}{p} \left(\frac{T}{T_0}\right)^n$$

Figure 4.7 Opposite. (a) The total water diffusivity coefficient D_θ that is defined by $D_\theta = D_{\theta v} + D_{\theta l}$ has been calculated from the changes of volumetric water content profiles with time in unsteady state experiments under isothermal condition. The clay sample was sodium bentonite (Kunigel-VI) 100 per cent with dry density of 1.8 mg/m³. The higher the temperature, the larger the water diffusivity coefficient. (b) The total water diffusivity coefficient D_θ that is defined by $D_\theta = D_{\theta v} + D_{\theta l}$ has been calculated from the changes of volumetric water content profiles with time in unsteady-state experiments under isothermal condition. The clay sample was sodium bentonite (Kunigel-VI) mixed with 30 per cent quartz sand with a dry density of 1.6 mg/m³. The higher the temperature, the larger the water diffusivity coefficient.

where D_0 is the molecular diffusion coefficient at a reference condition, p is the air pressure, p_0 is the reference pressure, T_0 is the reference temperature and n is a constant that is 2.3 for vapour (Rollins *et al.*, 1954).

4.2.5   *Water movement in swelling clays*

To study the effect of a volume change on the characteristics of water movement in a swelling clay, the test set-up shown in Figure 4.5 is used, with the restraining tie rods loosened to allow for unlimited swelling of the test sample. The wetting front advance into a natural clay loam (Ste. Rosalie clay), which has a small proportion of montmorillonite as shown in Figure 4.8. For a no-volume-change test conditon, the relationship for the rate of wetting front advance into the sample, vis-à-vis the square root of time, should be linear, as shown in the left-hand illustration in Figure 4.6 and also in Figure 4.8. The linearity characteristic comes from the use of the Boltzmann transform $\lambda = x/\sqrt{t} = \lambda(\theta)$ in the similarity solution technique invoked to reduce equation 4.7 to an ordinary differential equation:

$$\frac{\lambda}{2}\frac{d\theta}{d\lambda} = \frac{d}{d\lambda}\left(D\frac{d\theta}{d\lambda}\right) \tag{4.9}$$

The transform used requires that physical linearity is obtained between x and $\sqrt{t}$.

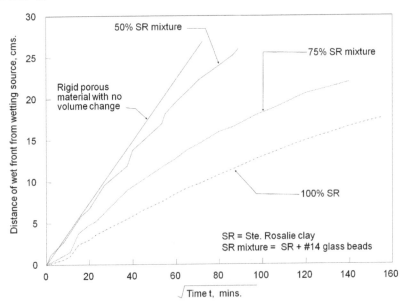

Figure 4.8 Influence of swelling clay content on rate of wetting front advance. The Ste. Rosalie clay is a clay loam with a small montmorillonite content.

The results shown in Figure 4.8 indicate that when free swelling of the test sample is allowed to occur, departure from linearity between wet-front advance and the square root of time occurs. As one would expect, the higher the swelling capability (potential) of the sample, the greater is this departure – as shown by the results for the use of mixtures of the clay with sterile glass beads. The significance of permitted swelling of the clay lies in the realization that in many clay barriers used to line landfills, some restricted swelling can occur. Prediction of the advance of any leachate front emanating from the landfill should take into account the swelling potential of the clay barrier, meaning that leachates should be used for 'water uptake' tests, and that some degree of swelling be allowed to occur during the uptake test.

When restrictions are placed on swelling volume change, the wetting-front profile becomes considerably affected, as shown in Figure 4.9. The comparison of the wetting-front profiles between free-swell and no-swell show that the rate of advance of the wet front is reduced significantly. The higher the initial density of the clay, the greater will be the reduction in rate of advance of the wet front – with the same type of swelling clay and with the condition of *no-volume-change*. For clay buffers that surround HLW canisters in deep underground repositories, the conditions of clay buffer placement restrict volume change to local changes within the buffer itself. Macroscopic volume change is minimal at best.

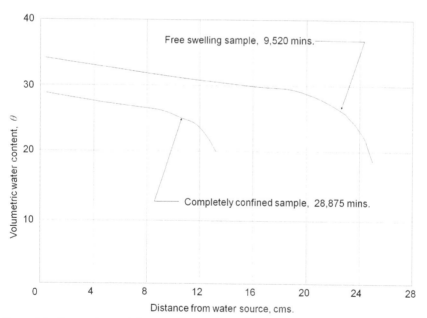

Figure 4.9 Comparison of free-swelling and confined sample wetting profiles. Both samples are similar except for a slightly lower initial density for the completely confined sample.

In situations of negative hydraulic heads, as shown, for example, by the experimental set-up in Figure 4.5, water uptake by clays and particularly by swelling clays is due to suction forces associated with the matric ψ_m and osmotic ψ_π potentials, as stated previously. The literature has used the terms *matric potential* and *matric suction* to mean the same thing. Hillel (1998) cites the definition of the *matric suction* in accord with the ISSS (International Soil Science Society) Terminology Committee, 'as the negative gauge pressure, relative to the external gas pressure on soil water, to which a solution identical in composition with the soil solution must be subjected in order to be in equilibrium through a porous membrane wall with the water in the soil'. By this definition, one assumes that if physical measurements of suction are obtained, as for example in the test set-up shown in Figure 4.5, these measurements are the matric suctions. Furthermore, one could further assume that the effect of the presence of solutes in the porewater will be integrated into the developed suction that one measures with a suction device, as, for example, a tensiometer or a pressure plate device previously discussed in Chapter 3. By extension of this reasoning, it has been generally argued that one could accept the matric potential ψ_m or the matric suction to be the total clay–water potential ψ (total suction).

It is not clear that the preceding reasoning, concerning acceptance of the matric potential to be the total clay–water potential, would apply to high-swelling clays, such as the smectites used in clay barriers and especially in clay buffers used to surround HLW canisters in repositories. The simple osmometer experiment shown in Figure 4.4 demonstrates that the solute potential (solute suction) is an active component of the total suction. As we have indicated in Chapter 3, the terms *solute potential* ψ_s and *osmotic potential* ψ_π are used interchangeably. One would expect that in low-swelling clay it is possible to ignore the contribution of the osmotic potential ψ_π to the total clay–water potential, without significant loss of accuracy. However, when swelling of the clay is a major issue, and especially when swelling is restrained, i.e. the swelling clay is confined, it is reasonable to argue that ψ_π cannot be ignored.

The relationship given in equation 4.7 does not apply when measurable volume change of the clay occurs during wetting. The material coordinate system used by Philip and Smiles (1969) provides a simple treatment of the volume change issue. The relationship between the spatial coordinate x and the material coordinate m is defined as $dm/dx = 1/(1 + e)$ where e is the void ratio. The Darcy model used in this treatment relates the velocity of water relative to the moving clay particles v_{ws} as:

$$v_{ws} = -k \frac{\partial \psi}{\partial x}$$

When the external pressure acts on the surface of clays, the total potential ψ includes the overburden potential Ω due to any surface load. In the system of three phases consisting of water, pore (gaseous) and solid, the continuity condition is described as:

$$\left[\frac{\partial}{\partial t}\left((1+e)\theta\right)\right]_m = -\frac{\partial v_{ws}}{\partial m} \tag{4.10}$$

It follows therefore (Philip and Smiles, 1969):

$$\frac{\partial}{\partial t}\left((1+e)\theta\right) = \frac{\partial}{\partial m}\left(\frac{k}{1+e}\frac{\partial \psi}{\partial m}\right)$$
$$= \frac{\partial}{\partial m}\left(D_m \frac{\partial e}{\partial m}\right) \tag{4.11}$$

where:

$$D_m = \frac{k}{1+e}\frac{d\psi}{de}$$

For water movement in a two-component system, i.e. a fully saturated state, the relation:

$$e = \frac{\theta}{1-\theta}$$

is determined in a fully saturated condition. Therefore, assuming that:

$$D(\psi) = k(\psi)\frac{\partial \psi}{\partial \theta}$$

then D_m reduces to the following form:

$$D_m = \frac{D(\psi)}{(1+e)^3} = (1-\theta)^3 D(\psi) \tag{4.12}$$

Using the water ratio $\vartheta = (1+e)\theta$ to refer to the volume of water per unit volume of solid, equation 4.11 reduces to:

$$\frac{\partial \vartheta}{\partial t} = \frac{\partial}{\partial m}\left(D_\vartheta \frac{\partial \vartheta}{\partial m}\right) \tag{4.13}$$

where:

$$D_\vartheta = \frac{k}{1+e}\frac{d\psi}{d\vartheta}$$

and ϑdm is the volume of water per unit cross-section of dm.

In a clay–water system where the initial water content of the clay is uniform, the water content at the surface in contact with incoming water will change as time progresses – because of swelling constraints at the far end. The initial and boundary condition are given as follows:

$$\vartheta = \vartheta_i = \text{const. } (dm/dx) > 0,\ t = 0$$

$$\vartheta = f(t)\ (dm/dx)_{x=0} = 0,\ t > 0$$

where ϑ is the initial water ratio.

4.2.6 Water and solids content profile in confined swelling clays

When water migrates into air-dried confined swelling clays, the changes of water content profiles with time are significantly different from that of non-swelling clays. The reasons are that (a) water in the liquid state and vapour move simultaneously in the clay mass; (b) the rate of liquid water movement is relatively low in comparison with the rate of vapour movement; and (c) vapour transfer, which precedes liquid water movement, reaches farther positions in the clay mass, condenses onto the surfaces of clay particles and rapidly increases the water content of the farther location. This situation is particularly significant in clay buffers used to embed HLW canisters in repositories located in deep geological formations, because of the presence of temperature gradients. This scenario will be discussed in detail in Chapter 6 when we discuss the impact of temperature, hydraulic, mechanical, chemical and biological factors on the short- and long-term performance of clay buffers and barriers.

The profiles of water content that are formed with the combined vapour–water movement – with vapour transfer ahead of water movement – will show ill-defined (obscure) wetting fronts. In the meantime, significant changes in the dry density will occur in the clay mass because of swelling resulting from transfer of both liquid and vapour – as mentioned in section

4.2.4. The dry density at the location adjacent to the water supply decreases markedly, whereas the part distant from the water supply increases gradually when the clay mass is completely confined at the boundaries of the mass as shown in Figure 4.10. The measurable dry density changes can be analysed using a concept of the flow of clay particles, as the changes are large enough to fall outside the bounds of mechanical deformation theory. The changes in dry density with time can be simulated using the clay particle diffusivity coefficient D_σ that has been calculated from the changes of solid profiles with time obtained in unsteady state experiments (Nakano *et al.*, 1986). In HLW repositories founded in geological rock formations, the presence of fissures and cracks will create the situation that would allow for penetration of the swelling clay. In such an event, the decreasing dry density at the location adjacent to the water supply will become more rapid when clay particles flow into the cracks of surrounding rocks.

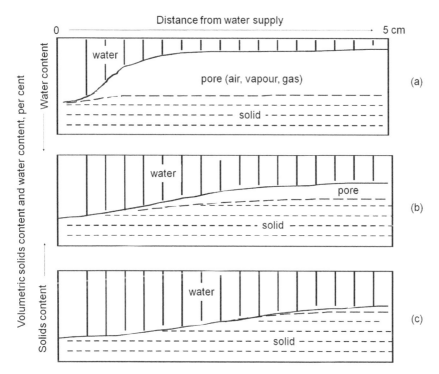

Figure 4.10 Changes in water content and solids content observed in a confined bentonite (Kunigel-VI) with initial dry density of 0.846 g/cm³ and initial water content of 0.078 g/cm³ (air-dried condition). The illustration shows the relative proportions of solid, water and pore air or vapour or gas after a defined period of water uptake. Beginning from the top, the diagram shows the distribution of solid, water and pore air–vapour–gas (a) after one day of water uptake and (b) after 26 days of water uptake and (c) the profile estimated at infinite time ($t = \infty$).

On the theory that clay particles will flow becasue of the gradients of the total water potential ψ, it is assumed that:

(a) $v_{cp} = -k_{cp} \dfrac{\partial \psi_{cp}}{\partial x}$

where subscript cp expresses the amount of clay particles;

(b) $\dfrac{\partial \psi_{cp}}{\partial x} = -\delta \dfrac{\partial \psi}{\partial x}$ where δ is a coefficient.

In combination with the continuity equation, one obtains:

$$\frac{\partial \sigma}{\partial t} = \frac{\partial}{\partial x}\left(D_\sigma(\sigma,\theta,D_\theta) \frac{\partial \sigma}{\partial x} \right) \tag{4.14}$$

where σ is the volumetric clay particle content, D_σ is the clay particle diffusivity coefficient, and D_θ is the total water diffusivity coefficient.
  The initial and boundary conditions are:

$\sigma = \sigma_i = $ const. at $x > 0$, $t = 0$

$\sigma = f(t)$ at $x = 0$, $t > 0$

where σ_i is the initial volumetric clay particle content.

4.3   Transport of porewater solutes

4.3.1   Solute transport in porewater

Generally speaking, porewater solute transport can be both by diffusion and by advective transfer depending on the magnitude of the temperature and external hydraulic gradients. In unsaturated clays, solute transport is directly affected by volumetric water content, as the transport process occurs in the porewater. Assuming that the flux of solute can be expressed by an equation of the Fickian type, the flux q_s is described as follows:

$$q_s = -D_s\theta \frac{\partial s}{\partial x} + q_w s \tag{4.15}$$

where s is the concentration of solute in the porewater, D_s is solute diffusivity coefficient, θ is volumetric water content, q_w is water flux, and $q_w s$ denotes advective solute transfer. As it is difficult to directly measure velocity v in the pores, an appropriate procedure is to use the volumetric flux to express the amount of solute and water (in volume) transferred per unit time per unit area of clay. In general, the velocity v is expressed by $v = q/\theta$.

The continuity equation of solute in clays is:

$$\frac{\partial(\theta \cdot s)}{\partial t} = -\frac{\partial q_s}{\partial x} - \frac{\rho_d}{\rho_w}\frac{\partial s^*}{\partial t} - \lambda \cdot \theta \cdot s \qquad (4.16)$$

where s^* is the concentration of solutes sorbed by the clay solids, ρ_d is dry density, ρ_w is the density of water, λ is a decay coefficient of dimension $1/t$ and $\lambda \cdot \theta \cdot s$ is the decay rate of radioactive solute.

Equations 4.15 and 4.16 yield the following general equation for solute transfer in unsaturated clays as follows:

$$\frac{\partial(\theta \cdot s)}{\partial t} = \frac{\partial}{\partial s}\left(D_s \theta \frac{\partial s}{\partial x}\right) - \frac{\partial(q_w \cdot s)}{\partial x} - \frac{\rho_d}{\rho_w}\frac{\partial s^*}{\partial t} - \lambda \cdot \theta \cdot s \qquad (4.17)$$

The first term in the right-hand side expresses the dispersive transport of solute. The second term expresses the contribution of advective transport. This can be ignored when advective transport is absent. To determine whether advective transport is to be accounted for in the equation, the Peclet number is used. The Peclet number is defined as the ratio of the contribution of advective transport to the contribution of molecular diffusion as follows: $P_c = v_L d/D_0$, where D_0 is the molecular diffusion coefficient of a specified solute in an infinite dilute solution, d is the average clay particle diameter, and v_L is the longitudinal flow velocity (advective flow). It is generally appropriate to consider transport of solutes as being diffusive when $P_c < 1$.

Figure 4.11 shows a solute transport diagram using information reported by Perkins and Johnston (1963). The information given in Figure 4.11 shows that for Peclet numbers smaller than 10^{-2}, i.e. $P_c < 10^{-2}$, the effects of advective velocities on the transport of solutes in the porewater may not be easily or readily discounted or ignored. The transition zone between dominantly diffusive transport and predominantly advective transport of the dissolved species occurs in the range of Peclet numbers between 10^{-2} and 10. When the Peclet number is greater than 10, i.e. $P_c > 10$, advection becomes the dominant mechanism of solute transport in the clay.

For natural clays, and especially for swelling clays, microstructure and interlayer water are two characteristic features that render solute transport modelling more complicated. The pore spaces in natural clays are not uniform, not only because of the irregular shapes and sizes of the clay particles, but also because of the presence and distribution of the microstructural units

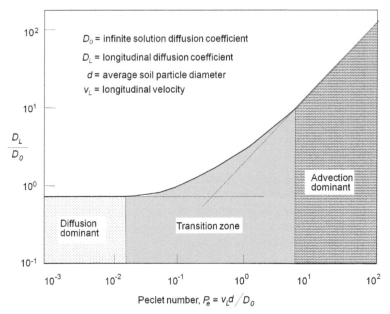

Figure 4.11   Diffusion and advection dominant flow regions for solutes in relation to Peclet number (adapted from Perkins and Johnston, 1963).

of varied sizes. They will have different species and concentrations of solutes conducting different characteristics of solute transport, and requiring attention to be paid to (a) the wide range of solutes diffusing within the microstructural units and (b) transport velocities within and between pores (Philip, 1968; Skopp and Warrick, 1974; Rao *et al.*, 1980). A very significant feature of this kind of microstructural feature is that some pore spaces could be nonconducting. Continuity between the pore spaces may not exist. Many micropores are too small and will not permit easy movement of solutes because of the prohibitive energy requirements. The disparity in sizes between the macro- and micropores in the microstructural units, in combination with their diffusive and transport characteristics, require consideration of the microstructural units as sources/sinks for solutes. A mechanistic model of porewater solute transport in such a medium is shown in Figure 4.12. Given the various kinds of transport within a representative volume, one needs to redefine the solute diffusivity coefficient to take into account the complex transport process.

Assuming that the microscopic advective flow of solute and the local diffusion solute transport in the stagnant region acting as sink/source could be described by a Fickian process, one could define solute diffusivity coefficient D_s in a manner suggested previously by Paissioura (1971), Rao *et al.* (1980) and Wagenet (1983):

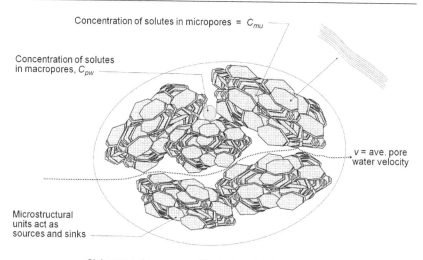

Concentration of solutes in micropores = C_{mu}

Concentration of solutes in macropores, C_{pw}

v = ave. pore water velocity

Microstructural units act as sources and sinks

Sink-source phenomena – diffusive transfer of solutes between restricted conducting (micropores) and conducting macropores

Figure 4.12 Sink-source phenomenon created by presence of microstructural units in diffusive flow of water and solutes through the clay–water system.

$$D_s = (D_m + D_h + D_{stag})$$ (4.18)

The coefficient D_s is used to represent the longitudinal diffusion–dispersion coefficient. This is a reflection of the advective velocity modification of the diffusive flow of solutes. This is then called the hydrodynamic dispersion coefficient sometimes. The term *longitudinal* is used in conjunction with this coefficient to signify flow in the direction of the advective velocity. The literature commonly refers to D_s as the *longitudinal diffusion coefficient*. D_m represents the molecular diffusive coefficient of the solute under consideration and is given as a function of volumetric water content, the infinite dilute solution diffusion coefficient D_0 and a tortuosity factor τ as shown in the next section (section 4.3.2). D_h represents the advective dispersive coefficient of solutes and is given by $D_h = \alpha v$, where α is a dispersivity parameter and v is the advective velocity. D_{stag} is the dispersion coefficient accounting for dispersion effects caused by diffusion of solutes from stagnant to mobile regions and, using the expression given by Paissioura (1971) and Rao *et al.* (1980), D_{stag} is given by:

$$D_{stag} = \frac{v^2 r^2 (1 - \phi)}{15 D_{es}}$$

where r is the average equivalent diameter of the clay particles, ϕ is the porewater fraction in the conducting region that is equivalent to volumetric water content θ and D_{es} is the effective diffusion coefficient in the stagnant region.

4.3.2   Macropore–micropore channels and molecular diffusive coefficient

The molecular diffusion of solutes in partly saturated swelling clays will be impacted by the types of micropores and macropores and their distribution, and by the properties of interlayer water. At low water contents, close to air-dried condition, solutes in the porewater of clays will move not only as film boundary transport, but also within the water-filled micropores. Continuity of film water, which will be formed by these hydration layers, will establish contact between adjoining particles in the microstructural units. These are in essence film boundaries. They provide the opportunity for diffusive transport of solutes. It is safe to say that in partly saturated clays, solute diffusion is the predominant mechanism of transport of porewater solutes – in film boundaries and in water-filled micropores of the microstructural units. Brownian activity of solutes in the porewater contributes significantly to solute transport in the film boundaries and in the interlayer water of partly saturated clays. In the absence of temperature and other driving force gradients, solute concentration differences between contiguous points in a partly saturated clay serve as the driving forces for diffusive transport of the solutes. By and large, the absence of straight-line diffusion or flow paths is one of the principal characteristics of clays. This is particularly important for partly saturated clays if one assumes that the molecular diffusive coefficient of a specific solute species given as D_m is simply related to its infinite dilute solution diffusion coefficient D_0 through a tortuosity factor τ, i.e. $D_m = D_0 \tau$. The tortuosity factor τ is generally taken as the ratio of the path length of connected pore channels conducting the solutes L_{pc} to the straight-line path length L_0. As the path length of connected pore channels is always greater than the straight-line path, i.e. $L_{pc} > L_0$, this ratio will be greater than one. A commonly used value for this factor is $\sqrt{2}$. For unsaturated swelling clays, the physico-chemical tortuosity factor $\tau^* = [L_{pc}/L_0]^2 \omega \chi$ was defined and used as $\tau = \tau^*$ by Bresler *et al.* (1982), where ω is a coefficient accounting for the effects of the reduction in water viscosity due to the presence of charged particles on water viscosity, and χ is the coefficient that accounts for the retarding effects of anion exclusion on flow in the vicinity of negatively charged particles. This relationship takes into consideration the effects of water content, the charge of particles and anion exclusion.

    The electrical resistance–conductance approach of Manheim and Waterman (1974) for estimation of the molecular diffusion of solutes in clays uses a formation factor f. This factor describes the ratio of the electrical resistance of a brine-filled porous medium R_p to that of the brine R_w occupying the same volume as the bulk porous medium;

$$f = \frac{R_p}{R_w} = \frac{1}{\varphi^n}$$

where n ranges from 2.5 to 5.4, and φ represents the ratio of fractional cross-sectional area available for conductance of electric current A_{pc}, and the gross cross-sectional area A_o, which corresponds to water content. With this approach, one obtains the molecular diffusive coefficient as follows: $D_m = D_0 \varphi^2$.

For smaller macropores and for micropores – in the range of nanometres – the relationship proposed by Renkin (1954) provides one with a model that includes consideration of tortuosity of pores and the boundary drag offered by the walls of the pore spaces. The following is the short list of approximations for D_m:

$$D_m = D_0 \tau \quad \text{simplified tortuosity model} \tag{4.19}$$

$$D_m = D_0 \varphi^2 \quad \text{Mannheim and Waterman (1974), Lerman (1979)} \tag{4.20}$$

$$D_m = D_0 \left(1 - \frac{r}{r_p}\right)^2 \left[1 - 2.10\frac{r}{r_p} + 2.09\left(\frac{r}{r_p}\right)^3 - 0.95\left(\frac{r}{r_p}\right)^5\right] \quad \text{Renkin (1954)}$$

$$\tag{4.21}$$

where r is the radius of the dissolved species of solutes, and r_p represents the radius of a typical pore space. Note that when r_p approaches r in size, $D_m \cong 0$.

For most, if not all, reported approaches used to determine molecular diffusive coefficients for estimation of solute transport, one still needs to determine the infinite dilute solution diffusion coefficient D_0. Considering the diffusion of solutes to be a Fickian process, the rate of diffusion of a specific species of solutes in a dilute solution J_D, will be given by the following simple relationship:

$$J_D = -D_0 \frac{dc}{dx}$$

where c and x represent the concentration of the solutes and the spatial distance, respectively. The studies of both Nernst (1888) and Einstein (1905) on the movement of suspended particles controlled by the osmotic forces in a solution provide us with the following relationships for D_0:

$$\text{Nernst–Einstein: } D_0 = \frac{uRT}{N} = u\kappa T \tag{4.22}$$

$$\text{Einstein–Stokes:} \quad D_0 = \frac{RT}{6\pi N\eta r} = 7.166 \times 10^{-21} \frac{T}{\eta r} \tag{4.23}$$

$$\text{Nernst:} \quad D_0 = \frac{RT\lambda}{F^2 |z|^2} = 8.928 \times 10^{-10} \frac{T\lambda}{|z|^2} \tag{4.24}$$

where u is the absolute mobility of a solute, R is the universal gas constant, T is the absolute temperature, N represents Avogadro's number, κ is Boltzmann's constant, λ is the conductivity of the target solute, r is the radius of a hydrated solute, η is the absolute viscosity of the fluid, z is the valence of the ion and F is Faraday's constant. Nernst, Einstein and Stokes show that the infinite solution diffusion D_0 coefficient is a product that includes consideration of such factors as ionic radius, absolute mobility of the ion, temperature, viscosity of the fluid medium, valence of the ion and equivalent limiting conductivity of the ion. Compiled values and discussions on various species of solutes and their corresponding D_0 under various conditions can be found in Li and Gregory (1974), Jost (1960) and Lerman (1979), amongst others, and experimental values for λ for many major ions at various temperatures can be found in Robinson and Stokes (1959). Calculations made using the Nernst–Einstein relation, represented by equation 4.22, show significant increases in the magnitude of D_0 with increasing temperature. As temperatures in clay buffers and barriers are expected to vary considerably over some time interval, proper recognition and accounting of the D_0 value are important.

4.3.3   Impact of tortuosity

Recognizing that tortuosity of channel paths impacts significantly on the hydraulic conductivity, we can account for the effects of tortuosity by modelling various path tortuosity scenarios. The following treatment has been adapted from the account given by Pusch and Yong (2006). In the discussion to follow, the three-dimensional flow models with code 3Dchan, developed by Neretnieks and Moreno (1993), have been adopted for calculations of the example scenarios. An orthogonal pattern of interconnected channels filled with permeable clay gels has been assumed to constitute the microstructure of the clay being examined, in the belief that this configuration is a reasonable representation of the general microstructural heterogeneity of the type of clay under study. Estimation of the size and frequency of the channels containing clay gels can be made along the guidelines suggested by Pusch *et al.* (2001) as follows: (a) the channels have a circular cross-section; and (b) the diameter of the widest channel for clay A is $50\,\mu$m (MX-80) (with a bulk density of $1570\,$kg/m³), the widest channel for clay B is $20\,$m (with a bulk density of $1850\,$kg/m³), and the widest channel for clay C is $5\,$m (with a bulk density of $2130\,$kg/m³) – all estimated from transmission electron microscopy. The size of the channels demonstrates a normal statistical distribution (Figure 4.13).

The channel size distributions shown in Table 4.2 have been obtained with the aid of these guidelines. Gel-filled voids with a diameter less than 0.1 μm are considered to be of no importance to the bulk conductivity.

The conceptual model and the computational code consider that to all intents and purposes, all permeating water will flow in the three-dimensional network of gel-filled channels depicted in Figure 4.14. The shape of the channels is characterized by their lengths, widths, apertures and transmissivities, which are all stochastic. The rest of the clay matrix is assumed to be porous, but largely impermeable. Calculation of the bulk hydraulic conductivity is made by assuming that a certain number (commonly taken to be six) of channels intersect at each node of the orthogonal network. Each channel in the network consists of a bundle of N capillaries with circular

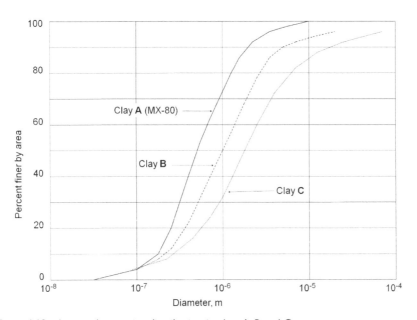

Figure 4.13 Assumed pore-size distribution in clays A, B and C.

Table 4.2 Number of differently sized channels per $250 \times 250 \, \mu m^2$ cross-section area representing the representative elementary volume

Bulk density (kg/m³)	Number of 20- to 50-μm channels	Number of 5- to 20-μm channels	Number of 1- to 5-μm channels
2130	0	0	135
1850	0	10	385
1570	2	85	950

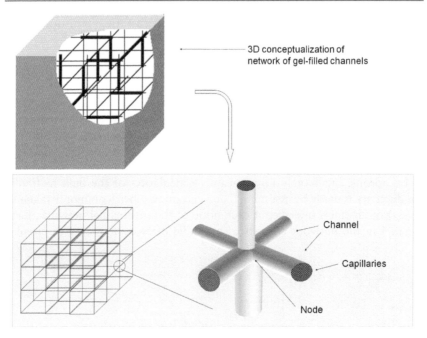

3D conceptualization of
network of gel-filled channels

Channel

Capillaries

Node

Figure 4.14 Three-dimensional conceptual model (top), and idealized channel network mapped as a cubic grid with gel-filled channels intersecting at a node in the grid (bottom).

cross-section, as shown in the lower portion of Figure 4.14. Note that the channels have different diameters. With the total porosity of the clay as the bounding condition, the number of channels with assumed length L with bundles of N capillaries with a diameter (d) must be proportional to the channel width and must obviously equal the total porosity of the clay. Following complete particle expansion, the voids filled with homogeneous clay gels are assumed to have a normal size distribution with the same intervals as in the two-dimensional model, i.e. 1–5 μm for the clay with 2130 kg/m³ density, 1–20 μm for the clay with 1850 kg/m³ density, and 1–50 μm for the clay with 1570 kg/m³ density.

Using the experimentally obtained hydraulic conductivity k values for the samples with bulk densities shown in the left-hand column of Table 4.3 for comparison with theoretically calculated values, Pusch and Yong (2006) find that the calculated bulk conductivity of MX-80 clay with the density 1570 kg/m³ at saturation is in the same order of magnitude as typical experimental data. However, the computational model significantly overpredicts the conductivity of the denser clays.

It appears that because the computational model does not take into account the fact that as the density becomes higher, variations in cross-sections between channels and also along the paths of individual channels, i.e. along L are likely to occur – thereby invalidating the assumption of a constant cross-section along length L. As might be expected, the narrowest parts, i.e. the smallest cross-sections, will control the transport of the fluid, and, as the conductivity of a capillary is an exponential function of the diameter, the net effect of constrictions imposed by the smaller cross-sections is more obvious for denser clays.

4.3.4   Coupling flow of water and solute

The flow of liquid water, vapour and solute will occur simultaneously in partly saturated swelling clays. In non-isothermal conditions, heat flow will be coupled with simultaneous mass transport. Mass and heat flow are non-equilibrium processes and are the result of driving forces that include water potential, concentration of solutes and temperature gradients. For instance, in the case of the solute concentration gradient, it could synchronously cause the diffusion flow of water associated with solute diffusion. For a temperature gradient, water diffusion and solute diffusion due to the temperature gradient occur simultaneously with heat transfer in clays. We will discuss these in detail in Chapter 6 when we address the problems of clay buffers and barriers embedding HLW canisters.

Although it is difficult to proportion or separate a fraction of each flow attributable to each of the driving forces in experiments, we can nevertheless develop a theoretical basis for these using phenomenological relationships. The contribution of each flow due to all driving forces to the total flow in the coupling flow phenomena using phenomenological relations can be written using the general flux and the driving force format as:

$$J_i = \sum_{k=1}^{n} L_{ik} X_k$$

Table 4.3 Hydraulic conductivity (k) of three clay types prepared by compacting MX-80 powder and saturating them with electrolyte-poor water

Bulk density of air-dry powder (kg/m^3)	Dry density (kg/m^3)	Density at saturation (kg/m^3)	n	Calculated k (m/s)	Experimental k (m/s)
2000	8010	2130	0.13	3×10^{-12}	2×10^{-14}
1500	1350	1850	0.20	1.3×10^{-11}	3×10^{-12}
1000	900	1570	0.47	2.4×10^{-10}	8×10^{-11}

n, porosity.

where $i = 1, 2 \ldots n$, J_i is flux of flow of i, X_k is driving force of k ($k = 1, 2 \ldots n$) and L_{ik} are the phenomenological coefficients.

Onsager's fundamental theorem states that the matrix of phenomenological coefficients L_{ik} becomes symmetric, i.e. $L_{ik} = L_{ki}$ ($i, k = 1, 2 \ldots n$) provided a proper choice is made for fluxes and forces. These identities are called the Onsager's reciprocal relations. For a non-equilibrium state, when flux occurs, the deviation ΔS of the entropy from its equilibrium value consequently changes in the system. The entropy production per unit time $\Delta S / \Delta t = \sigma$ is given in irreversible processes as follows (de Groot, 1961):

$$\frac{\Delta S}{\Delta t} = \sigma = \sum_i^n J_i X_i \tag{4.25}$$

Invoking a dissipation function Φ defined as $\Phi = T\sigma$, we obtain:

$$\Phi = J_s \cdot grad\left(-T\right) + \sum_1^n J_i \cdot grad\left(-\mu_i\right) \tag{4.26}$$

where J_s is the entropy flow, J_i signifies material flow and μ_i is the chemical potential of the ith matter. The dissipation function is also expressed by the sum of products of flows and forces, although the flow of entropy is applied instead of heat flow. In an isothermal state, the dissipation function is given as:

$$\Phi = \sum_1^n J_i \cdot grad\left(-\mu_i\right) \text{ because } gradT = 0.$$

We know that the phenomenological relationship:

$$J_i = \sum_{k=1}^n L_{ik} X_k$$

can be constructed between the flows and forces on water and solutes that are defined by this dissipation function, provided that the chemical potential of the clay solution can be separated into the chemical potential of water and solute. These phenomenological relations can be specified in the partly saturated state after rearrangement as equation 4.27. This takes into account (a) the chemical potential of water expressed in terms of the matric potential and (b) the chemical potential of solute expressed in relation to the concentration of solutes:

$$J_\theta = -L_{\theta\theta}\frac{\partial \psi_m}{\partial x} - L_{\theta s}\frac{\partial s}{\partial x}$$

$$J_s^* = -L_{s\theta}\frac{\partial \psi_m}{\partial x} - L_{ss}\frac{\partial s}{\partial x}$$

(4.27)

where J_θ is the flux of water and J_s^* is the solute flux without advective transfer of solutes, i.e. $J_s^* = J_s - q_w s$, ψ_m is the matric potential of water (see section 3.8.1) and s is the concentration of solute. $\partial \psi_m/\partial x$ and $\partial s/\partial x$ are the driving forces of water and solute, respectively. $L_{\theta\theta}$, $L_{\theta s}$, $L_{s\theta}$, and L_{ss} are the phenomenological coefficients. It requires $L_{\theta s} = L_{s\theta}$ to be made available.

Combining with the continuity law associated with adsorption and decay, we can obtain the expression for water and solute transport as follows:

$$\frac{\partial \theta}{\partial t} = \frac{\partial}{\partial x}\left\{\left(D_{\theta\theta} + D_{\theta s}\right)\frac{\partial \theta}{\partial x}\right\}$$

$$\frac{\partial(\theta \cdot s)}{\partial t} = \frac{\partial}{\partial x}\left\{\left(D_{s\theta} + D_{ss}\right)\frac{\partial s}{\partial x}\right\} - \frac{\partial(q_w s)}{\partial x} - \frac{\rho}{\rho_w}\frac{\partial s^*}{\partial t} - \lambda \cdot \theta \cdot s$$

(4.28)

where:

$$D_{\theta\theta} = L_{\theta\theta}\frac{\partial \psi_m}{\partial \theta}, \ D_{s\theta} = L_{s\theta}\frac{\partial \psi_m}{\partial s}, \ D_{ss} = D_s\theta \ \text{and} \ D_{\theta s} = L_{\theta s}\frac{\partial s}{\partial \theta}$$

are moisture, solute–moisture, solute, and moisture–solute diffusivity coefficients respectively.

The choice of the phenomenological driving forces and coefficients can be performed using several techniques. Elzahabi and Yong (1997) chose the osmotic potential ψ_π of the porewater as the driving force of solute and defined the phenomenological coefficients related to solute transport using van't Hoff law $\psi_\pi = RTs$. The functional forms for the phenomenological coefficients are determined on the basis of experimental information on the distribution of solutes along columns of test samples. Yong and Xu (1988) have provided a useful *identification technique* for evaluation of these phenomenological coefficients.

4.4   Concluding remarks

In this chapter, we build on the material discussed in the previous chapter relating to clay–water interactions and uptake of water. The primary focus of this chapter has been to look at the various interactions and reactions in the porewater of clays, with a view to determine how these will affect the interactions between the clay solids and contaminants when we come to the discussion in Chapter 5. We have learnt from the previous chapter about the basic macro- and microstructure of clays and the associated surface functional groups.

In this chapter, we are introduced to abiotic and biotic reactions in the porewater due to the presence of dissolved solutes. We want to know more about how these affect or impact on the uptake and movement of water in anhydrous and partly saturated clays. We are interested to know when the porewater is in a more-or-less immobile state, for example hydration water layer, and how this affects the transport of porewater solutes. Water movement in partly saturated soils occurs along film boundaries in clay pore spaces that are not completely filled with water. Pore channel flow occurs for those pore spaces that are completely filled with water. For Peclet numbers $P_e \gg 10^{-2}$ the effects of advective velocities on the transport of solutes in the porewater cannot be ignored.

Water and solute movement in swelling clays, especially when they are in a partly saturated state, presents challenges in modelling because of the importance of the different phenomena associated with water uptake from the anhydrous condition. The importance of pore channels, and tortuosity of channels, becomes more acute in the case of these kinds of clays.

Chapter 5

Contaminant–clay
interactions and impacts

5.1   Introduction

5.1.1   Contaminants and clay barriers

Interactions between contaminants and clay buffers and barriers used in high-level radioactive waste (HLW) and hazardous solid waste (HSW) containment occur when (a) leachates from waste landfills escape from their confining geomembrane liner system, thereby allowing the leachates to make direct contact with the underlying engineered clay barrier/liner system and substrata, and (b) corrosion products and fugitive radionuclides escaping from an encapsulated HLW canister come into direct contact with the surrounding engineered clay buffer system. Interactions between contaminants and clays also occur when (a) natural subsurface clay strata that are being used as contaminant-attenuating material react with incoming leachates and (b) clays that are used in specially designed permeable reactive barriers (PRBs) intercept leachates as their planned design function. In such instances, one relies on the assimilative capability of the clay. In the case of PRBs, clays and other attenuating aids are used to control, mitigate and even eliminate the transport of contaminants. These will be discussed in a later section in this chapter.

Even though source differences and circumstances leading to leachate and contaminant plume generation between waste landfills and HLW repositories may be different, commonality exists in the types of contaminants generated from these sources. In deep geological HLW repositories, such as those shown in Figures 2.6 and 2.7, exposure of the metal inserts in the canisters to water will eventually result in the production of corrosion products, i.e. from corrosion of the cast iron and/or copper inserts in the canister. Release of radionuclides from exposed fuel matrix if/when the canisters are breeched will add to the variety of contaminants making contact with the clay buffer surrounding the canisters. These include caesium, iodine, strontium, selenium, zirconium, technetium, palladium and tin.

The composition of a leachate emanating from a waste landfill is highly varied, as it is a product of dissolution products, compounds formed from chemical reactions resulting in complexation and speciation, and organic chemicals. It is highly varied, not only with respect to the source of the contaminants and processes generating the contaminants, but also with respect to the maturity of the waste in the landfill. Thus, it is often difficult to provide an exact or detailed listing of contaminants and their quantitative values. To demonstrate the variety and range of contaminants found in waste landfill leachates, we can consult the summary version of leachate site characterizing parameters and data (Table 5.1) that is often used as a checklist in determining the nature of leachates escaping from landfill liner systems containing various kinds of wastes.

5.1.2 Interactions and impacts

The nature of the interactions between contaminants and the constituents in a clay depends on the type of contaminant and on the physical and chemical properties of the clay. Figure 5.1 highlights the primary sets of issues that arise when contaminants interact with an engineered clay barrier, with attention to the impact of the interactions on the nature of the clay being exposed to the contaminants. The questions of major interest revolve around whether the interactions with the contaminants will degrade or otherwise impact negatively on the capability of the engineered clay buffer/barrier to fulfil its design function. More specifically, have the interactions and reactions occurring in the clay diminished the physical and chemical properties to the extent that the clay is no longer an effective buffer or barrier? How can we anticipate or predict what the changes will be?

The specific issues of concern include:

- *Nature of contaminants and of the clay used in the engineered clay buffer or barrier.* The concerns arise from requirements for engineered

Table 5.1 Typical characterization parameters and data for evaluation of the nature of waste leachates from landfills

Grouping/type	Characterization parameters and data
General	Appearance, pH, oxidation–reduction potential, conductivity, colour, turbidity, temperature
Organic chemicals	Phenols, chemical oxygen demand, total organic carbon, volatile acids, organic nitrogen, oil and grease, chlorinated hydrocarbons, tannins, lignins
Inorganic chemicals	Total bicarbonate, solids (TDS, TSS), volatile solids, chloride, phosphate, alkalinity and acidity, nitrite and nitrate, heavy metals, ammonia, cyanide, fluoride
Biological	Biological oxygen demand

TDS, total dissolved solids; TSS, total suspended solids.

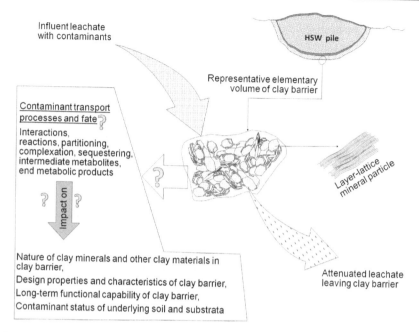

Influent leachate
with contaminants

HSW pile

Representative elementary
volume of clay barrier

Contaminant transport
processes and fate

Interactions,
reactions, partitioning,
complexation, sequestering,
intermediate metabolites,
end metabolic products

Impact on

Layer-lattice
mineral particle

Nature of clay minerals and other clay materials in
clay barrier,

Design properties and characteristics of clay barrier,

Long-term functional capability of clay barrier,

Contaminant status of underlying soil and substrata

Attenuated leachate
leaving clay barrier

Figure 5.1   Interaction of leachate with a representative elementary volume of a clay buffer/barrier with particular attention to the questions of transport and fate and their impact on the long-term stability of clay buffers and barriers.

clay barriers and/or buffers to assimilate contaminants in its immediate microenvironment so that the contaminants are not transported further downstream. What are contaminants? And how do we choose a clay that would fulfil the requirements for a competent engineered clay buffer/ barrier?

• *Alteration and transformation of clays in the engineered clay buffer/ barrier, and chemical alteration of contaminants in the clay.* These questions arise because of (a) possible dissolution and changes in original clay minerals to other kinds of minerals due to interactions with the various inorganic and organic chemicals constituting the contaminants and (b) chemical changes in the chemical contaminants into other forms of contaminants. What do we know about the stability of the clay minerals and other clay constituents in the clay buffer and barrier, and also about the chemicals constituting the contaminants? Should we be concerned with mineral transformation and degradation?

• *Evolution and long-term functionality of the clay buffer and barrier being affected and long-term hazardousness and toxicity of contaminants.* The concern for long-term functionality is founded on the fact that 'time' is a very important factor in the kinetics of abiotic and biotic chemical reactions. The properties and characteristics of the clay used in the buffers/barriers will evolve as time progresses, in the main on account of the various reactions occurring in the clay. The hazardousness and toxicity of contaminants will also change as time progresses because of the various chemical reactions. How well do we understand the long-term processes that contribute to the evolution of the clay and to the decrease or increase in hazardousness and toxicity of contaminants within the engineered clay buffer/barrier?

- *Nature and long-term performance of rocks, underlying soils and substrata used as natural barriers.* The source of this set of concerns arises from knowledge that these natural barriers could supply some chemical constituents to the clay buffer/barrier by their dissolution from interactions with contaminants and clays or owing to their performance in nature.
- *Nature of long-term performance of over-packs and containers containing wastes.* We need to be concerned with the likelihood that over-packs and containers will corrode and/or degrade. The result of the breakdown will be fugitive contaminants that will interact with the clay buffer or barrier and also with other contaminants already present in the buffer/barrier. The result could be degradation of the clay buffer/barrier and changes in the toxicity of the contaminants in the buffer/barrier.

5.2 Interactions and sorption mechanisms

5.2.1 Surface properties and interactions

We can group the factors influencing the mechanisms of interactions between contaminants and clay solids into three broad groups as follows:

- *Clay.* The type and distribution of clay fractions (type of clay minerals and other constituents in the clay); the properties of the clay fractions' surfaces; surface functional groups; microstructure, macrostructure, density, water content and degree of saturation.
- *Contaminants.* The types, distribution and concentration of contaminants in the porewater; the functional groups.
- *Clay–contaminant system.* The Eh and pH of the system; microorganisms; the local temperature and pressure environment.

The processes resulting in the transfer of ions, molecules and compounds from the porewater onto clay particles' surfaces involve molecular interactions, i.e. interactions between nuclei and electrons. The major types of interatomic bonds are:

- *ionic* – electron transfer between atoms that are subsequently held together by the opposite charge attraction of the ions formed;

- *covalent* – electrons are shared between two or more atomic nuclei, i.e. each atom provides one electron for the bond; when a bond is formed by the sharing of a pair of electrons provided by one atom, this is called *coordinate covalent bonding*;
- *hydrogen* – very strong intermolecular permanent dipole–permanent dipole attraction;
- *Coulombic* – ion–ion interaction;
- *van der Waals* – dipole–dipole (Keesom), dipole–induced dipole (Debye), and instantaneous dipole–dipole (London dispersion);
- *steric* – involves ion hydration surface adsorption.

The reactive surfaces of clay minerals are important factors in the control of contaminant–clay interaction mechanisms. Montmorillonites, with their 2:1 layer–lattice structure (see Figure 3.3) show siloxane-type surfaces on both bounding surface, compared with kaolinites with their 1:1 structure, which gives us a siloxane upper bounding surface and a gibbsite layer at the opposite bounding surface. Because of the structural arrangement of the silica tetrahedral sheet and the nature of the substitutions in the layers, siloxane-type surfaces are reactive surfaces. When surface silanol groups dominate, the surface will be hydrophilic. The surface silanol groups are weak acids. However, if strong H-bonding is established between silanol groups and neighbouring siloxane groups, the acidity will be decreased. In silanol surfaces, the OH groups on the silica surface become the centres of adsorption of the water molecules. If internal silanol groups are present, hydrogen bonding (with water) could exist between these internal groups, in addition to the bonding established by the external silanol groups.

The amount of silanol groups on the siloxane bounding surface depends upon the crystallinity of the interlinked SiO_4 tetrahedra. The regular structural arrangement of interlinked SiO_4 tetrahedra, with the silicon ions underlying the surface oxygen ions, results in cavities bounded by six oxygen ions in ditrigonal formation. If there is no substitution of the silica in the tetrahedral layer and lower valence ions in the octahedral layers, the surface may be considered to be free of any resultant charge. When replacement of the ions in the tetrahedral and octahedral layers by lower valence ions occurs through isomorphous substitution, one obtains resultant charges on the siloxane surface, thus rendering this a reactive surface.

When the edges of the layer lattice minerals are broken – as is the general case for most of the particles – hydrous oxide types of edge surfaces are obtained. The surfaces of the hydrous oxides of iron and aluminium will coordinate with hydroxyl groups which will protonate or deprotonate depending on the pH of the surrounding medium. Exposure of the Fe^{3+} and Al^{3+} on the surfaces promotes development of Lewis acid sites when single coordination occurs between the Fe^{3+} with the associated H_2O, i.e. Fe(III) CH_2O acts as a Lewis acid site.

5.2.2 Surface functional groups and interactions

In sorption interaction mechanisms involving short-range forces between clay particle surfaces and contaminant ions, the acid–base type of reactions will be predominant. The surface functional groups for clays and contaminants have been briefly discussed in Chapters 3 and 2 respectively. These groups are chemically reactive atoms or groups of atoms bound into the structure of a compound. The functional groups for most clay minerals are either acidic or basic. In contaminants, the nature of the functional groups that form the compound will influence the characteristics of the compound and its ability to 'bind' with clay minerals. For example, depending on how they are placed, the functional groups will influence the characteristics of organic compounds, and will thus contribute greatly in the determination of the mechanisms of accumulation, persistence and fate of these compounds in clay. The chemical properties of the functional groups will influence the surface acidity of the minerals in a clay. Surface acidity is very important in the adsorption of ionizable organic molecules by clays. It is the major factor in the adsorption by clays of amines, s-triazines, and amides and substituted ureas in which protonation takes place on the carbonyl group (Burchill and Hayes, 1980).

The nature of functional groups in organic molecules, shape, size, configuration, polarity, polarizability, and water solubility are important factors in the adsorption of organic chemicals by clays. Many organic molecules, for example amine, alcohol and carbonyl groups, are positively charged by protonation and are adsorbed onto the surfaces of clay minerals as part of the cation exchange process. As the cation-exchange capacity (CEC) of clay minerals differs, it is expected that there will be differences in their capacity to adsorb organic cations – not only because of the differences in CEC, but also because of the influence of the molecular weight of the organic cations. Because of their size and higher molecular weights, large organic cations are adsorbed more strongly than inorganic cations by clays (Morrill *et al.*, 1982).

Hydroxyl (OH) functional group

The *hydroxyl group* consists of a hydrogen atom and an oxygen atom bonded together. As we have learnt from Chapter 3, the hydroxyl group is the dominant reactive surface functional group for clay minerals, amorphous silicate minerals, metal oxides, oxyhydroxides and hydroxides. This group is also present in two broad classes of organic chemical compounds: (a) alcohols (e.g. methyl, ethyl, isopropyl and n-butyl) and (b) phenols [e.g. monohydric (aerosols) and polyhydric (obtained by oxidation of acclimated activated sludge: pyrocatechol, trihydroxybenzene)]. Alcohols are basically hydroxyl alkyl compounds (R-OH) with a carbon atom bonded to the hydroxyl group.

The more common ones are CH$_3$OH (methanol) and C$_2$H$_5$OH (ethanol). They are neutral in reaction, as the OH group does not ionize. Phenols, on the other hand, are compounds with a hydroxyl group that is attached directly to an aromatic ring. This is illustrated in Figure 5.2, along with the other functional groups.

The hydroxyl of alcohol can displace water molecules in the primary hydration shell of cations adsorbed onto clay minerals – depending on the polarizing power of the cation. The other mechanisms for adsorption of hydroxyl groups of alcohol are through hydrogen bonding and cation–dipole interactions. Most primary aliphatic alcohols form single-layer complexes on clay minerals, with their alkyl chain lying parallel to the surfaces of the clay particles. Some short-chain alcohols such as ethanol can form double-layer complexes with the clay minerals.

Other than the hydroxyl group the two main compound functional groups are (a) functional groups having a C–O bond (e.g. carboxyl, carbonyl, methoxyl, and ester groups) and (b) nitrogen-bonding functional groups (e.g. amine and nitrile groups). The aquisition of a positive or negative charge occurs through the process of dissociation of H$^+$ from or onto the functional groups, depending on the dissociation constant of each functional group and the pH. The compounds can thus be fixed or variable charged compounds, or a mixture of each.

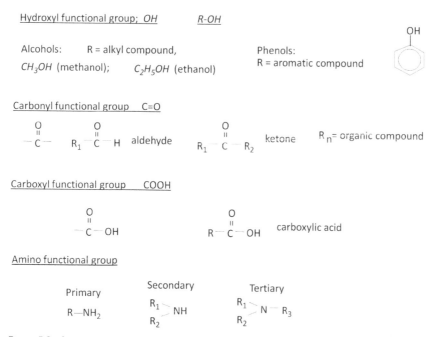

Figure 5.2   Some common surface functional groups for clay particles and organic chemical contaminants.

Strictly speaking, the *phenolic functional group*, which consists of a hydroxyl attached directly to a carbon atom of an aromatic ring, belongs as a subcategory of the hydroxyl functional group. The phenolic functional group compounds can combine with other compounds such as pesticides, alcohol and hydrocarbons to form new compounds, for example anthranilic acid, cinnamic acid, ferulic acids, gallic acid, *p*-hyroxybenzoic acid. The major types of phenolic compounds found in soils include pesticides, cyclic alcohols and napthols.

Carbonyl (C=O) functional group

The carbonyl (C=O) functional group consists of a carbon atom bonded to an oxygen atom by two pairs of electrons (double bond). The compounds that contain the *carbonyl functional group*, called carbonyl compounds, include aldehydes, ketones and carboxylic acids. Most carbonyl compounds have dipole moments because the electrons in the double bond are unsymmetrically shared. Even although they can accept protons, the stability of complexes formed between carbonyl groups and protons is very weak. The carbonyl groups of both aldehydes and ketones will determine their characteristics. Aldehydes are easily oxidized to the corresponding acids, whereas ketones are difficult to oxidize because of the absence of hydrogen attachment to the carbonyl group. The carbonyl group of ketones is adsorbed onto clay minerals, with the adsorption being a function of the nature of the exchangeable cation and hydration status of clay.

Acetone and nitrobenzene form double-layer complexes with the clay particles, with linkage through a water bridge to the cation of the exchange complex. Acetone is adsorbed both physically and by electrostatic interaction with montmorillonites. However, the electrostatic interaction between the cation and acetone is weakened when a hydration shell surrounds the cation.

Carboxyl (COOH) functional group

The carboxyl functional group is obtained through a combination of the carbonyl and hydroxyl groups into a single unit. Carboxylic acids are typical of the carboxyl group. The different types of carboxylic acids include (a) lower acids, which are liquids with an unpleasant odour (e.g. formic and acetic acids) and are miscible with water, and (b) higher acids, which are wax-like 'solids' (e.g. oleic acid) and are almost insoluble. Especially in clays containing expanding layer–lattice minerals, interaction of the carboxyl group of the organic acids (e.g. benzoic and acetic acids) with the minerals is either directly with the interlayer cation or through the formation of a hydrogen bond with the water molecules coordinated to the exchangeable cation. Water bridging is an important mechanism in the adsorption process, together with the polarizing power of the cation. In addition to coordination and hydrogen bonding, organic acids can be adsorbed through the formation of salts with the exchangeable cations. Their ability to donate hydrogen ions to form basic substances renders most carboxyl compounds acidic. However, in comparison with inorganic acids, they are weak acids.

Amino NH_2 functional group

The amino NH_2 functional group is found in primary amines. These are organic bases that form stable salts with strong acids. They may be aliphatic, aromatic or mixed, depending on the nature of the functional groups. By and large, aliphatic amines are stronger bases than ammonia, and aromatic amines are much weaker than aliphatic amines and ammonia. We should note that amines can protonate in clays and can replace inorganic cations from the clay complex by ion exchange. They can be adsorbed with their hydrocarbon chain perpendicular or parallel to the surfaces of clay minerals depending on their concentration. For example, ethylenediamine (EDA) is adsorbed onto montmorillonite by hydrogen bonding coordination and aniline is adsorbed onto clay particles by cation–water bridges.

5.2.3   Sorption mechanisms

The term *sorption* is used to refer to the many adsorption (interaction) processes that result in partitioning of the dissolved contaminant solutes in the porewater onto the surfaces of the clay minerals and other clay fractions. The dissolved solutes include ions, molecules and compounds. Because it is difficult to single out or separate the various adsorption processes that result in partitioning, one speaks of *sorption of contaminants* as a general reference to such processes as:

*   *physical adsorption (physisorption)* – occurs principally as a result of ion-exchange reactions and van der Waals forces;
*   *chemical adsorption (chemisorption)* – involves short-range chemical valence bonds.

Physical adsorption

Physical adsorption of contaminants occurs when contaminants in the porewater are attracted to the surfaces of the clay minerals in response to the charge deficiencies of the clay minerals. Cations and anions are specifically or non-specifically adsorbed by the clay minerals, depending on whether they interact in the diffuse ion layer or in the Stern layer. The counterions in the diffuse ion layer will reduce the potential ψ (see Figure 3.10 and equation 3.12). These counterions, which are generally referred to as *indifferent ions*, are *non-specific*, in that although they will reduce the magnitude of ψ, they do not reverse the sign of ψ. The term *non-specific adsorption* is used to refer to the fact that the counterions are held primarily by electrostatic forces. This has been referred to as outer-sphere surface complexation of ions by the functional groups associated with the clay particles (see section 3.6.3).

The adsorption of most alkali and alkaline earth cations by clay minerals is a good example of non-specific adsorption. Determination of the spatial distribution of the non-specifically adsorbed counterions in the direction perpendicular to mineral surface can be obtained on the same basis as determined by the diffuse double-layer (DDL) model. If a cation is considered as a point charge, as assumed in the Gouy–Chapman model (Chapter 3), the adsorption of cations would be related to their valence, crystalline and hydrated radii. All else being equal, Coulomb's law tells us that cations with the smaller hydrated size or large crystalline size would be preferentially adsorbed. Replacement of exchangeable cations involves cations associated with the negative charge sites on clay minerals through largely electrostatic forces. Ion exchange reactions occur in the various clay constituents, i.e. clay minerals and non-clay mineral fractions.

The position of adsorbed cations on clay mineral surfaces is called the adsorption binding site of clay minerals for cations. In other words, the adsorption binding sites refer to the position of exchangeable cations on clay minerals. Three-dimensional situations of adsorbed cations in the interlayer water, i.e. the horizontal and vertical distribution of counterions on mineral surface, can be demonstrated by molecular dynamics simulations using the interaction potential energy for a pair of atoms. The horizontal distribution of adsorbed cations on clay minerals will designate the adsorption binding site. Chang *et al.* (1995) indicated that the adsorption atoms will hover and roam much more freely over the siloxane surface. Smith (1998) conceptually summarized and demonstrated that there will be five adsorption binding sites on a montmorillonite particle surface: the binding sites are situated over (1) a tetrahedral-layer Si atom; (2) a tetrahedral-layer Al atom; (3) the hexagonal cavity that is adjacent to a tetrahedral-layer Al atom; (4) the hexagonal cavity; and (5) the octahedral-layer Mg atom. The adsorption binding sites will thereby be identified as the position at which moving cations will stay frequently for a long time period on the clay mineral–water interface. Figure 5.3 shows the trajectory of Cs atoms in the interlayer water that was simulated using equations 3.8–3.10 and Table 3.3 (Nakano and Kawamura, 2006). The positions where the trajectories concentrate and form black lumps can be presumed to be the adsorption binding sites. One can determine that there are three kinds of adsorption binding sites: (a) over a centre of hexagonal cavity; (b) over a corner oxygen atom in hexagonal cavity; and (c) over a side of tetrahedron on the plane of the 2:1 sheet.

For the many organic molecules that are positively charged by protonation, it is expected that these molecules will be adsorbed on the surfaces of

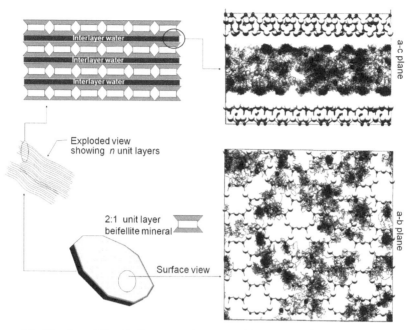

Figure 5.3   The thread-like thin lines are trajectories of the roaming Cs atoms (a) in an a–c plane, which is the vertical section of the 2:1 unit layer particle, and (b) in an a–b plane (surface plane) for a hydrated caesium beidellite with 7.5 H_2O, which is equivalent to two or three layers of water. The tetrahedral-layer O atoms on the mineral surface are shaded circles and the tetrahedral-layer Si atoms are small black circles. Trajectories concentrate in the positions that are (a) over a centre of hexagonal cavity, (b) over a corner oxygen atom in hexagonal cavity and (c) over a side of tetrahedron on the plane of the 2:1 sheet.

clay minerals, depending on the cation exchange capacity of the clay. The most probable sources of protons in the protonation of organic compounds are (a) exchangeable H^+ occupying cation exchange sites; (b) water associated with metal cations; and (c) proton transfer from another cation species already adsorbed at the clay mineral surface.

Specific adsorption

Specific adsorption refers to the situation where ions are adsorbed onto clay particle surfaces by forces other than those associated with the electric potential within the Stern layer. The ions involved, which are generally referred to as specific ions, have the ability to influence the sign of ψ. Sposito (1984) refers to specific adsorption as the effects of inner-sphere surface complexation of the ions in solution by the surface functional groups associated with the clay fractions. Arnold (1978) indicates that any tendency for cations to be specifically adsorbed in the inner part of the double layer (Stern layer) will lower the point of zero charge (*pzc*). Specific adsorption of anions on the other hand will tend to shift the *pzc* to a higher value.

Chemical adsorption

Chemical adsorption, sometimes referred to as *chemisorption,* refers to high affinity, specific adsorption that occurs in the inner Helmholtz layer through covalent bonding. The cations penetrate the coordination shell of the structural atom and are bonded by covalent bonds via O and OH groups to the structural cations – with valence forces that are of the type that bind atoms to form chemical compounds of definite shapes and energies. The chemisorbed ions have the ability to influence the sign of ψ, and are referred to as *potential determining ions (pdis)*. It is not easy to distinguish from electrostatic positive adsorption, except for the fact that considerably higher adsorption energies are obtained in chemical adsorption. Reactions can be either endothermic or exothermic, and usually involve activation energies in the process of adsorption, i.e. the energy barrier between the molecule being adsorbed and the surfaces of the clay particles must be surmounted if a reaction is to occur. Strong chemical bond formation is often associated with high exothermic heat of reaction. By and large, the first layer is chemically bonded to the surface and additional layers are held by van der Waals forces. The three principal types of chemical bonds between atoms include (a) ionic bonds, when electron transfer between atoms results in an electrostatic attraction between resulting oppositely charged ions; (b) covalent bonds, when there is more or less equal sharing of electrons; and (c) coordinate covalent bonds, when the shared electrons originate only from one partner.

5.3   Partitioning of inorganic contaminants

Partitioning of contaminants refers to the phenomenon when contaminants in the porewater are transferred from the porewater onto the surfaces of the minerals. The phenomenon applies to all kinds of soils, and in the case of clays used for engineered clay buffers and barriers, assessment of partitioning of contaminants is an important aspect of determining the competency of the clay chosen for the barrier. The processes involved include both chemical and physical processes, and are identified as chemical and physical mass transfer, respectively. Determination of partitioning of contaminants is commonly undertaken through laboratory testing using two types of tests: (a) batch equilibrium adsorption isotherm tests and (b) column or cell leaching tests. Although these two tests share a common objective, i.e. to physically 'measure' the concentration or quantity of contaminants adsorbed by the clay minerals of a candidate clay, the means whereby measurements are made are completely different (Figure 5.4). Because of the differences in the laboratory test methods and also in the methods of preparation of the candidate clay, direct and unfiltered comparison of the adsorption capacities determined by these two types of tests cannot be obtained easily.

Batch equilibrium tests are commonly used to determine the adsorption characteristics of candidate clay for specific inorganic contaminants. As will be discussed in the next section, with respect to clays used for buffers and barriers, the clay particles are dispersed in a solution containing the contaminants of interest. One expects that all the clay particles' surfaces (i.e. surfaces of clay minerals and other clay fractions) are in contact with the contaminants in the solution. In contrast, column leaching tests are designed to determine the adsorption characteristics of a candidate compact clay to specific leachates that contain contaminants of interest. One therefore relies on contaminant leachate flow through the compact clay to establish interaction between the contaminants and clay particles and microstructural units. The proportion of clay particles' surfaces exposed to interaction with the contaminants in the porewater can vary anywhere from 15 to 70 per cent, depending on the type of clay, density and microstructure.

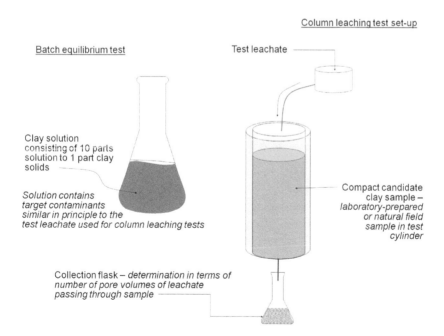

Figure 5.4 Batch equilibrium test for determination of adsorption isotherms and column leaching test set-up, showing the differences in test protocol and test sample.

5.3.1   Batch equilibrium tests and adsorption isotherms

All else being equal, sorption of an inorganic chemical contaminant species from the porewater onto the clay fractions or aggregates is a function of (a) the composition of the clay and (b) the various kinds and distribution of contaminants in the porewater. The sorption capability or potential of a clay for a particular chemical contaminant species is determined by studying the characteristics of adsorption by the clay for that particular contaminant species under specified conditions, i.e. ideal or under certain chemical mix species conditions, etc. The common procedure for determining adsorption isotherms is through batch equilibrium testing of candidate clays with contaminants, and with the use of replicate samples and varying concentrations of contaminants, one obtains characteristic adsorption isotherm information as shown in the top right-hand diagram in Figure 5.5.

Adsorption isotherms are the characteristic adsorption curves relating the adsorption of individual target contaminants by candidate clays in relation to the available target contaminants. The test procedure is not restricted to clays, and can be applied as a procedure to study the adsorption characteristics of all kinds of soils. The batch experiments usually utilize the clay as a clay suspension, as shown in Figure 5.5, with increasing concentrations of the chemical species to be sorbed in the different batches of clay suspension.

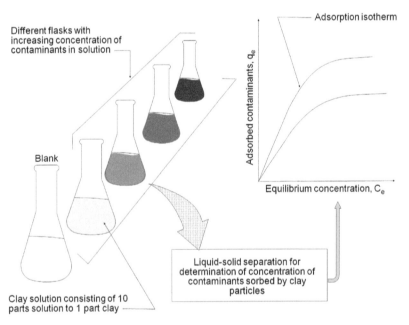

Figure 5.5 General procedure for determination of adsorption isotherm using batch equilibrium testing procedure.

The curve obtained in a graphical portrayal of the relationship between the equilibrium concentration of contaminants in solution and the equilibrium concentration of contaminants sorbed by the clay minerals or clay aggregates (microstructural units) is defined as the *adsorption isotherm* of the clay. A typical set of adsorption isotherms, representative of the adsorption characteristics of clays and other kinds of soils, ranging from high to low sorption capability, is shown in Figure 5.6. The sorbed concentration of contaminants, i.e. the contaminants attracted to the surfaces of the clay particles' surfaces, is denoted as s^*. We keep saying *clay particles' surfaces* because it must be remembered that most clays do not exist as pure 100 per cent clay minerals. It is important to continually recognize that other non-clay mineral fractions form part of any clay, as has been described in Chapter 3. The relationship for the *constant adsorption* curve given as $s^* = k_1 s$ in the figure relates the sorbed concentration of solutes s^* to the equilibrium concentration of solutes in solution s via a constant k_1. The obvious characteristic of this linear relationship lies in the fact that one predicts limitless adsorption of contaminants by the candidate clay being tested. This is highly unlikely, as one would need an infinite number of adsorption sites associated with the clay fractions. The Freundlich and Langmuir adsorption isotherm models on the other hand are more realistic isotherms than the constant adsorption model. It is important to remember that these models are *equilibrium sorp-*

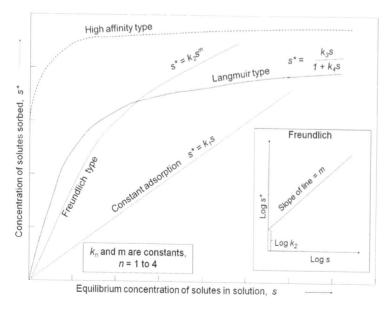

Figure 5.6 Characteristic adsorption isotherms obtained from batch equilibrium tests of different types of soils. Right-hand inset shows the log-plot format for the often-used Freundlich relationship for determination of the constants k_2 and m.

tion models. Non-equilibrium sorption models will be discussed when we address the topic of transport and fate of contaminants (section 5.5).

If one takes the often-used Freundlich relationship shown in the figure, i.e. $s^* = k_2 s^m$, and expresses this in a logarithmic form, one obtains log $s^* = \log k_2 + m \log s$. This is shown in a graphic form in the bottom right-hand portion of Figure 5.6, where the Freundlich constants k_2 and m can be easily determined as the intercept on the ordinate and the slope of the line respectively. It is important to note that the concentration of contaminants or solutes in the solution (i.e. the parameter used to describe the variation along the abscissa) has been expressed as either the equilibrium concentration of solutes (contaminants) or the initial concentration of solutes. Regardless of which form is used for the abscissa, it is important to note that the units used for sorbed concentration (ordinate) and the concentration of solutes in the solution should be consistent with each other. Figure 5.7 shows the adsorption isotherms for an illitic clay in relation to Pb as a solute (contaminant) species. Also shown in the graph is the pH of the clay solution in relation to the concentration of solutes in the solution. The pH of the solution decreases as the amount of Pb (in nitrate form) is added to the solution. The initially high pH at low Pb input concentrations is the result of the high pH of the illitic clay.

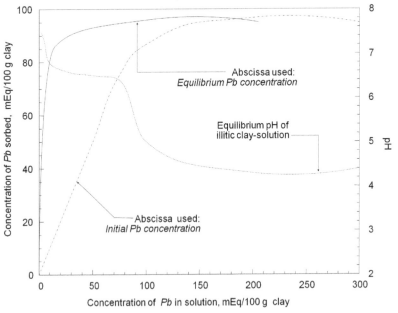

Figure 5.7 Pb adsorption isotherms for an illitic clay, showing two types of isotherms depending on what one chooses to use as the abcissa. For the equilibrium Pb concentration in solution abscissa, the isotherm obtained mimics a high-affinity type isotherm, and for the initial Pb concentration, the isotherm obtained is better approximated as a Freundlich-type adsorption isotherm.

It is important to take note that if the abscissa is plotted in terms of the concentration of solutes in the leachate solution used, as opposed to the equilibrium concentration (i.e. concentration remaining in solution), we will obtain two different adsorption isotherm curves, as demonstrated in the graphical plot shown in Figure 5.7. This is particularly significant, not only because of the marked differences in the characteristic shapes of the isotherms, but also because slope k_d of the chosen curve is used as the distribution coefficient in the 'diffusion–dispersion' contaminant transport relationship, as will be evident in the discussion later in this chapter.

The units used for sorbed concentration (ordinates) of solutes and the concentration of solutes (abscissae) must be consistent with each other. This is important in the subsequent use of the isotherm for determination of the distribution coefficient k_d. We can control the pH of each batch of clay solution by adding buffering agents. However, when we choose to do so, we must be aware of the fact that the adsorption characteristics can also change – the amount and the nature of which will be dependent on both the composition of the clay and the solute used in the solution. It is important to note also that the adsorption isotherms are likely to be different when multiple species of solutes are used in the solution. These, together with the types of conjugate ions used for the solutes, will also contribute to the characterization of shape of the adsorption isotherm. Finally, should the clay contain such fractions as oxides of aluminium and other types of amorphous materials, their higher specific surfaces and types of functional groups will have considerable influence on the character of the adsorption isotherms.

5.3.2   Column leaching tests

Sorption profiles

There are at least two classes of column leaching tests – distinguished not only by their length–diameter (L/D) ratio, but also by the intent of the test itself. In the first class of columns, these have L/D ratios larger than unity. The length of these columns vary according to the need to determine the nature of contaminant sorption profiles. The literature shows columns with diameters of 5 cm and L/D ratios varying between 1 and 2, but also with L/D ratios as high as 3 or more. These are generally referred to as long leaching columns, and are specifically designed to determine (a) the contaminant sorption profile in relation to the nature of leachate input and (b) the total amount of input leachate required to exhaust the sorption or attenuating capability of the clay being tested (Figure 5.8).

Unlike batch equilibrium testing, column leaching tests use compact clay samples obtained as either laboratory-prepared samples or as samples retrieved from the field. In that regard, one is testing a compact clay to determine its capability to sorb or attenuate contaminants. This type of

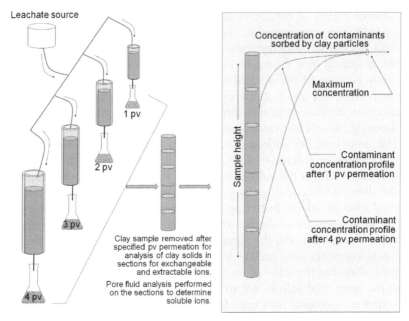

Figure 5.8 General procedure for leaching column tests for determination of sorption of contaminants by clay solids. Replicate samples are permeated with leachate and samples tested after permeation of one to four to six pore volumes (*pvs*) of test leachate. Samples are removed after the predetermined permeation and sectioned. Sectioned samples are analysed for concentration and species of sorbed contaminants, and for chemistry of porewater.

information is important in the decision-making process for choosing the proper clay for an engineered clay barrier or for its use as a contaminant-attenuating barrier. For long column leaching tests using replicate samples as shown in Figure 5.8, various quantities of a test leachate, measured in terms of numbers of pore volumes (*pvs*) are leached into individual clay samples. Removal of each test sample after the end of the specified leachate permeation allows one to section the sample for analysis of the porewater in each section for soluble ions. For the solids, laboratory analyses of the ions associated with the clay minerals allows one to determine the concentration of exchangeable and extractable ions. The sorption characteristic profiles shown in the right-hand part Figure 5.8 illustrate a typical set of contaminant concentration profiles (in relation to length of the test sample), which include both exchangeable and extractable ions – identified as sorption curves.

Because the test samples are compact clay samples, the total surface area of minerals (clay solids or microstructural units) interacting with the contaminants in the influent leachate is less than that in batch equilibrium tests for an equivalent amount of clay solids. Everything else being equal, i.e. clay and leachate composition being unchanged, the primary physical interaction factors involved in characterization of the sorption profile of a clay are (a) nature and distribution of clay solids and microstructural units and (b) density of the clay. The nature and distribution of clay solids and microstructural units control the amount of particle surface areas exposed for interaction with the contaminants in the leachate. The density of the clay impacts on the nature of the macropores and the tortuosity of flow path. The resultant total exposed surface areas would be considerably less than the sum total of individual mineral surface areas – as would be the case of batch equilibrium testing.

The choice of whether one uses batch equilibrium or column leaching tests depends on what questions need to be answered. Adsorption isotherms obtained from batch equilibrium testing are useful for providing information concerning the adsorption potential of a candidate clay for a target contaminant species. Because one uses a totally dispersed clay in the test solution, in theory all the clay solids are available for interaction with the target contaminant. However, unless actual confirmation of complete dispersion of clay solids or particles in the solution is available, one cannot discount the fact that for smectitic clays, microstructural units may constitute some of the *dispersed solids*. This is an important point to consider when one seeks to establish a standard base for comparison of sorption capability between different soils and/or different target contaminants. In contrast, the use of compact samples in leaching column or cell tests permits one to obtain an appreciation of *field conditions*. The assumption is made that leaching of a representative compact clay sample in the test column provides one with an appreciation of the *real field condition*. A basic flaw in this type of reasoning is one's inability to properly replicate field hydraulic conditions in laboratory leaching columns. As hydraulic gradients used in leaching column tests or conductivity cells will far exceed field hydraulic gradients, a proper set of procedures and analyses is required to take into account the role and effect of hydraulic gradients in characterization of the leaching profiles. What is generally overlooked in the conduct of these tests and analyses of results is the importance of *reaction kinetics* in contaminant–clay interaction. There is a direct relationship between hydraulic heads and advective velocities in contaminant transport and contaminant–clay reaction times.

Breakthrough curves

Sampling and analysis of the leachate output from the leaching columns, collected in the beakers shown in Figure 5.8, will provide one with information on the concentration of contaminants in the output leachate in relation to the number of throughput leachate *pv*s. Expressing this throughput concentration in relation to the initial concentration used as input, and relating this to the number of throughput *pv*s as shown in Figure 5.9 allows one

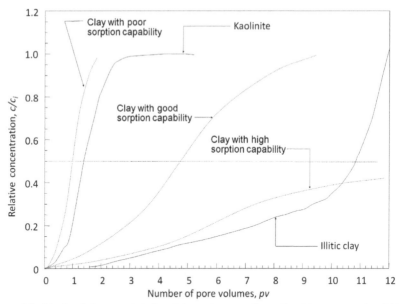

Figure 5.9   Idealized characteristic breakthrough curves for soils with poor, good and high sorption capabilities. For comparison, actual Pb breakthrough curves from column testing with L/D ratio of 2 for a kaolinite and an illitic clay are shown. c represents the concentration of contaminants in the collection beaker and c_i represents the initial concentration of the influent leachate.

to determine sorption capability of the soil. We define the *breakthrough point* as the pore volume throughput identified by the 50 per cent relative concentration c/c_i, where c is the concentration of contaminants and c_i is the concentration of contaminants in the influent leachate. The curves in the diagram are called *breakthrough curves*. Three idealized types of breakthrough curves are shown in the figure: poor, good and high clay sorption capabilities. For comparison, the breakthrough curves for a Pb contaminant for a kaolinite and an illitic clay are shown.

5.3.3   Partitioning of heavy metals

Partitioning of heavy metals in a clay barrier includes the physical and chemical adsorption processes described in section 5.2.3. In addition, as partitioning refers to removal of dissolved solutes (contaminants) from the porewater, at least two other mechanisms need to be considered: speciation/complexation and precipitation. All of these, from physical adsorption to precipitation, fall under the category of interactions of contaminants with clays in a clay–water system. Having said all that, it is important to realize that it is not easy to distinguish the various processes responsible for the removal, as the various sorption mechanisms and precipitation all result in the removal of heavy metals from the porewater.

The more common heavy metals found in wastes in landfills, chemical waste leachates and sludges, include: lead (Pb), cadmium (Cd), copper (Cu), chromium (Cr), nickel (Ni), iron (Fe), mercury (Hg) and zinc (Zn). The interactions of heavy metals with the clay fractions in engineered clay buffers and barriers constitute the problem of major concern in containment of HLW and HSW. This does not take away the fact that inorganic contaminants such as alkali, alkaline-earth metals and transition metals are also pertinent contaminants that will find their way into clay buffers and barriers. Alkali and alkaline-earth metals are elements of groups I and II in the periodic table, such as Li, Na and K – with Na and K being very abundant in nature. Other alkali metals in group IA (Rb, Cs and Fr) are less commonly found in nature. The metals in group II include Be, Mg, Ca, Sr, Ba and Ra – with the more common ones such as Mg and Ca being strong reducing agents. These metals react well with many non-metals. In general, alkali metals are strong reducing agents and are never found in the elemental state, as they will react well with all non-metals. Although the metals with atomic numbers that are higher than that of Sr (atomic number 38) are classified as *heavy metals*, we have indicated in Chapter 2 that common usage includes elements with atomic numbers greater than 20 as heavy metals. A summary recounting these has been given in section 2.3.3.

Complexation and speciation

Speciation, in the context of a clay–water system, refers to the formation of complexes of metallic ions and ligands in the porewater. In a clay–water system this means that there is a competition for sorption of the metallic ions (e.g. heavy metals) between ligands and the clay minerals. Metallic ions are generally coordinated, i.e. chemically bound to water molecules, and in their hydrated form they exist as $M(H_2O)_x^{n+}$, where M^{n+} denotes metallic ions such as Cu^{2+} or Zn^{2+}. The water molecules that form the coordinating complex are the ligands. *Ligands* are anions that can form coordinating compounds with metal ions. The characteristic feature of ligands is their free pairs of electrons. The metal ion M^{n+} occupies a position known as the central atom, and the number of ligands attached to a central metal ion is defined as the *coordination number.* The coordination number of a metal ion is the same regardless of the type or nature of ligand. Replacement of water as a ligand for M^{n+} occurs through replacement of the water molecules bonded to the M^{n+}. Outer sphere complexes or ion pairs formed between metals and complexant ligands are generally weak electrostatic interactions because one or both of the charged species retains their hydration shell. Strong associations between the metals and complexant ligands are obtained in inner-sphere complexes through covalent bonding between the metal ions and the ligands.

Precipitation

Precipitation of metals as hydroxides and carbonates will occur when the ionic activity of heavy metal solutes exceeds their respective solubility products, generally resulting in the formation of new substances. This means that alkaline conditions are favourable conditions for precipitation of heavy metals. Precipitates can also be attached to clay particles from a two-stage process involving nucleation and particle growth. Several critical factors are involved in precipitation of heavy metals. These include (a) the pH of the clay–water system; (b) the nature of heavy metal contaminants, such as concentration, species, and whether one has single and/or multiple species of heavy metals; (c) the nature of inorganic and organic ligands in the porewater; and (d) the precipitation pH of each and every heavy metal species in the porewater.

In the case of multi-species heavy metal contaminants, the presence of other heavy metals affects the precipitation behaviour of individual species. Precipitation data from tests reported by MacDonald (1994) for heavy metals in a nitrate solution are shown in Figure 5.10. The curves shown in the bottom of the figure report on the precipitation behaviour of the three

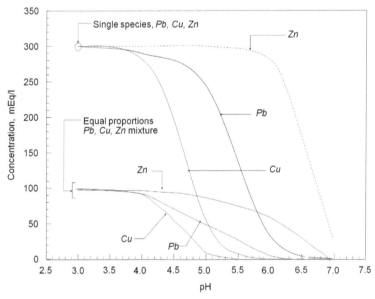

Figure 5.10 Precipitation of heavy metals Pb, Cu and Zn in aqueous solution. Bottom curves are precipitation of individual metal from a mixture of Pb, Cu and Zn in equal proportions – to attain final total metal nitrate solution concentration of 300 mEq/L. Top curves are for single solutions of individual metals at 300 mEq/L concentration (interpreted from data reported by MacDonald, 1994).

heavy metals in a mixture with equal proportions of each of the metals. The top curves depict the precipitation behaviour of individual species of the heavy metals. Precipitation of any of the metals is a gradual and ongoing process. It is not always easy, nor is it entirely possible, to specifically determine the pH at which the actual precipitation occurs (i.e. precipitation pH) in a solution containing multiple heavy metals. One can only compare the point at which *onset of precipitation* occurs. It is difficult to comment further at this stage, except to say that precipitation behaviour of individual species of metals is affected by other ions in solution. In a clay–water system, in which other ions and reactive clay minerals exist, one would expect that the precipitation phenomenon will become more complicated.

5.3.4 Preferential sorption of heavy metals – selectivity

Preferential sorption of heavy metals occurs in metals–clay interaction when multi-species heavy metals are present in the porewater of the clay–water system. By this we mean that there is a degree of selectivity in sorption of heavy metals in clays, i.e. there is a preference in sorption of specific species of heavy metals by clays. All else being equal, one observes that for a particular clay, sorption preference is conditioned by the species of heavy metals and their concentrations. In addition, the types of inorganic and organic ligands present in the porewater are seen to exert influence on sorption preference. Preference in sorption of metal species is called *selectivity*. Selectivity in metal sorption differs between any two clays because of the impact of the nature and distribution of the reactive mineral surfaces available in the clay. Table 5.2 gives an example of preferential sorption of some metal species by three clays.

The table shows that the selectivity order – reported in the literature – depends on the type of clay and the pH environment. In the case of divalent metal ions, it has been reported that when the concentrations applied to a clay are the same, a correlation between effective ionic size and selectivity order may be expected (Elliott *et al.*, 1986). The ease of exchange, i.e. the strength with which metallic ions of equal charge are held within the clay matrix, is generally inversely proportional to the hydrated radii, or proportional to the unhydrated radii (Bohn, 1979). If one predicts a selectivity order on the basis of unhydrated radii, one should obtain $Pb^{2+}(0.120\,nm)$ $> Cd^{2+}(0.097\,nm) > Zn^{2+}(0.074\,nm) > Cu^{2+}(0.072\,nm)$, where the measurements in parentheses refer to the unhydrated radii. Yong and Phadungchewit (1993) show a general selectivity order to be $Pb > Cu > Zn > Cd$ – somewhat different from the ranking based on unhydrated radii. However, the Yong and Phadungchewit ranking scheme agrees well with the ranking based on the *pk* of the first hydrolysis product of the metals – using *k* as the equilibrium constant when $n = 1$ for the reaction in the relationship: $M^{2+}(aq) + nH_2O \leftrightarrow M(OH)_n^{2-n} + nH^+$.

Table 5.2 Sorption preference for heavy metals by kaolinite, illite and montmorillonite

Clay	Sorption preference	Reference
Kaolinite (pH 3.5–6)	Pb > Ca > Cu > Mg > Zn > Cd	Farrah and Pickering (1977)
Kaolinite (pH 5.5–7.5)	Cd > Zn > Ni	Puls and Bohn (1988)
Illite (pH 4–6)	Pb > Cu > Zn > Cd	Yong and Phadungchewit (1993)
Illite (pH 3.5–6)	Pb > Cu > Zn > Ca > Cd > Mg	Farrah and Pickering (1977)
Montmorillonite (pH ≈ 4)	Pb > Cu > Zn > Cd	Yong and Phadungchewit (1993)
Montmorillonite (pH ≈ 5)	Pb > Cu > Cd ≈ Zn	Yong and Phadungchewit (1993)
Montmorillonite (pH ≈ 6)	Pb > Cu > Zn > Cd	Yong and Phadungchewit (1993)
Montmorillonite (pH 3.5–6)	Ca > Pb > Cu > Mg > Cd > Zn	Farrah and Pickering (1977)
Montmorillonite (pH 5.5–7.5)	Cd = Zn > Ni	Puls and Bohn (1988)

With this ranking scheme, one obtains a selectivity as follows – Pb(6.2) > Cu(8.0) > Zn(9.0) > Cd(10.1) – where the numbers in the brackets refer to the *pk* values.

5.3.5   *Distribution of partitioned heavy metals*

The term *distribution* used in the context of partitioned heavy metals refers to the proportion of heavy metals (transferred from the porewater) retained/sorbed by each type of clay constituent (clay fraction). *Distribution* is essentially a reflection of the different sorption capabilities of the various clay fractions. Knowledge of the distribution of partitioned heavy metals is useful, as it informs us about the contribution to the partitioning process by the individual clay fractions in a clay. A general procedure for determination of the distribution of partitioned heavy metal contaminants in a clay is the use of a selective sequential extraction technique commonly known as the *selective sequential extraction (SSE)* technique.

Selective sequential extraction technique and analysis

Application of the SSE technique for removal of sorbed heavy metals from individual constituent clay fractions requires the use of chemical reagents, chosen for their capability in selectively destroying the bonds established between heavy metal contaminants and specific individual clay fractions. A judicious choice of chemical reagents for selective bond destruction is key to the success of the technique. A proper application of the selective sequential extraction technique requires the chemical reagents to release the heavy metal contaminants from specific clay fraction by destroying the bonds binding the heavy metals to the target clay fraction. Two particular points need to be noted:

- Aggressive chemical reagents will not only destroy the bonds, but also threaten the integrity of individual clay fractions. One should use extractant reagents that have a history of use in routine clay analyses. These are available and are classified as concentrated inert electrolytes, weak acids, reducing agents, complexing agents, oxidizing agents and strong acids.
- A chemical reagent chosen to destroy the bond between heavy metals and a target clay fraction may have a small unintended collateral destructive effect on the bonds associated with another clay fraction. To avoid or to minimize this, the sequence of application of the extractant reagents should start with the least aggressive extractant.

Because of the dependence on specific laboratory techniques and choice of extractants, quantitative results reporting on the 'measurement' of distribution of the partitioned heavy metal must be considered as *operationally defined*. Although these measurements should be considered to be more qualitative than quantitative, they are nevertheless useful in that they provide a good insight into the distribution of the partitioned heavy metals. Table 5.3 provides a short summary of a variety of reagents used by different researchers in application of the SSE technique.

The numbers 1 to 5 at the beginning of each column in Table 5.3 refer to the sequence of extraction. With the SSE technique, one obtains five different heavy metal bonding categories, which are operationally defined as groups, categories or phases, with *phases* being the more common terminology used. The SSE technique is shown in Figure 5.11, and a sample of the type of information obtained is shown in Figure 5.12.

The results of the analyses are described in terms of metal–clay fraction associated groups as follows:

- *Exchangeable metals (second column in Table 5.3)*. The metals extracted through the use of neutral salts as ion-displacing extractants, such as $MgCl_2$, $CaCl_2$, KNO_3 and $NaNO_3$, are considered to be exchangeable metals. The literature sometimes refers to these released metals as 'in the exchangeable phase', i.e. they are considered to be non-specifically adsorbed and ion exchangeable and can be replaced by competing cations. The clay fractions associated with these exchangeable metals are the clay minerals, organics and amorphous materials. If the salt solutions are applied at neutral pH, one would expect that, at most, only minimal dissolution of carbonates would result. Other types of salts such as NH_4Cl and NH_4OAc, may dissolve considerable amounts of compounds such as $CaCO_3$, $MgCO_3$, $BaCO_3$ and $MgSO_4$. Extractants such as $CaSO_4$ and NH_4OAc can cause some dissolution of manganese oxyhydrates and metal oxide coatings.

Table 5.3 Summary of some extractants used in SSE technique (adapted from Yong, 2001)

Reference	Exchangeable	Bound to carbonates	Bound to Fe–Mn oxides	Bound to organic material	Residual
Tessier et al. (1979)	(1) $MgCl_2$	(2) NaOH/HOAc	(3) $NH_2OH.HCl$ in 25% HOAc	(4) H_2O/ HNO_3 + NH_4OAc	(5) HF + $HClO_4$
Chester and Hughes (1967)	(1) NH_3OHCl + CH_3OOH	(2) NH_3OHCl + CH_3COOH	(3) NH_3OHCl + CH_3COOH		
Chang et al. (1984)	(1) KNO_3	(4) Na_2EDTA		(3) NaOH	(5) HNO_3 (70–80°C)
Emmerich et al. (1982)	1-KNO_3	(4) Na_2EDTA		(3) NaOH	(5) HNO_3
Gibson and Farmer (1986)	(1) CH_3COONH_4, pH 7	(2) CH_3COONa, pH 5	(3, 4) Hydroxylammonium + HNO_3/acetic acid	(5) H_2O_2 + HNO_3 (85°C)	(6) Aqua regia + HF + boric acid
Yanful et al. (1988)	(1) $MgCl_2$ + Ag thiourea	(2) CH_3COONa + CH_3COOH	(3) $NH_2OH.HCl$	(4) + sulphides, H_2O_2 + HNO_3	(5) HNO_3 + $HClO_4$ + HF
Clevenger (1990)	(1) $MgCl_2$	(2) NaOAc/HOAc		(3) HNO_3/H_2O_2	(4) HNO_3 (boiled)
Belzile et al. (1989)	(1) $MgCl_2$	(2) CH_3COONa/ $NH_2OH.HCl/HNO_3$, room temperature	(3) Mn oxide, $NH_2OH.HCl$/ HNO_3, NH_4OAC/HNO_3	(4) + Sulph. H_2O_2/ HNO_3, NH_4OAc/ HNO_3	
Guy et al. (1978)	(1) (exch. + adsor. + organic) $CaCl_2$ + CH_3COOH + K-pyrophosphate	(4) (carb. + adsor. + Fe-Mn nodules) NH_3OHCl + CH_3COOH	(2) (metal oxides + org.) H_2O_2 + diothinite + bromoethanol		
Engler et al. (1977)	(1) (exch. + adsorb.) NH_4OAc	(2) $NH_2OH.HCl$	(3) $NH_2OH.HCl$	(3) H_2O_2/HNO_3	(4) $Na_2S_2O_4$/HF/ HNO_3
Yong et al. (1993)	(1) KNO_3	(2) NaOAc, pH 5	(3) $NH_2OH.HCl$	(4) H_2O_2 (three steps)	(5) $HF/HClO_4$ + HCl

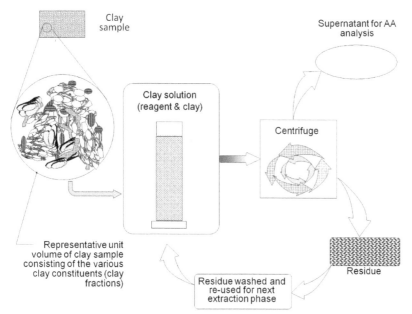

Clay
sample

Supernatant for AA
analysis

Clay solution
(reagent & clay)

Centrifuge

Representative unit
volume of clay sample
consisting of the various
clay constituents (clay
fractions)

Residue washed and
re-used for next
extraction phase

Residue

Figure 5.11 General laboratory procedure for implementation of SSE procedure in determination of partitioning of heavy metals onto specific clay fractions.

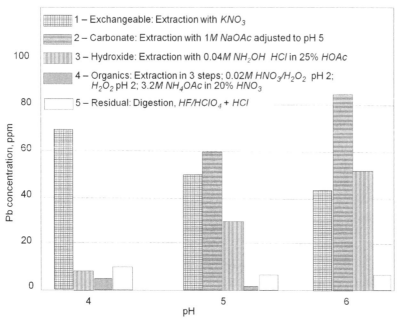

1 – Exchangeable: Extraction with KNO_3

2 – Carbonate: Extraction with $1M$ *NaOAc* adjusted to pH 5

3 – Hydroxide: Extraction with $0.04M$ *NH_2OH HCl* in 25% *HOAc*

4 – Organics: Extraction in 3 steps; $0.02M$ *HNO_3/H_2O_2* pH 2; *H_2O_2* pH 2; $3.2M$ *NH_4OAc* in 20% *HNO_3*

5 – Residual: Digestion, *$HF/HClO_4$ + HCl*

Figure 5.12 Distribution of retained Pb concentration for illite using SSE. The various sequential extraction steps, from 1 to 5, are shown at the top part of the diagram (adapted from Yong and Mulligan, 2003).

- *Metals associated with carbonates (third column in Table 5.3).* The metals precipitated or co-precipitated as natural carbonates can be released by application of acidified acetate as the extractant. A solution of 1 mol/L HOAc–NaOAc (pH 5) is generally sufficient to dissolve calcite and dolomite to release the metals bound to them without dissolving organic matter, oxides or clay mineral particle surfaces.
- *Metals associated with oxides (fourth column in Table 5.3).* The metals released in this sequence of extractant treatment include both metal contaminants attached to amorphous or metals from poorly crystallized metal oxides, such as Fe, Al and Mn oxides.
- *Metals associated with organic matter (fifth column in Table 5.3).* The general technique used to release metals bound to organic matter as a result of oxidation of the organic matter is to use oxidants at levels well below their (organic matter) solubilities. As the binding mechanisms for metals in association with organic matter include complexation, adsorption, and chelation, it is expected that some overlapping effects will be obtained with those methods designed to release exchangeable cations.
- *Metals contained in the residual fraction (sixth column in Table 5.3).* The amount of metal contained in the residual fraction is generally not considered to be significantly large. The metals are contained within the lattice of silicate minerals, and can become available only after digestion with strong acids at elevated temperatures. Determination of the metal associated with this fraction is important in completing mass balance calculations.

5.3.6   Application

There are several answers or responses to some important questions relating to the design, material selection, and construction of engineered clay barriers that require informed knowledge of the several aspects of partitioning of contaminants in the clays used for the buffers and barriers. Many of these will be discussed in the various sections in this chapter. One of the most important aspects of engineered clay buffers and barriers is the ability of the clay to function as a contaminant sorption and attenuating buffer/barrier. Some of the particular questions and areas of immediate concern are given below:

- What are the specific sorption and attenuation qualities (properties and characteristics) of the variety of constituents that constitute the clay needed to construct the buffer/barrier?
- Partitioning studies using batch equilibrium and column leaching tests provide initial assessment of the available candidate clays, and SSE studies

should provide qualitative reports on how the various constituents or fractions in the clay contribute to the desired qualities.

- What can we learn about the transport and fate of contaminants in the chosen clays, and how can we develop the appropriate predictive/ analytical models?
- Partition or distribution coefficients can be determined from both batch equilibrium and column leaching tests. However, as will be discussed in the next section, the selection of an appropriate set of distribution functions is an exercise that demands proper understanding of partitioning behaviour.

5.4   Partitioning of organic chemical contaminants

5.4.1   Adsorption isotherms

Partitioning of organic chemical contaminants is described in a manner similar to that used to describe the partitioning of heavy metals. As in the case of partitioning of inorganic contaminants, the procedures and techniques that are used to determine partitioning of organic chemical contaminants apply to all kinds of soils. As the soils being discussed in this book are clays, we will consider the partitioning of organic chemical contaminants specifically in relation to clays. The distribution coefficient, identified as k_d, refers to the ratio of the concentration of organic chemical contaminants retained by the clay fractions to the concentration of chemical contaminants in the porewater (aqueous phase). This means that $C_s = k_d C_w$, where C_s refers to concentration of the organic chemical contaminants retained by the clay fraction, and C_w refers to the concentration remaining in the aqueous phase respectively. The term *aqueous phase* is most often used instead of porewater when chemicals are involved in clays. Although the partitioning of organic chemical contaminants generally involves many processes, the distribution coefficient k_d is usually obtained using batch equilibrium procedures similar to those used to obtain adsorption isotherms for inorganic contaminants. Clay-suspension tests with target organic chemical contaminants and specified (or actual) clay fractions can yield three types of adsorption isotherms that describe the partitioning behaviour of organic chemicals. These types have often been characterized by the Freundlich isotherm relationship identified in Figure 5.6.

Denoting C_s and C_w as the organic chemical retained by the clay fractions and remaining in the aqueous phase respectively – to avoid confusion with the inorganic solutes isotherms used previously – the Freundlich isotherm for organic chemical contaminants will be given as $C_s = k_1 C_w{}^n$. As before, k_1 and n are the Freundlich parameters. When $n = 1$, a linear relationship between C_s and C_w is obtained. When $n < 1$, the retained organic chemical decreases proportionately as the available organic chemical increases. This would indicate that all available retention processes are being exhausted, and when $n \ll 1$, one would expect the retention capacity to be fully taxed. The reverse is true when $n > 1$, i.e. the retention capacity of the clay keeps increasing as more organic chemicals are retained. This can happen in situations where initial retention of the chemicals by a clay will disrupt the clay microstructural units, thus releasing more reactive particle surfaces for interaction with the organic chemicals.

Studies using paddy soils in Japan (Kibe *et al.*, 2000), with organic contents ranging from 1.2 to 6.8 per cent, SSA from 6.2 to 34 m²/g, and CEC from 12 to 45 mEq/100 dry soil, were conducted in interaction with different herbicides such as esprocarb ($C_{15}H_{23}NOS$), pretilachlor ($C_{17}H_{26}ClNO_2$), simetryn ($C_8H_{15}N_5S$) and thiobencarb ($C_{12}H_{16}ClNOS$). The Freundlich-type relationships obtained had n values ranging from 1.0 to 1.6 (Kibe *et al.*, 2000). Experiments conducted using batch equilibium tests with aniline (C_6H_7N) and trichloroethylene (TCE, C_2HCl_3) in interaction with 6 per cent bentonite produced Freundlich-type adsorption isotherm results after 7 and 10 days for both chemicals. Other types of isotherms have been obtained with different types of organic chemicals and soils – not all of them easily 'fitted' by a Freundlich relationship. For example, in a different set of studies, Langmuir-type adsorption isotherms were found for tetracycline ($C_{22}H_{24}N_2O_8$).

Organic molecules are generally varied in nature, i.e. they have varied size, shape, molecular weight, etc. By and large, these molecules demonstrate less polar characteristics than water, and their water solubility will influence or control their partitioning. In addition, their water solubility will also influence the various processes such as oxidation–reduction, hydrolysis and biodegradation, which result in their transformation. Figure 5.13 demonstrates the above as a case in point. The water solubility of naphthalene ($C_{10}H_8$) and 2-methylnaphthalene ($C_{11}H_{10}$) is 30 mg/L and 25 mg/L respectively. As these water solubilities are very similar, their isotherms also demonstrate close similarity. However, in the case of the 2-naphthol ($C_{10}H_8O$) – an intermediate along the pathway of reaction sequence of naphthalene by *Cunninghamella elegans* – whose water solubility is about 750 mg/L, one obtains a significantly different isotherm. An organic chemical with higher water solubility will have a greater amount of that chemical remaining in the aqueous phase, in contrast with another organic chemical with a much lower water solubility. Adsorption studies involving petroleum hydrocarbons (PHCs) and surfaces of clay particles show that adsorption occurs only when the water solubility of the PHCs is exceeded, and that the hydrocarbons are accommodated in the micellar form.

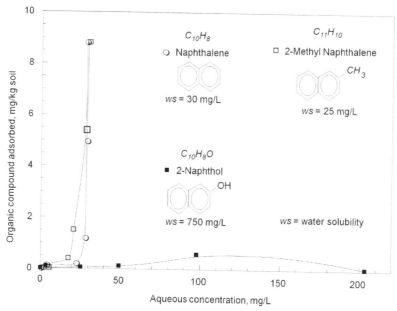

Figure 5.13   Adsorption isotherms for naphthalene, 2-methyl naphthalene and 2-naphthol with kaolinite.

5.4.2   Equilibrium partitioning – octanol–water partition coefficient, k_{ow}

Partitioning of organic chemicals in a clay–water system is determined in terms of the relative fugacity of the organic compound in an organic solvent and in water. If we define C_{os} as the concentration of an organic chemical in an organic solvent, and C_w as the concentration of that organic chemical in water, the ratio C_{os}/C_w is defined as k_{os}, the equilibrium partitioning coefficient. Because of its very low solubility in water, n-octanol is commonly used as the organic solvent, and the partition coefficient determined with this solvent is called the *octanol–water partition coefficient k_{ow}*. As n-octanol is amphiphilic, i.e. part lipophilic and part hydrophilic, it can accommodate organic chemicals with the various kinds of functional groups. The dissolution of n-octanol in water is about eight octanol molecules to 100,000 water molecules in an aqueous phase, i.e. a ratio of about 1:12,000 (Schwarzenbach *et al.*, 1993). As water-saturated n-octanol has a molar volume of 0.121 L/mol compared with 0.16 L/mol for pure n-octanol, this close similarity allows us to ignore the effect of the water volume on the molar volume of the organic phase.

The k_{ow} octanol–water partition coefficient has been widely adopted in studies of the environmental fate of organic chemicals. The literature shows that good correlations have been obtained between solubilities of organic compounds and their n-octanol–water partition coefficient k_{ow}. It has also been found to be sufficiently correlated not only to water solubility, but also to clay sorption coefficients. The relationship for the n-octanol–water partition coefficient k_{ow} in terms of the solubility S has been reported by Chiou *et al.* (1982) as:

$$\log k_{ow} = 4.5 - 0.75 \log S \text{ (ppm)} \tag{5.1}$$

Organic chemicals with k_{ow} values of less than 10 generally have high water solubilities and small adsorption coefficients, and are considered to be relatively hydrophilic. On the other hand, organic chemicals with a k_{ow} values greater than 10^4 generally have low water solubilities. They are very hydrophobic and are not very water soluble. The nature and amount of organic matter in a soil has a significant influence on the value of k_{ow}. A correlation of sorption can be obtained with the proportion of organic carbon in the soil. The coefficient that describes the organic carbon content in soil organic matter, k_{oc}, is in essence a measure of the hydrophobicity of the chemical contaminant, and can be determined as follows (Olsen and Davis, 1990):

$$k_{oc} = \frac{k_d}{f_{oc}} = 1.724 k_{om} \tag{5.2}$$

where f_{oc} is the organic carbon content in the organic matter in the clay–water system, k_d is the distribution coefficient discussed in the previous subsection and k_{om} is the distribution coefficient for the organic matter. Values for k_{om} and k_{oc} for a number of organic chemicals can be found in the various handbooks dealing with environmental data for such chemicals. Because of competing sorption sites from other clay fractions, clays with organic matter content of less than 1 per cent by weight can give high values for k_{oc}. It has been suggested that the relationship given as equation 5.2 should not be used when $f_{oc} < 1$. McCarty *et al.* (1981) give a critical minimum level for the organic carbon content f_{oc-cr} as:

$$f_{oc-cr} = \frac{SSA}{200(k_{ow})^{0.84}} \tag{5.3}$$

where SSA denotes the specific surface area of the soil.

Several relationships between f_{oc} and k_{ow} have been proposed in the literature, for example:

$$\log k_{oc} = \log k_{ow} - 0.21$$
$$\log k_{oc} = 1.029 \log k_{ow} - 0.18 \tag{5.4}$$
$$\log k_{oc} = 1.06 \log k_{ow} - 0.68$$

These have been obtained in studies involving chemicals ranging from polycyclic aromatic hydrocarbons (PAHs) to polychlorinated biphenyls (PCBs) and pesticides. Figure 5.14 shows the relationship in terms of log k_{oc} and k_{ow} for many of the chemicals. The values used for k_{ow} are in the mid-range of the numerous results reported from many different studies. The k_{oc} values shown in the figure have been obtained either as reported measured values or as reported computed values using log k_{oc}–log k_{ow} relationships [e.g. Karickhoff *et al.*, (1979); Kenaga and Goring (1980); Rao and Davidson (1980); Schwarzenbach and Westall (1981); Verscheuren (1983); Montgomery and Wellkom (1991)]. The approximate relationship shown by the line in Figure 5.14 is the third relationship given in equation 5.4. With this type of relationship, one has the opportunity to anticipate partitioning of a large number of organic chemical compounds.

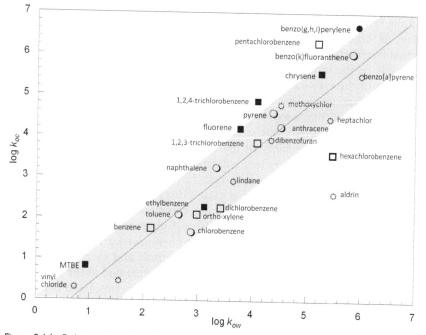

Figure 5.14   Relationship of log k_{oc} and log k_{ow} for several organic chemical compounds. The shaded band encompasses a large majority of the chemical compounds.

5.5   Transport and fate of contaminants

5.5.1   *Basic considerations*

The primary vehicle for transport of inorganic contaminants and organic chemicals in the subsurface soil is water, i.e. water is the principal carrier of contaminants in the subsurface. Water exists in the ground, as porewater in subsurface soils and rocks, and as aquifer water. Water entering the soil subsurface can come as the aqueous phase of leachates generated from wastepiles or as direct infiltration and seepage from rainfall or snowfall or other water discharge systems. A secondary vehicle for transport of organic chemicals could be the liquid chemicals themselves, discharged illegally into the ground, or as accidental discharge of liquid chemicals. In the case of organic chemical contaminants found in the subsurface environment, volatilization provides another means for transport of chemical contaminants.

The fate of contaminants depends to a large extent on (a) hydrodynamic forces responsible for production of advective flow of leachates, etc.; (b) physico-chemical processes resulting in the partitioning mechanisms (described previously) controlling the transport of these contaminants; (c) biogeochemical reactions and transformations occurring in the clay buffer/barrier; and (d) redox transformations. The outcome of all of these processes will impact directly on the ultimate fate of these contaminants. It must be noted that in addition to the impact of the factors described, from (a) to (d), on the fate of contaminants, the impact on the nature and properties of the clays in the engineered clay buffers and barriers cannot be overlooked. In actual fact, there is considerable concern over the detrimental impacts caused not only by the interactions from contaminants on the stability and design function of the clay barriers/buffers, but also by long-term biogeochemical processes that are not necessarily provoked by contaminants entering the system. In short, one needs to be concerned with evolution of the clays in clay buffers and barriers due to short- and long-term processes leading to a reduction in the capability of the clays to fulfil their design function in the engineered clay buffers and barriers. These concerns are addressed in Chapter 8 in the discussion of long-term buffer/barrier performance.

With respect to the engineered clay buffers and barriers that are used for containment of HLW and HSW, some commonality exists in the overall contaminant transport phenomenon. What distinguishes between the HLW and HSW clay buffer/barrier–contaminant phenomena are (a) the presence of a significant heat source lasting over hundreds and even thousands of years; (b) swelling pressures of some magnitude generated in the clay buffer as a result of buffer–water interaction; (c) the need for consideration of long-term buffer integrity because of biogeochemical reactions and transformations impacting on the nature of the buffer, i.e. evolutionary processes; and (d) radiological considerations. In comparison with the clays used for HSW engineered clay barriers, the clay in the buffers used for HLW confinement consists primarily of smectites, with a montmorillonite content greater than 50 per cent. In contrast, the clays used in the engineered clay barriers for HSW liner containment are generally natural clays with good assimilative capacity and CEC, SSA or sand–bentonite mixtures.

Transport and fate – assessment and prediction

There exist several kinds and several levels of sophistication of transport and fate analytical computer models. With respect to clay buffers and barriers, transport and fate models can provide analyses and evaluation of performance characteristics of contaminant plumes in transport through the clay buffer/barrier. The type, level of sophistication, and validity of any of these models are related to (a) the model developer's (hereafter referred to as the *developer*) depth of understanding and appreciation of the system, problem and/or processes to be modelled; (b) the developer's ability to provide an accurate, conceptualized model of actual system performance, phenomena and processes; (c) purpose, i.e. scoping studies, assessment, analysis, prediction, design, etc.; (d) types and availability of input information; and (e) end-user requirements and preferences. More and more, one finds that they are required elements in planning and design of engineered clay buffer/ barrier containment systems, and in assessment of effectiveness of remediation technologies. As predictive tools, these models are essential tools in the decision-making process for regulatory agencies and practitioners in (a) assessment and evaluation of attenuation competence of engineered clay barriers or clay substrate; (b) prediction of continued progress of contaminant plumes; and (c) risk and safety assessment and management.

It is not the purpose of this section or this book to deal with the extensive development of transport and fate models. The reader should consult more specialized textbooks and dedicated journals dealing with transport modelling for detailed treatment of the analytical and numerical techniques. In recent years, there has been some considerable development of software transport models – many of them being available commercially and also through regulatory agencies. Instead, the focus on transport and fate models in this section is towards the capability of the models to fully accommodate or account for the phenomena (results) resulting from the various interactions between contaminants and clay minerals. In this section, we will be concerned with the main aspects of the phenomena involved in the transport and fate of inorganic and organic chemical contaminants, particularly in relation to the buffer/barrier containment problem and generally in relation to subsurface soils. Most of these considerations will pertain directly to HSW containment, and elements of these will also apply to HLW repository containment. The extra aspects introduced by HLW repository containment, mentioned previously, will be discussed in the next chapter (Chapter 6).

5.5.2   *Heavy metal contaminants*

We consider the transport and fate of metals in general, and heavy metals in particular in this section, as these are the inorganic contaminants that are most likely to be found in leachates emanating from waste landfills and as corrosion products from repository canisters containing HLW. Aside from the hydrodynamics of contaminant leachate flow that impact directly on the advective flow of the leachate, two important pieces of information needed in formulating the analytical relationships describing contaminant transport in clays are (a) the diffusive properties or capabilities of the contaminants and (b) the partitioning of these contaminants in their transport through the clay buffer/barrier system. Although most (if not all) transport relationships and models address soils as the transport medium, our concern in this book is directed towards clays used as engineered buffers and barriers. In that respect, the transport mechanisms pay attention to processes invoked during transport in clays. The commonly cited one-dimensional relationship used to describe contaminant transport in clays under the saturated condition takes into account (a) molecular diffusion and mass flow of the contaminants; (b) contaminant partitioning; and (c) storage, in the form given as follows:

$$\frac{\partial c}{\partial t} = D_L \frac{\partial^2 c}{\partial x^2} - v \frac{\partial c}{\partial x} - \frac{\rho}{n\rho_w} \frac{\partial c^*}{\partial t} + S \tag{5.5}$$

where c refers to the concentration of contaminants of concern, $t =$ time, D_L is the diffusion coefficient, v is the advective velocity, x is the spatial coordinate, ρ is the bulk density of the clay, ρ_w is the density of water, n is the porosity of the clay, c^* refers to the concentration of contaminants retained by the clay (as generally determined from batch equilibrium adsorption isotherms), and S is the storage term that accounts for *sources* and *sinks* in the clay–contaminant system. In clays, where microstructural units are features that dominate the macroscopic structure of the clay, the storage term S and the retention parameter c^* are impacted by the presence and distribution of these microstructural units – through the properties of these units.

Clay microstructural units serve as sources s_i and sinks s_j for solutes in general and metal contaminants in particular (see Figure 4.12). Taking into account all the sources from $i = 1$ to $i = a$ and sinks from $j = 1$ to $j = b$, and the solutes and/or metal contaminants involved in the sources/sinks, the storage term S can be written as:

$$S = \sum_{i=1}^{a} s_i - \sum_{j=1}^{b} s_j \tag{5.6}$$

An alternative way of handling the storage phenomena is to consider them in terms of their effect on the overall transport of solutes/metals. This is done by incorporating the diffusion of solutes/metals from stagnant regions in the micropores of microstructural units to mobile regions in the macropores – as has been discussed in section 4.3.1, with respect to the diffusion of solutes in partly saturated clays.

In HLW repositories, consideration of transport of fugitive radionuclides in the clay buffers and barriers is required. In such a case, equation 5.5 can be written as:

$$\frac{\partial c}{\partial t} = D_L \frac{\partial^2 c}{\partial x^2} - v\frac{\partial c}{\partial x} - \frac{\rho}{n\rho_w}\frac{\partial c^*}{\partial t} + \left(\frac{\partial c}{\partial t}\right)_{decay} + S \qquad (5.7)$$

where the added term in equation 5.7 dealing with the radioactive decay of the contaminant, takes the following form:

$$\left(\frac{\partial c}{\partial t}\right)_{decay} = -\frac{\ln 2}{\lambda}c$$

In this case, the half-life of the radionuclide is denoted as λ.

The c^* term in equation 5.5, which refers to the concentration of contaminants retained by the clay solids, i.e. the contaminants removed from the pore fluid, can be expressed in terms of the influent concentration, for example using adsorption isotherms such as those shown in Figure 5.6. The adsorption isotherms obtained from batch equilibrium testing (section 5.3.1) are the results of measurements of *equilibrium* sorption, and if they are used for transport and partition modelling/analyses, account must be given to the requirement for the rate of sorption of the contaminants to be far in excess of the transport time in a clay buffer/barrier. This means to say that when equilibrium adsorption isotherms such as the constant, Freundlich or Langmuir relationships are used, it is important to ensure that equilibrium in contaminant partitioning is attained during transport of any of the contaminants through the buffer/barrier. Choosing a constant adsorption isotherm (see Figure 5.6), and denoting k_1 as k_d, the distribution coefficient, one obtains $c^* = k_d c$. Accordingly, with this substitution into equation 5.5, the following is obtained:

$$\left(1+\frac{\rho}{n\rho_w}k_d\right)\frac{\partial c}{\partial t} = D_L \frac{\partial^2 c}{\partial x^2} - v\frac{\partial c}{\partial x} + S \qquad (5.8)$$

The first bracketed term in equation 5.8 is generally referred to as the *retardation factor R*, i.e.:

$$R = \left[1 + \frac{\rho}{n\rho_w} k_d \right]$$

This allows one to write the final equation as:

$$R\frac{\partial c}{\partial t} = D_L \frac{\partial^2 c}{\partial x^2} - v \frac{\partial c}{\partial x} + S \tag{5.9}$$

The constant adsorption isotherm can be faulted because it allows for infinite adsorption of contaminants – a situation that is not realistic. A more popular choice of a characteristic adsorption isotherm is the Freundlich type of isotherm. Using the relationship shown in Figure 5.6, and substituting c^* for s^* and using k_{Fd} to indicate that this is the Freundlich distribution coefficient instead of k_2, we obtain $c^* = k_{Fd} c^m$. Substituting this relationship into equation 5.5 gives us:

$$\frac{\partial c}{\partial t} = D_L \frac{\partial^2 c}{\partial x^2} - v \frac{\partial c}{\partial x} - \frac{\rho}{n\rho_w} \frac{\partial \left(k_{Fd} c^m \right)}{\partial t} + S \tag{5.10}$$

The retardation factor R used in equation 5.9 now becomes:

$$R = \left[1 + \frac{\rho k_{Fd} m c^{m-1}}{n\rho_w} \right]$$

When contaminant sorption equilibrium (by the clay fractions) cannot be obtained during transport through an engineered clay buffer/barrier, it would be inappropriate to use these types of equilibrium models. This situation arises when other internal forces in an engineered clay buffer/barrier interfere with or compete with the clay fractions for the incoming contaminants. Precipitation, dissolution, hydrolysis, acid–base reactions and complexation, to name a few, are some of the interactions/reactions that will impact on sorption and the partitioning process. In such situations, one should use kinetic models that describe the rate of sorption of contaminants by the clay minerals. Kinetic sorption models can be either *irreversible* or *reversible first-, second-, third-order, etc.* kinetic sorption models.

Irreversible kinetic sorption models are used in situations when contaminants sorbed by the buffer/barrier clay minerals are tightly bound to the minerals and cannot be displaced by further incoming leachate contaminants, i.e. the sorbed contaminants are immobile. Examples of sorption mechanisms resulting in sorbed immobile metals in relation to various clays are shown in Table 5.4.

Table 5.4 Heavy metal retention by some clay minerals (adapted from data in Bolt, 1979)

Clay mineral	Chemisorption	Chemisorption at edges	Complex adsorption	Lattice penetration[a]
Montmorillonite	Co, Cu, Zn		Co, Cu, Zn	Co, Zn
Kaolinite			Cu, Zn	Zn
Hectorite			Zn	Zn
Brucite			Zn	Zn
Vermiculite			Co, Zn	Zn
Illite	Zn	Zn, Cd, Cu, Pb		
Phlogopite			Co	
Nontronite			Co	

Note
a   Lattice penetration = lattice penetration and embedding in hexagonal cavities.

The retention mechanisms shown in Table 5.4 do not include ion exchange because this retention mechanism allows for detachment and mobility of the ion-exchanged metals. The retention mechanisms shown in the table include specific adsorption, chemisorptions involving hydroxyl groups from broken bonds in the clay minerals, formation of metal–ion complexes, and precipitation as hydroxides or insoluble salts. These are not expected to be sensitive to actions leading to detachment of the sorbed metals. Studies on Ni^{2+} adsorption by montmorillonite reported by Xu *et al.* (2008) showed that adsorption equilibrium (of the metal) was achieved in a matter of hours. This evidence, coupled with the results obtained in subsequent desorption experiments that showed failure to detach previously sorbed metals, indicated that chemical adsorption and surface complexation were responsible for retention of metal – as opposed to physical adsorption. The standard value of enthalpy registered in the adsorption process indicated that adsorption was endothermic, a similar conclusion reached previously in the study on the sorption kinetics of caesium and strontium ions on a zeolite reported by El-Rahman *et al.* (2006).

Reversible kinetic sorption models allow one to account for detachment and mobility of previously sorbed contaminants by incoming leachate contaminants. The basic concept of reversible kinetic sorption deals with the rates of adsorption of contaminants and rates of desorption of previously sorbed contaminants. There are several conceptual scenarios that one needs to consider in structuring a reversible kinetic sorption model. These relate to the question of *degree of reversibility*. Figure 5.15 shows the results of sorption–desorption (of contaminants) kinetics.

Considerations of the rate of desorption of sorbed contaminants require one to determine (a) the sorption characteristics of the particular contaminant with the clay chosen for use in the engineered clay buffer and barrier system; (b) the overall chemical composition of the contaminating leachate;

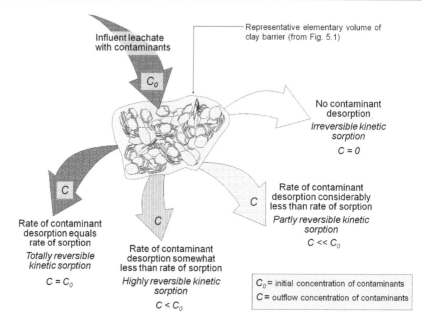

Figure 5.15 Contaminant leachate outflow results due to irreversible, partly reversible and totally reversible kinetic sorption of contaminants.

(c) the hydrodynamic aspects governing leachate penetration into the clay buffer/barrier; and (d) the assimilative capacity of the clay. The assimilative capability of a clay in a buffer/barrier system establishes the ultimate capacity of the clay to retain the incoming contaminants. The *carrying capacity* of the clay defines the capacity point beyond which no further contaminants can be sorbed or retained. This is an important consideration since one cannot expect the clay to retain incoming contaminants ad infinitum. With this in mind, let us consider the implications of the various scenarios posed in Figure 5.15. The bottom left arrow illustrates a totally reversible kinetic sorption effect. Obviously, this not a 'design' situation, i.e. one does not choose a clay that would not function as a contaminant attenuating barrier.

The *highly reversible kinetic sorption model* assumes that the rate of desorption of previously retained contaminants is somewhat less than the rate of sorption. The discharge leachate flow from a clay buffer/barrier with this kind of situation will show a concentration of contaminants C less than the entry concentration C_0, i.e. $C < C_0$. Continued inflow of the contaminant leachate will eventually fully tax the carrying capacity of the clay constituting the clay buffer/barrier, at which time the discharge concentration $C = C_0$. The *partly reversible kinetic sorption model* situation is similar in many respects to the highly reversible kinetic sorption situation with the exception that a greater concentration of contaminants is retained in the sorption–desorption process, resulting in early discharge concentrations $C << C_0$. The outcome of the higher level of sorption in the sorption–desorption process is that one would most likely reach the full carrying capacity of the clay (in the clay barrier) more rapidly, thereby resulting in an earlier arrival of discharge $C = C_0$. The ability to determine or obtain the proper sorption rate and desorption rate constants for individual species of contaminants in relation to the type of clay being penetrated by the contaminant leachate is the key to a successful prediction of the transport of contaminants in a clay.

A possible kinetic sorption–desorption model

There is no partitioning model that fits all situations. The choice of model to be used, to fit the third term on the right-hand side of equation 5.7, i.e. $\partial c^*/\partial t$, depends on one's appreciation of the sorption–desorption processes that attend the situation at hand. There are some simple rules that can be followed, and some simple experiments that can be conducted that would provide one with an appreciation of the kinds of partitioning processes that might be operative in the transport of metal contaminants in a clay. The basic question that needs to be answered is 'Is the time taken for equilibrium partitioning in batch equilibrium experiments achieved well within the time frame for transport of the contaminant leachate through the clay buffer/barrier system?' If the answer is 'yes' then it is a matter of selection of an appropriate equilibrium adsorption isotherm. If the answer to the basic question is 'no' then one needs to determine the contaminant adsorption and desorption rates. The general desorption-type experiments consist of batch tests and/or column leaching experiments using de-ionized water as the 'desorbing' fluid. The important point to be made here is that the sorption and desorption types of test must be similar in character, i.e. if one uses batch tests to determine adsorption isotherms, it follows that the same kinds of tests must be used to determine desorption isotherms. The same holds for column leaching tests. Assuming a linear kinetic desorption–adsorption relationship for the third term in equation 5.7 as follows:

$$\frac{\partial c^*}{\partial t} = k_{des}c - k_{sorb}c^*$$

(5.11)

We can now write equation 5.7 in the following form:

$$\frac{\partial c}{\partial t} = D_L \frac{\partial^2 c}{\partial x^2} - v\frac{\partial c}{\partial x} - \frac{\rho}{n\rho_w}\left(k_{des}c - k_{sorb}c^*\right)$$

(5.12)

where k_{des} and k_{sorb} are time-rate constants. For non-linear kinetic relationships, one has a choice of k_{des}, k_{sorb}, c and c^* to describe the following (a) non-linear desorption and linear sorption, (b) non-linear desorption and non-linear sorption and (c) linear desorption and non-linear sorption. Options available include bilinear and multilinear desorption and sorption models.

Desorption tests

There is no universal agreement or consensus on the conduct or technique or procedure for desorption experiments or tests. The two candidate procedures include (a) batch experiments using de-ionized water as the suspension fluid and (b) column leaching tests, also using de-ionized water as the influent fluid. In both cases, previously contaminated clay is used, either as the material for the clay suspension with de-ionized water, or as the compact clay in the column destined for leaching with de-ionized water. As we have stated previously, it is important to maintain consistency between sorption and desorption tests, i.e. the same experimental procedures must be used.

Experiments have been conducted where solutions containing specified species and concentrations of solutes have been used as the suspending or leaching fluid. Determination of the species and concentrations of solutes will be a decision based on scenarios contemplated as likely events. Experiments have also been conducted using a 'cyclical approach', i.e. conducting a sorption test that is immediately followed by a desorption test without removal of the test sample. In other words, these are sorption–desorption experiments – either batch or column leaching types. Regardless of the types of tests chosen and the kinds of test parameters, it is important to obtain a record of time frames involved in the adsorption and desorption processes. The procedures are tedious but essential if one seeks to apply non-linear kinetic adsorption–desorption models to describe partitioning of contaminants during transport through a clay buffer/barrier system.

Physico-chemical considerations in transport

Figure 5.16 shows a lumped parameter approach that takes into account the many physical and chemical reactions involved in transport of contaminants in a clay. These parameters are portrayed as F forces, which, together with a chemico-osmotic flow component portraying the results of physico-chemical reactions, provide the basis for development of a relevant analytical model. The influent contaminant concentration of solutes c_i is greater than the concentration c at the outflow end, and the chemico-osmotic flow shown at the right-hand side of the diagram occurs as a result of this difference in concentration of solutes. The chemico-osmotic effect becomes more significant with greater surface-active clay particles in combination with lower the porosities and clay permeability.

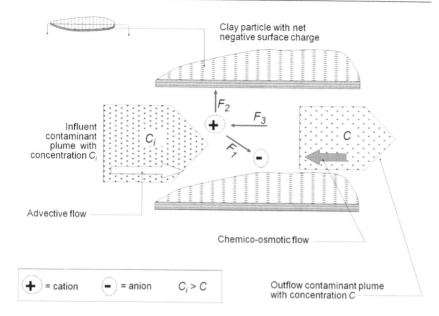

Figure 5.16 Physico-chemical forces and interacting ions in transport of contaminants through a clay. F_1, forces dealing with interaction between ions; F_2, forces of attraction due to presence of negative surface charge on clay particles; F_3, fluid drag forces.

The three sets of thermodynamic F forces acting on the cation shown in the diagram are as follow:

- F_1: *forces due to interaction with other ions.* The mobilities of the various ions in the clay–water system will be different because of their differing sizes and activities. This leads to forces of interaction between ions, and if the ions are oppositely charged, these forces of interactions will have great significance. Ionic interaction forces affect the activity coefficients of the ions, and consequently their diffusion coefficients. A net effect of these interactions is acceleration of diffusion of ions with lower mobilities, and a reduction in diffusion of other ions.
- F_2: attractive forces attributable to the reactive surfaces of clay particles and the drag forces of the pore fluid. Increasing the concentration of cations in the porewater increases cations in the diffuse double layer, and, as ionic concentration increases, overall activity decreases. This results in decreased diffusivity of the ions.
- F_3: drag forces exerted by the porewater on the diffusive movement of the contaminant solutes. This can be described with a modified Stokesian relationship as $F_3 = 6\pi\eta r v_c$, where η is the viscosity of the porewater, r is the radius of the moving contaminant solute, and v_c is the velocity of

the moving contaminant solute. The drag force increases with increasing porewater viscosity. The viscosity, which is a function of the concentration of solute ions is expressed as (Jones and Dole, 1929) $\eta = \eta_o(1 + a\sqrt{c} + bc)$, where η_o represents the viscosity of the pure porewater (solvent), and a and b are coefficients to be determined from experiments. The term $a\sqrt{c}$ accounts for the retardation forces due to electrostatic interaction of the contaminant ions, and the term bc accounts for the interactions between the porewater and the contaminant ions.

In the simplest first-order analysis, the thermodynamic forces F_1 and F_2 act on the contaminant ions, and are primarily described as functions of the concentration of contaminants in the porewater. Higher orders of interaction, such as chemical and biological, are not considered in the first-order analysis. As a first approximation, we consider the diffusion coefficient of each solute component as a function of its own concentration in the porewater. For higher orders of approximation and higher-order analyses, one could account for multi-species ions, ligands, speciation and complexation. For the system shown in Figure 5.16, the general relationships between the rates of flow (fluxes J) and the thermodynamic forces (F) responsible for the fluxes can be described by a power series. In a near-equilibrium state, it is sufficient to use the first power in the series, i.e. $J_i = \Sigma L_{ij} F_j$, where L_{ij} denotes the phenomenological coefficients. The relationships obtained from the second postulate of irreversible thermodynamics for a one-dimensional case (Onsager, 1931) gives us:

$$J_w = L_{pp} \frac{\partial \psi_p}{\partial x} + L_{pc} \frac{\partial \psi_c}{\partial x}$$

$$J_c = L_{cp} \frac{\partial \psi_p}{\partial x} + L_{cc} \frac{\partial \psi_c}{\partial x}$$

(5.13)

where J_w is the fluid flux, J_c is the contaminant solute flux relative to water, $L_{pp}, L_{pc}, L_{cp}, L_{cc}$ represent the phenomenological coefficients and:

$$\frac{\partial \psi_p}{\partial x}, \frac{\partial \psi_c}{\partial x}$$

are the thermodynamic forces due to the gradient of porewater pressure u and the gradient of concentration c respectively, and are described as follows:

$$\frac{\partial \psi_p}{\partial x} = V_w \frac{\partial(-u)}{\partial x}$$

$$\frac{\partial \psi_c}{\partial x} = \frac{RT}{c} \frac{\partial(-c)}{\partial x}$$

(5.14)

where V_w is the water molar volume, R is the gas constant and T is the absolute temperature. From equations 5.13 and 5.14 we obtain:

$$J_w = L_{pp} V_w \frac{\partial(-u)}{\partial x} + L_{pc} \frac{RT}{c} \frac{\partial(-c)}{\partial x}$$
$$J_c = L_{cp} V_w \frac{\partial(-u)}{\partial x} + L_{cc} \frac{RT}{c} \frac{\partial(-c)}{\partial x}$$

$$(5.15)$$

L_{pp} can be determined from the first relationship in equation 5.15 using Darcy's relationship for the situation when $\partial(-c)/\partial x = 0$. This gives us: $L_{pp} = k/\rho_w V_w n$, where k is Darcy's coefficient (permeability or water conductivity), ρ_w is the specific weight of water and n is the porosity. Similarly, L_{cc} can be determined from the second relationship in equation 5.15 when $\partial(-u)/\partial x = 0$ using Fick's first law. This gives us:

$$L_{cc} = \frac{c}{RT} D$$

where D is the diffusion coefficient.

To take into account the physico-chemical processes, we can define the coefficient of osmosis k_π as:

$$k_\pi = \frac{V_w RT}{c} L_{pc}$$

This coefficient can be determined through parameter estimation from experimental results. We can further define a coefficient of ionic restriction k_{ir} as follows:

$$k_{ir} = \frac{n V_w}{c} L_{cp}$$

The third postulate of irreversible thermodynamics (Onsager, 1931) allows us to evaluate k_{ir} in terms of k_π. Expressing the pore pressure u in terms of the hydraulic head h as follows: $u = \rho_w h$, we are now in a position to write equation 5.15 as follows:

$$J_w = \frac{\rho_w k}{V_w n} \frac{\partial(-h)}{\partial x} + \frac{k_\pi}{V_w} \frac{\partial(-c)}{\partial x}$$
$$J_c = \frac{\rho_w c k_{ir}}{n} \frac{\partial(-h)}{\partial x} + D \frac{\partial(-c)}{\partial x}$$

$$(5.16)$$

where the second term on the right-hand side of the first relationship in equation 5.16 demonstrates the chemico-osmotic effect responsible for deviations from Darcy's law. The first term on the right-hand side of the second relationship in equation 5.16 represents the restrictions of movement of the ions relative to that of the porewater, as a function of the effect of negatively charged surfaces of the clay particles. Substituting the first relationship in equation 5.16 into the relationship $V_x = J_w V_w$ gives us:

$$V_x = \frac{\rho_w k}{n} \frac{\partial(-h)}{\partial x} + k_\pi \frac{\partial c}{\partial x} \tag{5.17}$$

We can obtain the flux of solute and ion relative to fixed coordinates from the two relationships shown in equation 5.16 to substitute into the following mass conservation for contaminant ions or solutes:

$$\frac{\partial c}{\partial t} + \frac{\partial J_c^x}{\partial x} \pm \frac{\rho_s}{n} \frac{\partial c^*}{\partial t} = 0 \tag{5.18}$$

where ρ_s is the dry density of the clay, c^* is the concentration of contaminants sorbed or desorbed, and J_c^x is the contaminant flux relative to fixed coordinates. Following this procedure, Yong and Samani (1987) have obtained the final relationship as:

$$\frac{\partial c}{\partial t} + \left(\frac{k_{ir}}{k} + 1\right) V_x \frac{\partial c}{\partial x} + \frac{k_\pi k_{ir}}{2k} \frac{\partial^2 c}{\partial x^2} = \frac{\partial}{\partial x}\left(D \frac{\partial c}{\partial x}\right) \pm \frac{\rho_s}{n} \frac{\partial c^*}{\partial t} \tag{5.19}$$

As with the other relationships and other analytical computer models dealing with the prediction/analysis of contaminant transport in the ground, one notes that these relationships are valid for continua that are homogeneous and uniform. This means to say that the point values for advective velocity, sorbed contaminant concentration and porosity are considered to be representative of the continua values. For a clay where there are at least two separate phases present, and especially where the microstructural units constituting the overall clay structure are highly variable in size and properties, one needs to deal with a representative elementary volume in the formulation of the various relationships. Thus, we would have $c = <c> + c'$, $V_x = <V_x> + V_x'$ and $n = <n> + n'$, where $<c>$, $<V_x>$ and $<n>$ are the mean values and c', V_x' and n' are the fluctuating components.

5.5.3   *Organic chemical contaminants*

The problem of considerable interest with respect to organic chemical contaminants in the leachate stream is the fate of the chemicals remaining (retained) within the clay buffer/barrier system. The continued presence of inorganic and organic contaminants in the system constitutes a problem that requires attention because of the possible health impacts on biotic receptors, and also because of possible degradative impacts on the integrity, stability and overall design function of the buffer/barrier system. We will turn our attention to the many processes that impact on the fate of organic chemical compounds as contaminants in retained in the ground and especially in the clay buffer/barrier system.

Transformations and persistence

The continued presence of contaminants in the subsurface soil and in the clay buffer/barrier system is defined as the *persistence* of the particular contaminant. Although the persistence of heavy metals refers to the continued presence of the metals in the system, the persistence of organic chemical contaminants presents a more complicated picture. This is because of potential transformations into intermediate and final products that can be more dangerous than the original chemical contaminant. These transformations are conversions of the original organic chemical contaminant into one or more resultant products by processes that can be abiotic, biotic, or a combination of both of these.

Organic chemical compounds transformed as a result of biotic activities are classed as *intermediate products* along the path towards complete mineralization. These are essentially *products of degradation*. In contrast, transformed products obtained as a result of abiotic processes do not classify as intermediate products. *Abiotic* transformations occur without the mediation of microorganisms and are influenced by the physico-chemical properties of both the organic chemical compound itself and of the clay. They are transformation processes that can occur with or without net electron transfer. Chemical reactions in a clay organic chemical system include hydrolysis, formation of a double bond by removal of adjacent groups, and oxidation–reduction reactions. It is not easy to distinguish between products of transformation due solely to abiotic reactions and those due to biotic activities. To a very large extent, this is because many of the transformed products are obtained from a combination of the two processes, i.e. an initial abiotic reaction could easily lay the groundwork for subsequent biotic activities that will lead to the transformed product. Or an initial set of biotic activities could lead to the subsequent chemical reactions and final transformed product.

Biotic transformation processes are biologically mediated transformation reactions, and include associated chemical reactions. The distinguishing feature between abiotic and biotic transformation products is that abiotic transformation products are generally other kinds of organic compounds, whereas biotic processes lead to mineralization of organic chemical compounds as the transformation product. Complete conversion of some organic chemical compounds to CO_2 and H_2O (i.e. mineralization) may not be achieved. However, the intermediate products obtained during this process point towards complete mineralization. Biotic transformation processes under aerobic conditions are oxidative, and the various processes involved include hydroxylation, epoxidation, and substitution of OH groups on molecules. On the other hand, anaerobic biotic transformation processes are most likely reductive processes, and could include hydrogenolysis, H^+ substitution for Cl^- on molecules, and dihaloelimination (McCarty and Semprini, 1994).

We define *persistent organic chemical contaminants* as those organic chemical contaminants that are resistant to conversion by abiotic and/or biotic transformation processes. The literature commonly refers to these as persistent organic chemical pollutants (POPs), and also as *recalcitrant organic chemical pollutants*. The persistence of organic chemical pollutants in the ground and in the clay buffer/barrier system depends on at least three factors: (a) the physico-chemical properties of the chemical compound itself; (b) the physico-chemical properties of the ground and the interacting clay; and (c) the microorganisms in the ground and interacting clay. We will discuss these factors in the next section, under the heading 'Natural attenuation of contaminants'.

Hydrolysis reactions

We refer to *non-reductive chemical reactions* as those reactions that involve attacks by nucleophiles on electrophiles. Hydrolysis is a specific instance of nucleophilic attack on an electrophile. The term *neutral hydrolysis* is often used to refer to nucleophilic attack by H_2O, to distinguish this from acid- and base-catalysed hydrolysis, when catalytic activity is accomplished by the H^+ and OH^- ions respectively. *Nucleophiles* are electron-rich reagents (nucleus-liking species) containing an unshared pair of electrons, and are generally negatively charged. Because of their 'nucleus-liking' nature, they are 'positive charge liking'. Common inorganic nucleophiles include HCO_3^-, ClO_4^-, NO_3^-, SO_4^{2-}, Cl^-, HS^-, OH^- and H_2O. On the other hand, *electrophiles* have electron-deficient (electron-liking species) reaction sites that form bonds by accepting electron pairs from nucleophiles. They are generally positively charged and, because they are 'electron liking', they are also 'negative charge liking'.

Hydrolysis reactions are chemical reactions between an organic chemicals and water. The water molecule or OH^- ion replaces groups of atoms (or another atom) in the organic chemical, and a new covalent bond with the

OH⁻ ion is formed. No change in the oxidation state of the organic molecule is involved in the transformation. The products of hydrolysis reactions are generally compounds that are more polar in comparison with the original chemical compound, and will therefore have different properties.

Clay-catalysed hydrolysis reactions associated with the surface acidity of clay minerals affect the kinetics of hydrolysis by affecting the hydrolysis half-lives of the reacting organic chemicals. The surface acidity of kaolinites derives from the surface hydroxyls on the octahedral layer of the mineral particles. Surface acidity in the case of montmorillonites is due to isomorphous substitution and to interlamellar cations. The layer of water molecules next to the charged lamellar sheet is strongly polarized, resulting in the loss of protons. Figure 5.17 shows the effect of moisture content on the acidity of a kaolinite. Surface acidity is reduced dramatically as the moisture content of a clay is increased, thereby decreasing the catalytic activity. The charge and nature of the cations affect the degree of catalytic activity, as they impact directly on the polarizing power and the degree of dissociation of the water in the inner Helmholtz plane. We would expect the surface acidity of montmorillonites to increase as we increase the valency of the exchangeable cations. Measurements on surface acidity of many clay minerals have shown that these can be at least anywhere from two to four units lower than that of bulk water (Mortland, 1970; Frenkel, 1974).

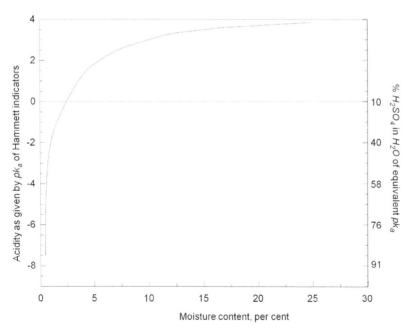

Figure 5.17 Effect of moisture content on surface acidity of kaolinite (adapted from Solomon and Murray, 1972).

Oxidation–reduction reactions

Oxidation–reduction reactions occurring in clays as a result of interactions with organic chemical contaminants can result in transformations. In contrast with transformations occurring through nucleophilic replacement reactions where no net transfer of electrons occurs, electron transfer occurs in oxidation–reduction (redox) reactions. *Oxidation* refers to a removal of electrons from the subject of interest, and *reduction* refers to the process whereby the electron acceptor or *oxidant* gains electrons from an electron donor (*reductant*). By gaining electrons, a loss in positive valence by the subject of interest results and the process is called a reduction. Redox reactions occur abiotically and also under biotic conditions. Because it is difficult to exclude direct involvement of any or some microbial activity in any of the reactions, it is not easy to distinguish between the two. This leads one to ask whether there is a critical requirement to distinguish between the two, as redox conditions are more likely than not to be the product of factors that include microbiological processes. The number of functional groups of organic chemical pollutants that can be oxidized or reduced under abiotic conditions is considerably smaller than those under biotic conditions (Schwarzenbach *et al.*, 1993). Quantification of specific reaction rates is difficult because of the many kinds of chemical reactions involving organic chemicals, microorganisms and the many different constituents in the clay. The scarcity of kinetic data makes it difficult to provide for quantitative calculations of redox reactions. The preceding notwithstanding, it is nevertheless instructive and informative to obtain a qualitative or descriptive appreciation of these reactions.

Clays are essentially electrophiles. They function well as electron acceptors (oxidizing agents or oxidants). The structural elements of clay minerals such as Al, Fe and Cu can transfer electrons to the surface-adsorbed oxygen of the clay minerals, which can be released as hydroperoxyl radicals (–OOH). These will serve as electron acceptors and can abstract electrons from the organic chemical contaminants. Figure 5.18 shows the results of oxidation of 2,6-dimethylphenol by Al^{3+}- and Fe^{3+}-saturated montmorillonites with pH values varying between 2 and 12, as reported by Yong *et al.* (1997).

The possible mechanisms for oxidizing the dimethphenol shown in the figure include the following elements: (a) the structural elements of the montmorillonite clay (Fe and Al) and (b) the partially coordinated aluminium on the edges of the clay minerals because of alumina sheet hydrolysis at low pH values. These function as Lewis acids that can accept electrons from aromatic compounds. The results shown in the figure indicate that more effective oxidation of the 2,6-dimethylphenol is obtained by the iron montmorillonite – presumably because of the greater oxidizing capability of the Fe(III). The intermediate product formed is a 2,6-dimethylphenol dimer of mass 242, as shown by the *degree of abundance* on the ordinate of the graph.

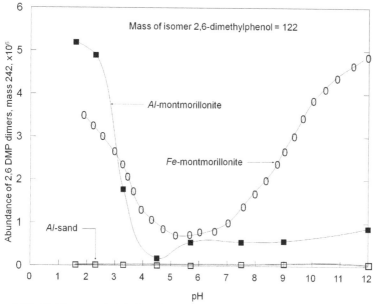

Figure 5.18   Oxidation of 2,6-dimethylphenol by aluminium montmorillonite, iron montmorillonite and aluminium sand (data from Yong et al., 1997).

The metabolic process in *biologically mediated redox* reactions is generally *catabolic* (i.e. energy releasing). This means that a transfer of electrons from the organic carbon will occur, resulting thereby in the oxidation of the organic chemical. Common electron acceptors in the clay are oxygen, nitrates, sulphates, Fe^{3+}, Mn^{4+} and other trace metals. The activities of microorganisms that result in transformation of organic chemical contaminants can also transform the clays used for engineered clay buffers/barriers. A more detailed discussion of the transformation of organic chemical contaminants due to microbiological activities will be found a later section (in this chapter) when we deal with the attenuation of contaminants in the clay buffer/barrier and in the ground. The transformation of the solid constituents in the clay (clay particles) due to microbiological activities is of considerable importance in the long-term integrity and stability of engineered clay buffers and barriers. This subject will be discussed in detail in a later chapter dealing with the evolution of clays in the buffer/barrier systems for HLW and HSW.

5.6   Contaminant impact on clay behaviour

The design function of clay buffers and barriers in engineered clay barrier systems for containment of HLW and HSW is generally developed with information on (a) the nature (type and composition) of the clay to be used and (b) the properties and characteristics determined from laboratory tests. In all instances, the various pieces of information obtained are current, i.e. they are 'today's information'. Over the life of the buffer/barrier system, changes in the nature of the clay and its properties/characteristics will occur as a result of chemical reactions due to leachate interaction, thermal, hydraulic and mechanical stresses, microbial actions, and natural microenvironmental forces/fluxes. Most of these changes and transformations will be discussed in a later chapter dealing with the long-term problem facing clay barrier containment systems, i.e. evolution of the clay in these barriers. In this section, we are concerned with the impact of contaminants on the design function of the clay in the engineered clay barrier system. A good example can be found in Figure 3.1, featuring the impact of exposure of a natural lacustrian clay (under a creep load) to a leachate ($0.25 \, N \, Na_2SiO_3 \cdot 9H_2O$) after 12,615 minutes of initial creep. The considerable increase in creep deformation resulting from the interaction between the clay and the leachate testifies to a change in the clay internal resistance mechanisms.

The impacts accruing from changes and transformations include (a) morphology and composition of the clay minerals; (b) assimilative capacity; (c) microstructural features; and (d) physical, transmission and mechanic properties. The underlying principal feature linking all of the impacts is the change in physical properties of the clay, resulting from the changes and alterations in the microstructural features and their effect on its physical properties. The various impacts are seen in Figure 5.19. We choose *microstructural features* and *physical properties* at the fundamental factors, as these are the building blocks that are instrumental in 'producing' the chemical

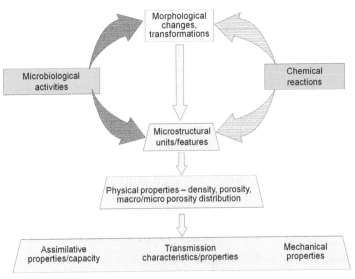

Figure 5.19 Impacts from chemical reactions and microbiological activities on clay in engineered clay buffer/barrier systems. Impacts from natural microenvironmental forces/stresses are not included in the diagram.

(assimilative potential), transmission (heat, hydraulic, stress) and mechanical (constitutive) performances of the clay

The clay mineral morphological changes in MX-80 smectite used for the clay buffer-barrier system in HLW confinement shown in Figure 5.20 provides a good example of the dissolution processes associated with hydrothermal treatment of the (montmorillonite) mineral. According to Pusch *et al.* (2007), substitution of the Al^{3+} by Fe^{3+} in the octahedral layer caused higher lattice stresses due to the larger ion radius of Fe^{3+} in comparison with Al^{3+}, thus reducing the Fe-enriched montmorillonite. Changes in the chemical composition of the montmorillonite mineral and the likely transformation of the mineral will be discussed in detail in the next chapter. The end result of these changes is a restructuring of the microstructures. The significant effect of this restructuring would be seen in terms of the changes in the proportions and distribution of the macro- and micropores in the clay. These changes, in turn, will significantly affect the physical properties of the clay. The major properties of interest for clay buffer and barrier systems are density, porosity and pore-channel connectivity in both the microstructural units and the overall macrostructure. The sink/storage role of microstructural units and their influence and control of the transport performance of contaminants cannot be overstated.

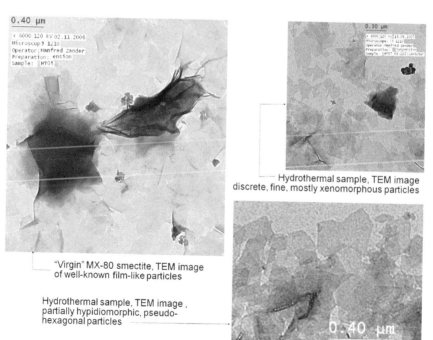

Hydrothermal sample, TEM image
discrete, fine, mostly xenomorphous particles

"Virgin" MX-80 smectite, TEM image
of well-known film-like particles

Hydrothermal sample, TEM image ,
partially hypidiomorphic, pseudohexagonal particles

Figure 5.20 Transmission electron microscopy (TEM) images of MX-80 smectite in 'virgin' state and after 5 years' exposure to HLW repository-like conditions (hydrothermal samples shown on the right) (adapted from Pusch *et al.*, 2007).

The effect of heavy metal contaminants in swelling clays used for engineered clay barrier systems on their rheological behaviour can be assessed using the familiar DDL model. Although direct accord between calculated and measured values may not be achieved – because of the ideal assumptions required for model calculations – calculated values nevertheless serve to provide one with a view of 'how the system will respond'. Calculations can be performed to determine the changes in osmotic equilibrium resulting from varying concentrations of heavy metals in the clay, using the van't Hoff relationship for overlapping diffuse ion layers – written simply as $\Pi = RTc$, where Π is the osmotic pressure, R is the gas constant, T is the absolute temperature and c is the concentration of ions. A three-dimensional graphical relationship linking the void ratio and osmotic pressure with the consolidation pressure for a series of tests on a smectite with varying proportions of lead as the heavy contaminant is shown in Figure 5.21.

The DDL potential ψ distribution (equation 3.12), developed in section 3.7.3, can be used to calculate the pressures between interacting particles with metal contaminants in the porewater. These can be compared with actual laboratory consolidation tests in which the porewater is contaminated with varying concentrations of heavy metal. The results of measured and calculated values are shown in Figure 5.22. The deviations between calculated and measured values are attributable to the ideal assumptions

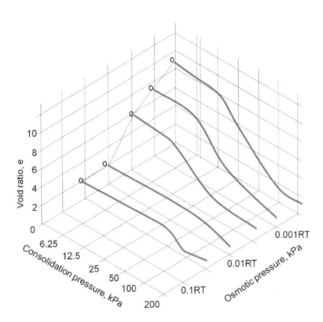

Figure 5.21 Variation of void ratio e, consolidation pressure and osmotic pressure (calculated) with Pb²⁺ in the porewater of the bentonite samples (adapted from Ouhadi et al., 2006). R, gas constant; T, temperature (K).

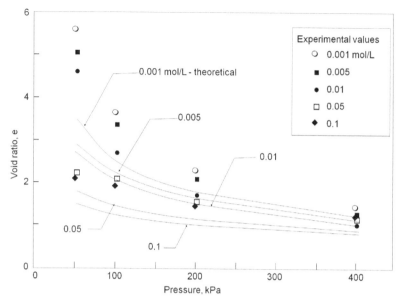

Figure 5.22 Experimental and theoretical consolidation curves for lead-contaminated bentonite. Solid lines are theoretically calculated values. Experimental points show the e-log *p* values in relation to the concentration of lead used in the form of Pb(NO$_3$)$_2$ (adapted from Ouhadi *et al.*, 2006).

invoked to simplify calculations. The essence of the results shown in all three figures (Figures 5.20–5.22) is that changes and/or alterations to the basic nature (morphology and structure) of the clay in the barrier system will, without doubt, impact on the performance of the clay. The introduction and continued presence of contaminants in the clay will also affect the performance properties of the clay through their effect on particle/microstructure interactions. DDL-type calculations can provide some hint as to what might happen, as shown in Figures 5.21 and 5.22. How severely the impacts would be, and to what degree these impacts will degrade the design performance, can only be answered by conducting test simulations.

5.7   Contaminant attenuation

The assimilative and buffering capacity of the clay used for buffer/barrier systems are of paramount importance in the success of these buffer/barrier systems. These are the basic components of the contaminant attenuating capacity of clays. The processes include dilution, partitioning of contaminants, and transformations, and involve a range of mechanisms and processes such as physical, chemical and biologically mediated reactions, and combinations of all of these. A listing of the major processes would include:

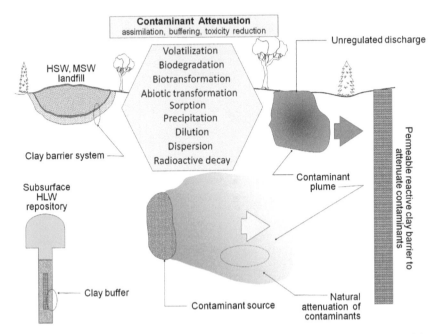

Figure 5.23 Illustration of the many applications of the contaminant attenuation capability of clays: (a) as engineered buffer/barrier systems, (b) as in their natural state in the ground (bottom centre of figure) and (c) as constructed reactive clay barriers to attenuate contaminants as it intercepts the advance of contaminant plumes (right-hand side).

- volatilization;
- biodegradation;
- biotransformation;
- abiotic transformation;
- sorption;
- precipitation;
- dilution;
- dispersion; and
- radioactive decay.

Not all of the above apply to all organic chemical and inorganic contaminants. In general, one could say that all of the processes, except precipitation and radioactive decay, are available in attenuation of organic chemical contaminants. In that same vein, one could expect that the processes available for attenuation of inorganic contaminants, including radionuclides, include biotransformation through to radioactive decay. Figure 5.23 shows the various applications of the capability for attenuation of contaminants. We have discussed the assimilative capacity of clays in the first part of this chapter. In this section, we turn our attention to the attenuation capability of the clay, not only in the buffer/barrier system, but also as the supporting sub-base (in the ground).

The attenuating capability of clays for the engineered buffer and barrier systems shown in the left-hand portion of the figure is, by and large, not the only consideration in the selection of candidate clays. Chapters 3 and 4 have discussed the characteristics and properties – other than attenuating capability – of clays necessary to fulfil the buffer/barrier role for these containment systems. The discussions in the first part of this chapter has shown that the contaminant attenuation capability of clays adds another dimension to the engineered buffer/barrier system. A good example of this added dimension is shown in the right-hand diagram of Figure 5.23 – the permeable reactive barrier (PRB). In most instances, the PRB consists of a clay with added constituents that include a range of oxidants and reductants, chelating agents, catalysts, microorganisms, etc. This will be discussed further later in this section.

5.7.1 *Natural attenuation of contaminants*

The illustration shown in the bottom middle part of Figure 5.23 is an example of the inherent capability of a clay (and *soils* in general) to attenuate contaminants. In this discussion of natural attenuation, we will continue to use the general term *soil*, as, in most instances, the ground material involved consists of clay mixed with other soil types. The literature has coined the term *natural attenuation* to refer to the attenuation of contaminants in soils by natural processes that serve to reduce the toxicity of the contaminants and/or the concentration of the contaminants. A measure of the recognized attenuation capability of soils (in general) is the approach taken by the US EPA in using natural attenuation as a potential tool for remediation of contaminated sites – subject to conditions favourable for the attenuating processes to function well. This approach has been called *monitored natural attenuation (MNA)* and has been defined in US EPA Science Advisory Board (2001) to mean 'a remediation approach based on understanding and quantitatively documenting naturally occurring processes at a contaminated site that protect humans and ecological receptors from unacceptable risks of exposure to hazardous contaminants'. The report further indicates that this is 'a knowledge-based remedy because instead of imposing active controls, as in engineered remedies, scientific and engineering knowledge is used to understand and document *naturally* occurring processes to clearly establish a causal link'.

Because of its *remedy by natural processes* procedure, one needs to obtain a proper understanding of the many principles involved in the natural attenuation processes that contribute to the end result. Monitoring of the contaminant plume at various positions away from the source is a key element of the use of MNA. Tracking of contaminants is essential if MNA is to be applied. Historically, the pollutants tracked have primarily been the organic chemicals including for example chlorinated solvents [perchloroethylene (PCE), trichloroethylene (TCE) and dichloroethylene (DCE)], and hydrocarbons such as benzene, toluene, ethyl benzene, and xylene (BTEX). The role of clay composition and potential interactions, such as processes involving electron transfer and partitioning of contaminants, are key elements in the success of MNA application. For natural attention of metals and radionuclides to occur, the primary attenuating mechanisms of sorption and dilutions must be available. One expects that in the case of radioactive materials, the daughter nuclides that will be produced must be assessed to determine whether they are more hazardous than the parent.

When active controls or agents are introduced into a soil to render attenuation more effective, this is called *enhanced natural attenuation (ENA)*. ENA refers to the situation in which, for example, nutrient packages are added to a soil system to permit enhanced biodegradation to occur, or catalysts are added to a soil to permit chemical reactions to occur more effectively. ENA could include biostimulation and/or bioaugmentation.

5.7.2   *Permeable reactive clay barriers*

The PRB shown in Figure 5.23 has been used in contaminant plume management strategies. These PRBs, or treatment walls, rely on (a) the natural or enhanced attenuation capability of the soil placed in the treatment wall; (b) proper control of the transport time of the contaminant plume through the treatment wall, i.e. the residence time of a specific volume of contaminant plume in the treatment wall must be long enough for the treatment processes to be effective; and (c) directing the contaminant plumes or contaminated groundwater to flow into and through the wall. A proper characterization of the hydrological setting is essential for locating the PRB. The *funnel-gate* technique is one of the more common techniques used to guide a contaminant plume to the intercepting reactive wall. The funnel, which is constructed or placed in the contaminated plume site, consists of confining boundaries of impermeable material (e.g. sheet pile walls), which narrow towards the funnel mouth where the reactive wall is located.

Some of the major contaminant removal and immobilization processes in the treatment wall include the following:

- *inorganic contaminants* – sorption, precipitation, substitution, transformation, complexation, oxidation and reduction;
- *organic chemical contaminants* – sorption, abiotic transformation, biotransformation, abiotic degradation, biodegradation.

Enhancement of the attenuation capability of the clay (and soils in general) in the reactive walls is obtained with the aid of reagents and compounds. These include the previously mentioned range of oxidants and reductants, chelating agents, catalysts and microorganisms. Other enhancement agents include zero-valent metals, zeolite, reactive clays, ferrous hydroxides, carbonates and sulphates, ferric oxides and oxyhydroxides, activated carbon and alumina, nutrients, phosphates and organic materials. Successful use of treatment walls reflects a proper knowledge of the nature of the contaminants and the choice of clay, reagents and compounds, together with manipulation of the pH–pE microenvironment used in the treatment walls.

5.8   Concluding remarks

The basic subjects covered in this chapter have all been pointing towards establishing the fundamental mechanisms and processes that endow clays with contaminant attenuation capability – through discussions of the many processes that collectively provide a clay with the capability to assimilate contaminants. The general term *sorption* has been used to include the results of transfer of contaminant solutes from the porewater to the clay minerals. Whether the transfer is due to actual adsorption of the contaminants onto the clay solids, or formation of a new phase through precipitation, the term *sorption* is meant to denote retention of contaminants in the clay.

Knowledge of the surface properties of both the clay minerals and the contaminants is important, as this will tell us what to expect with regard to the partitioning of contaminants. We need to be aware of the demand for equilibrium conditions when applying the results obtained from batch equilibrium tests. Specifically, although adsorption isotherms are very useful pieces of information, we need to match them with conditions in the field. When necessary, we should consider kinetic adsorption models. Chemical mass transfer is mainly responsible for partitioning of contaminants and for transport of and fate of contaminants.

The processes of sorption of heavy metals generally result in a drop in the pH of the immediate clay particles' environment – with the decrease of pH being a function of the species and concentration of the heavy metals. Release of hydrogen ions from metal–proton exchange reactions on the particles' surface sites, and from hydrolysis and precipitation of the metals in the porewater, is responsible for the pH drop. The surface functional groups, such as the carbonyl compounds, aldehyde, ketones and carboxylic acids, have dipole moments. The electrons in their double bonds are unsymmetrically shared, and although they can accept protons, the stability of complexes between carbonyl groups and protons is weak. Interactions between contaminants and layer–lattice clay particles occur either directly with interlayer cations or through formation of hydrogen bonds with water molecules coordinated to exchangeable cations.

The various sorption processes that contribute to bonding between organic chemical contaminants clay particles include partitioning (hydrophobic bonding) and accumulation – through adsorption mechanisms involving the clay minerals and other soil particulates such as carbonates, and amorphous materials. The more prominent properties affecting the fate of organic molecules by soil fractions include (a) clay particles – surface area, surface chemistry, surface charge (density, distribution and origin), surface acidity, CEC, exchangeable ions, microstructural units and their distribution; (b) organic chemicals – functional groups, structure, charge, size, shape flexibility, polarity, water solubility, polarizability, partitioning and equilibrium constants; and (c) clay–water system – microbial community, energy sources, temperature, inorganic/organic ligands available, pH, pE, salinity and physical gradients (fluxes).

We conclude our discussions by looking at how the attenuating capability of soils in general, and clays in particular, is used outside the engineered clay buffer/barrier systems utilizeded in containment of HLW, HSW and municipal solid waste (MSW). The use of the *monitored natural attenuation* technique as a remediation tool and the use of permeable reactive barriers not only testify to the attenuating capability of clays, but also remind us that much needs to be learnt about how these clays mature and evolve with time – in the HLW and HSW containment environment.

Chapter 6

Thermal, hydraulic, mechanical, chemical and biological processes

6.1 Processes initiated in clay buffer/barrier

6.1.1 Initial considerations

There are several processes initiated in engineered clay buffers and barriers used for containment of high-level radioactive waste (HLW), hazardous solid waste (HSW) and municipal solid waste (MSW). These processes are brought about because of (a) site specificities, such as containment constraints and exposure to groundwater in the case of deep geological repositories; (b) design requirements and functions, such as chemical and physical buffering requirements; and (c) nature and properties of wastepile and HLW canister. The processes that are initiated in the clays are due to forces, reactions, activities, and THMCB phenomena [i.e. thermal (T), hydraulic (H), mechanical (M), chemical (C), and biological (B)].

It is useful to recognize that, for completeness in accounting for impact factors on short- and long-term clay buffer–barrier performance, one should include the possibility that radiation-induced effects (R) may play a role in the long-term performance of clay buffers/barriers. It has been argued, however, that the extent of radiation-induced influence on long-term clay buffer/barrier performance would be small to insignificant. As these (radiation-induced) effects have yet to be properly defined, investigated and elucidated, the discussions in this chapter will confine themselves to the impacts associated with THMCB phenomena.

The significance and extent to which any of these processes apply to HLW, HSW and MSW depend on (a) the specific nature of the waste materials and substances; (b) how stable these materials are in the course of time; (c) whether external forces (i.e. external to the containment system) can act on the engineered clay buffer/barrier system; and (d) the nature and quantity of fugitive contaminants, and dissolution and corrosion products released if the primary containment system (e.g. canister, double-membrane liner system) is breached.

Figure 6.1 provides a capsule view of the impact of the individual THMCB factors and their fluxes on the engineered clay buffers/barriers used

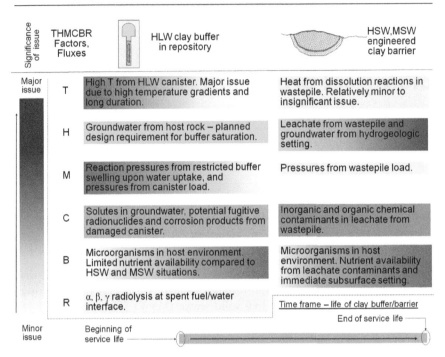

Figure 6.1 THMCBR factors/fluxes impact on engineered clay buffer/barrier used for containment of HLW, HSW and MSW. The gradient shading used to indicate level of significance of issue raised by the individual THMCBR factors is shown on the left. The service life of the engineered clay buffer or barrier is shown at the bottom right of the illustration. Note that the service life is the design service life and that the service life of the HSW and MSW clay buffer/barrier systems is considerably shorter than the design service life for HLW.

in containment of HLW, HSW and MSW. For completeness in presentation of the various impacts, we also show the potential impact of radiation-induced processes. The intensity of the dark background shading associated with the respective texts relating to each factor indicates the seriousness of the issue presented by each factor (see left-hand side of the diagram). The beginning (left-hand portion) of the background shading indicates the beginning period of the planned (design) service life of the clay buffer/barrier, and the right-hand end of the background shading indicates the ending period of the planned service life (see 'time frame' at bottom of figure). We expect the design service life of the HSW and MSW engineered clay buffer/barrier systems to be considerably shorter than the design service life of the clay buffers used in the repositories designed to contain HLW. The THMCBR factors, forces and issues of concern include:

- *Temperature and thermal fluxes (T).* We are concerned with (a) high heat source as in the case of containment of HLW canisters in deep underground repositories and (b) heat generated in dissolution processes in HSW and MSW, and frost effects on covers for landfills. In comparison with the high heat source problem in HLW containment, the issue of temperature or heat fluxes in HSW and MSW landfills is relatively minor.

- *Hydraulic issues (H)*. Water availability as groundwater from the host rock is a big concern in HLW containment. Saturation of the clay buffer in HLW containment systems is a planned design consideration. Water interaction with clay barriers for HSW and MSW will be due to (a) rainfall and snowmelt; (b) leachate leaking through geomembrane liners overlying the clay barrier component, and also from surrounding groundwater (if a portion of the engineered barrier system is located in the saturated subsurface); and (c) groundwater flow as a function of the hydrogeological setting of the containment site.

- *Mechanical forces (M)*. Principal concerns in HLW are (a) swelling pressures developed from water uptake of the smectite buffer and (b) canister weight. To a lesser extent, one is concerned with the lateral pressures from host rock and overburden pressures acting on the HLW canister. For HSW and MSW, the pressures from the wastepile appear to be the principal issue. These are not expected to be significantly larger than normal overburden pressures. Frost heave pressures for landfill sites in the north, acting on cover systems, may present problems.

- *Chemical issues (C)*. For HLW repositories, solutes in the groundwater (Na^+, K^+, Mg^{2+}, Ca^{2+}, etc.) and groundwater pH provide the initial sets of reactions and interactions with clay. At a later stage, if and when the HLW canister faces corrosion distress, corrosion products and fugitive radionuclides will add to the suite of interacting solutes. For HSW and MSW landfill containment systems, contaminants escaping from geomembrane liners will constitute the suite of chemicals for interaction with the clay buffer/barriers.

- *Biological activities (B)*. For all the waste containment systems (HLW, HSW and MSW), the shorter-term concerns are with respect to biologically mediated chemical reactions, and the longer-term concerns are those dealing with transformation and alteration of the clay minerals.

- *Radiation-induced processes (R)*. One expects that the major concerns with radiation-induced processes are those associated with α, β and γ radiolysis at the spent fuel–water interface in the repository containment of HLW.

6.1.2   *Containment factors*

The previous chapter dealt with the various mechanisms and processes involved in interactions between contaminants and clays, with specific interest focussed on clays used for engineered clay buffers/barriers. The discussions developed in Chapter 5 are relevant to situations that apply to engineered clay buffer/barrier containment of HLW, HSW and MSW. Although all the THMCB factors impact to some degree on the engineered clay buffer/barrier systems for HLW, HSW and MSW containment, Figure 6.1 shows that there are processes that are more specific to HLW containment – requiring separate treatment in analysis and discussion. These are tied into at least three important factors, namely (a) the nature of the waste being contained, i.e. continued long-term heat source and long term radiation threat; (b) the method of HLW containment and deep geological disposal site specificities; and (c) the extremely long-term performance requirement for the containment clay buffer/barrier system – at least 100,000 years. We can say that although *site specificities* is an issue that confronts all waste disposal facilities, the repository format used in deep geological disposal presents elements that require special attention. In addition, the 'extremely' long-term design performance expectations of the buffer containment system require extra attention to effects of long-term chemical processes and biological activities on the nature and properties of the clay used to provide the design function or capability of the buffer/barrier system. Another significant factor is the heat producing element in HLW containment requiring attention to coupled processes of (a) heat and mass transport; (b) abiotically and biotically driven chemical processes; and (c) a combination of both (a) and (b) in the clay buffer system and the attendant consequences.

In this chapter, we will focus our attention on the processes activated in the clay buffer system surrounding the HLW canister. It is convenient to group these processes according to their time-effective intensities. Thermal, hydraulic, mechanical and chemical reactions can be grouped as significant processes in maturation of the clay in the buffer system. Section 6.4 provides a descriptive discussion of *maturation, evolution* and *ageing*. Chemical reactions and clay mineral transformations, together with outcomes of biological and radiation-induced processes, can be considered to fall within the class of relevant long-term processes – classifying more as those processes that contribute directly to the long-term evolution of clay, i.e. ageing of the clay. There are no hard and fast rules that say that one process or set of reactions belongs in one or the other group. All of the processes and reactions grouped as maturation (short term) and ageing (long term) are operative throughout the life of the HLW repository system, albeit with more or less intensity and relative relevancy. The various processes and reactions are shown in Figure 6.2. Details on the various processes have been discussed previously in Chapters 4 and 5. The magnitude or intensity of the impacts of the various processes on the buffer containment system for a deep geological repository containment of HLW varies considerably over the long time span required for safe HLW containment.

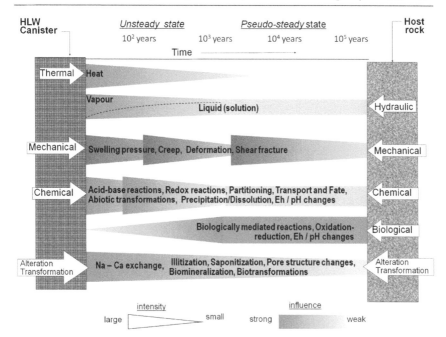

Figure 6.2   Source, intensity and influence of THMCB processes in maturation of clay buffer surrounding HLW canister in deep geological disposal. The location of the arrows indicate the source of the force/flux/phenomenon. For example, the thermal arrow located in the HLW canister shows that the heat flux is generated by the canister heat source.

6.2   HLW near-field processes

In the previous section and in Chapter 2, it has been indicated that highly radioactive waste (HLW) produces heat in conjunction with radioactive decay and beta/gamma radiations (see Figures 2.3 and 2.4). Figure 6.3 shows that some significant heat and radioactivity still remain after 10^5 and 10^6 years respectively. The disposal strategies have concentrated on containment of the HLW in repositories founded deep underground. Section 2.4 has provided some commonly described concepts for deep geological disposal of the high-level radioactive wastes, together with some of the more noticeable impacts on the near-field components. These concepts call for canisters to be placed in boreholes or tunnels bored at a depth of several hundred meters in crystalline rock, salt, argillaceous rock or plastic clay.

The near-field processes that occur, in the early maturation stage, in the buffer embedding the HLW canisters are the result of (a) the heat source from the HLW canister; (b) the type of material chosen as the buffer; (c) the method or requirements for buffer placement; and (d) groundwater access from fissures, cracks, etc. in the host rock. All HLW disposal concepts face the common problem of secure placement and functioning of a buffer as the

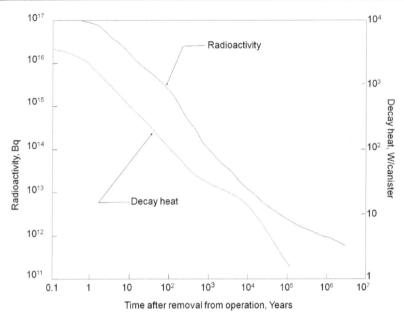

Figure 6.3   Level of radioactivity of spent nuclear fuel with a burn-up of 38 MW d/kg U (megawatt days per kg of uranium) in relation to time after reactor operation together with reduction in decay heat in the spent-fuel canisters. Note that the decay heat ordinate is on the right side of the diagram (adapted from information given in SKB, 1999; Hedin, 1997).

embedment material, i.e. a buffer that would protect the HLW canister and isolate the canister. The consensus choice of a swelling clay as the buffer material between the HLW canister and the host rock (see Figures 2.6 and 2.7) is predicated on the following system requirements:

- very tight embedment of the canisters to minimize water flow around them;
- necessary properties to retard and attenuate transport of corrosion-promoting ions from the rock and migration of fugitive radionuclides;
- ductile embedment of the canisters to attenuate rock stresses caused by tectonics and rock fall.

As the clay is generally placed in a semi-dry state, the following factors relating to post placement of the clay buffer require consideration:

- groundwater from the surrounding host rock provides the source-water for water uptake and eventual saturation of the initially semi-dry swelling clay buffer;
- swelling of the clay upon water uptake will result in development of swelling pressures because of the confining boundaries;

- the canister heat source, together with groundwater uptake (from the opposite direction – see Figure 2.8), provides the situation for development of complex coupled heat–mass transport in the buffer.

The active processes in the buffer, resulting from the primary factors listed in the preceding can be categorized into four simple groups:

- T (thermal – dealing with heat transfer);
- H (hydraulic – dealing with fluid mass transfer);
- M (mechanical – dealing with the pressures developed as a result of swelling of the buffer clay);
- C (chemical – dealing with chemical reactions in the clay–water system).

The C factor can be considered to comprise two sets of activities: (a) one that is related to reactions between solutes in the groundwater brought into the system via fissures, cracks, etc. in the host rock and (b) another that includes reactions with fugitive radionuclides and HLW canister corrosion products, and long-term reactions that lead to chemical transformation and alteration of the minerals in the clay. The first set, which we will call the *reaction set*, can be considered to be activities in the short term, and the second set, which can be called the *long-term set*, deals with processes that are long term in nature.

6.3   Clay buffer maturation

Clay buffer maturation refers to the evolution of the clay as a result of its time-related interactions and reactions with chemical, mechanical and hydraulic forces, and biological activities. The term *fully matured* is commonly used to refer to the state where a substance, material or biotic species has reached its full capability or potential. With respect to clay buffers and barriers used for containment of HLW, HSW and MSW, there are significant differences in the meaning of (a) fully matured and (b) the time taken to reach full maturity. Engineered clay barrier/liners that are utilized for the last barrier component in multi-barrier systems for containment of HSW and MSW, in conjunction with geomembranes and other types of synthetic membranes, will not generally be 'called into action' until ('if' and 'when') leakage of leachates through the overlying membranes occurs. Design and emplacement requirements for engineered clay barrier/liners are structured to ensure that these barriers/liners will reach their full potential when called into action.

Clay buffers used to surround and embed HLW canisters are emplaced in a semi-dry condition. As the full potential of the clay to function as an engineered buffer requires water saturation, the clay buffer is not expected to reach full maturity until this occurs. The impact of evolution of the clay buffer during its maturation and in the period beyond full maturity can be highly detrimental to the integrity and design function of the clay buffer, unless proper knowledge is obtained and design precautions taken. This section discusses many of the issues pertinent to this particular HLW containment problem.

6.3.1   Clay buffer maturation and the THMC processes

We define *maturation* of the clay buffer surrounding HLW canisters as the evolution of the clay in the first 100 years and more, during which time (a) the buffer becomes fully saturated and (b) the effects of long-term processes have not begun to be manifested. This time period exceeds most design lives of engineered clay barriers used in multi-barrier systems for HSW and MSW landfills. The discussions in Chapter 5 cover most of the issues concerning contaminant–clay interactions during the maturation period for HSW and MSW engineered clay liners. For this discussion, where the maturation period is something in the order of 100 years or more, it is the clay buffer/ barrier system used to embed HLW canisters in repositories that is of major concern.

There is no hard and fast rule or signpost that tells us when this maturation period is over, and when post-maturation ageing sets in. It is an arbitrarily designated period, which allows one to conduct analyses and assessments within a relatively simple set of controls and/or processes, such as the THM processes in the buffer surrounding the HLW canister, summarized in Figure 6.4. The left-hand illustration in the figure shows the basic elements involved, and the right-hand illustration shows an idealized three-phase (solids, water and gaseous) representation of the buffer and the major processes initiated by the presence of the hot canister and water availability from the host rock.

The buffer/barrier system requiring intimate contact with the HLW canister with a certain degree of flexibility in the buffer mass is met if the initially semi-dry clay buffer is allowed to expand in order to fill in the void spaces and establish firm contact with the canister and host rock face. The expandability and self-sealing potential of the clay must be robust, i.e. these properties must continue to exist for periods as long as tens to hundreds of thousands of years. The early evolution of the buffer is important, particularly with respect to the wetting rate. However, the impact of temperature in the several hundred or 1000-year-long hydrothermal period is believed to be even more important for survival of the buffer. All of these functions require that mineralogical changes should be relatively insignificant in this long time period, and that at least 50 per cent of the smectite remains unaltered after the required isolation time. These longer-term evolutionary aspects (chemical and biological) are discussed in a later section in this chapter.

We use the Swedish concept KBS-3V for underground storage of HLW, which is one of many concepts for HLW isolation, as an example to provide

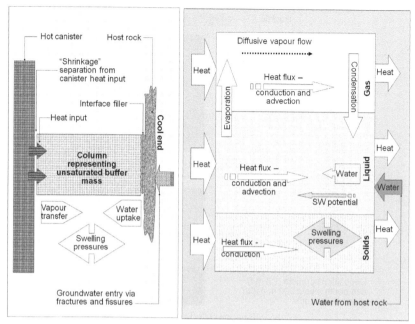

Figure 6.4   THM processes in the buffer surrounding (embedding) a HLW canister in a typical repository set-up in deep geological disposal/management. The left-hand illustration depicts the clay buffer as a column with the major elements of the THM processes. The right-hand illustration is a three-phase (solid, liquid, gaseous) representation of the clay, showing a simplified view of the THM actions in the respective phases. Note that chemical processes are not shown in the diagram and that SW potential = clay–water potential. Engineering usage identifies this as equivalent to clay suction.

a more specific focus for the discussion on the THM aspects of early matura-tion of the clay that constitutes the buffer. The HLW canisters are embedded in blocks of highly compacted sodium smectite, identified as MX-80, and placed in bored deposition holes of 1.8 m diameter and 8 m depth, in crystal-line rock at about 500 m depth (Figure 6.5). Figure 2.6 gives an illustration of the scheme of deposition holes spaced about 6 m apart in the underground tunnel. The HLW canisters produce energy that is about 1000 W at the time of placement, with an estimated heat decay time to equilibrium of about 3000 years. The maximum temperature at the canister surface is expected to be about 95°C and the temperature gradient across the buffer will become vanishingly small after about 1000 years.

In the early maturation stages, one could expect the following changes to occur in the compacted smectite blocks (Pusch and Yong, 2006):

•   Thermally induced redistribution of the initial porewater will occur, resulting in desiccation in the hottest part of the buffer and wetting of

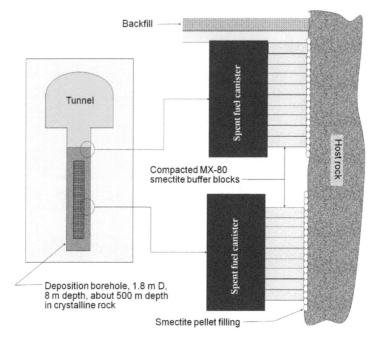

Figure 6.5 KBS-3V concept, showing deposition borehole and smectite (MX-80) buffer-blocks surrounding a HLW canister. Note that pellets filling the space between the buffer blocks are smectite. See Figure 2.6 for a diagram showing the location of the borehole/tunnel scheme.

the colder outer part of the buffer. Some shrinkage separation of the clay at the canister–buffer interface is likely, as shown in Figure 6.4.

- Access to groundwater from the host rock will allow maturation of the smectite pellet fill at the interface between the host rock and the smectite buffer blocks forming the buffer. Water saturation of the smectite pellet fill will lead to homogenization and consolidation of the buffer component under the swelling pressure exerted by the hydrating and expanding MX-80 buffer blocks.

- Uptake of water from the rock and backfill will occur, leading to hydration of the clay buffer. Swelling pressures are generated because of the volume change phenomenon associated with water uptake by the layer silicate MX-80.

- Expansion of the buffer results in an eventual tight contact between canister and buffer and rock. Note that the expansion will also occur in the canister-supporting portion of the buffer, i.e. in the buffer portion below the canister, which will cause an upward displacement of the canisters and the overlying backfill.

- Chemical processes will occur within the buffer and at the contact of buffer and canisters. A most important process is that salt, primarily Ca,

Cl and SO_4, will migrate with the water that migrates from the rock; precipitation in the hot part of the buffer will produce solid or brine NaCl and gypsum, thereby contributing to corrosion of the canisters; at some later time, the likelihood of corrosion products emanating from the canister will become more omnipresent.

Interface pellet wetting from rock groundwater and subsequent wetting of the smectite buffer is shown in Figure 6.6. Water availability to the interface pellet/gel serves as the source water to the buffer. This means that all of the stages depicted in the figure overlap each other. The attainment of a gel-like mass from the pellet configuration is not a sudden transformation. The transition stage is a process of wetting of the pellets, and it is not necessary for the pellets to become totally saturated or assume a total gel-like state before water can be drawn from them into the smectite buffer blocks. Using the engineering concept of soil suction to illustrate the sequence of actions, one could say that water from the interface pellet/gel will be drawn into the

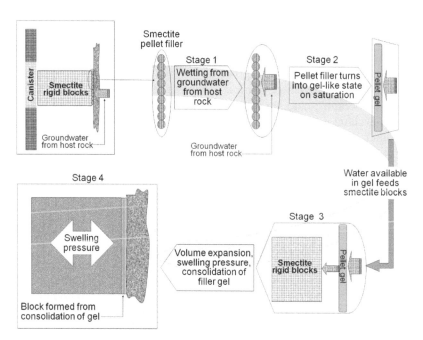

Figure 6.6   Different stages of interface smectite pellet wetting and buffer block saturation. Note that wetting of pellet filler in stage 1 continues throughout the life of the repository – or as long as groundwater in the host rock is available. The various stages overlap each other, i.e. stage 2 overlaps stage 1, and the process in stage 3 will automatically begin as soon as the soil suction is sufficient to draw water in from the pellet gel (which does not have to be completely formed or completely saturated). Similarly, the process in stage 4 will commence in conjunction with water uptake in the smectite rigid blocks in stage 3.

initially semi-dry buffer as soon as the clay suction in the buffer is capable of doing so. The discussion so far, concerning interface pellet wetting, has not included the influence of a thermal gradient in the buffer containment system.

How long it would take for water saturation of the buffer to occur depends also on the availability of water from the host rock. For a richly water-bearing crystalline rock, one could anticipate from simple analytical modelling procedures a period of a decade or so for the buffer to achieve near full to full saturation. However, in tight crystalline rocks one could anticipate full saturation in several decades. In argillaceous rocks, this full saturation time could extend to hundreds and thousands of years. The ability of the surrounding rock to supply the buffer with water is a key element in the design and performance assessment of buffer containment barriers. One should be careful to note that long-term physico-chemical processes and biological activities over the long containment period could lead to significant changes in one's anticipation of saturation time and buffer performance, especially in the time period of high temperature gradients.

The various processes associated with the THMCB factors are coupled in a complex manner – and not altogether in a constant relationship. Leaving aside the C and B factors for the time being, an example of the complicated picture is the 'couplings' in the processes of T–H, H–M and M–T. Because of the H factor, the movement of heat, vapour and liquid are in opposing directions. Temperature- and water content-dependent development of volume-change/swelling pressures are all, in one fashion or another, interdependent phenomena. This is demonstrated in the right-hand side of Figure 6.4. The various processes, including chemical and biological, operate with different intensities and at differing timescales, as shown in Figure 6.2. The important point to note from Figure 6.2 is the simple fact that all of the detailed processes evoked in THMCB and R are available, and are most likely operative, in the clay buffer–canister system.

The temperature and swelling pressure results of the clay buffer for the full-scale 5-year field experiment in the SKB underground laboratory (with the MX-80 clay buffer) are shown in Figure 6.7. In the canister set-up, similar to that shown in Figure 6.5, pressure cells were located at canister mid-height. The initial low swelling pressure gradient is the outcome of (a) initial phase of hydration of the pellets at the interface; (b) compression of the hydrated pellets due to swelling of the initially semi-dry MX-80 against the hydrated pellets; and (c) closure of gaps between the MX-80 blocks used to construct the buffer.

The fact that all the processes have time-scales of intensity and influence makes the problem of anticipating, predicting and hence assessment of system performance very difficult. Added to this is the fact that much is lacking in detailed factual information of system and component behaviours. Take, for instance, the pellets in the 50-mm water-filled gap between the

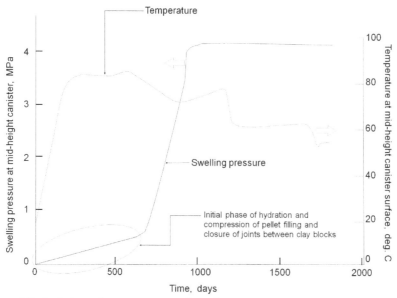

Figure 6.7 Evolution of temperature and swelling pressure at canister mid-height in full-scale 5-year experiment in SKB underground laboratory using MX-80 bentonite as the buffer. The stepwise reduction in temperature is the result of power failure of heater elements. The temperature ordinate scale is on the right and the swelling pressure ordinate is on the left (adapted from Pusch *et al.*, 2007).

smectite blocks and the rock face – shown as smectite pellet filling in Figure 6.6. Conceptually, one expects that when the interface filler is exposed to groundwater emanating from the host rock, hydration of the pellets will cause disintegration of the pellets resulting in the formation of a soft clay gel. This has been demonstrated in controlled laboratory experiments. The availability of water from the filler gel provides the opportunity for water uptake by the dense smectite blocks that serve as a buffer between the canister and the host rock. Volume-change swelling pressure of the clay blocks will consolidate the gel, resulting in a clay medium that will be as dense as the expanded blocks. In the longer-term picture, it is reasonable to expect that microbes can enter the pellet filling and survive and multiply until the density of the filling reaches about 1600 kg/m³. This brings in the B factor, which will be discussed in the later section dealing with long-term processes.

6.3.2   *Temperature and T (thermal) factor*

Of all the THMCB factors associated with deep geological HLW containment, we consider the T factor to be the most influential. This is because temperature plays a significant role not only in the types and rates of chemical reactions in a clay, but it also exercises its influence on biological processes. Temperature gradients in a clay create a set of reactions in the clay that are dependent not only on the type of clay, but also on the minimum clay temperature (or the mean temperature) in that temperature gradient. The various transport processes involved are complex in nature, not only because of the various heat transfer mechanisms involved, but also because of the heterogeneity of the clay – a property that gives rise to *point-dependent* properties of a clay. If we add in the varied nature of the various reactions, transformations, alterations, etc. that occur, we can add *time-dependent* to the complex picture that governs heat transport processes. In short, *time-variable* and *point-variable* properties of a clay make the prediction of transport of solutes in a clay somewhat difficult.

The terms *thermal energy* and *thermal energy transfer*, which have often been used in the literature to refer to *heat* and *heat transfer*, are, strictly speaking, not correct usage of the terms. In actual fact, thermal energy refers to that portion of the energy content of a material (in this case, clay) due to heat input into the material. Confusion arises when we wish to view it in thermodynamic terms, as thermal energy is the *internal energy* of a material or system in thermodynamic equilibrium as a result of its temperature. The internal or thermal energy of a system, such as a clay system (clay solids/particles, water and air), consists of at least five different energy components as follows: sensible, latent, chemical, nuclear and interaction. If the term *thermal energy* is to be used, one needs to first specify the a priori intrinsic energy content of the material.

The thermal properties of interest in a clay are (a) thermal conductivity, (b) thermal diffusivity and (c) heat capacity. Clay thermal properties vary considerably depending on the type of clay and especially on its water content and dry density (or microstructure and particle packing). With respect to the subject of our discussion, i.e. HLW smectite buffer/barrier systems, it follows that the rate of heat and mass transfer, volume change swelling and swelling pressures in the smectite buffer/barrier would also depend on the type of clay and its water content and dry density, other than the temperature and boundary conditions governing the system.

Heat transfer

There are three principal modes of heat transfer in clays: radiation, convection and conduction. In addition to these, we have an additional mode of heat transfer called *latent heat transfer*. This is particularly pertinent in the smectite buffer embedding HLW canisters. In clay buffer/barrier systems, radiation and convection play relatively minor to insignificant roles in the transfer of heat – to a very large extent because of the high density and the low water contents of the clay. In the initial stages of water uptake in a partly saturated clay, vapour transfer occurs in the continuous pore spaces and heat transfer is considered to be principally by convection. After the initial stages of wetting and water uptake, we can consider the primary mode of heat transfer to be by conduction.

The temperature in an infinitesimal volume of a clay changes when heat flows into or out of the volume. The increase and/or decrease in temperature are given by a relation that states that an amount of heat that flows into or out of the volume is equivalent to the product of the changes of temperature and heat capacity of the clay. In terms of the heat conservation law, this is expressed as follows:

$$c\rho \frac{\partial T}{\partial t} = -\frac{\partial q_h}{\partial x} + L \tag{6.1}$$

where T is the temperature, q_h is the flux of heat, L is a heat source or heat sink, c is the specific heat capacity and ρ is the mass of the clay. We can consider the product $c\rho$ as the heat capacity C, and L to take into account *latent heat transfer* and convection. The heat flux due to conduction can be expressed in terms of conjugate force–flux relationships using Fourier's empirical law:

$$q_h = -k_h \frac{\partial T}{\partial x}$$

where k_h is the thermal conductivity. Substituting this into equation 6.1, we obtain:

$$C\frac{\partial T}{\partial t} = \frac{\partial}{\partial x}\left(k_h \frac{\partial T}{\partial x}\right) \pm L_T(x,t) \tag{6.2}$$

The latent heat transfer term L_T refers to the sink or source term L in equation 6.1, in as much as convection of vapour and liquid water can be considered as small and safely ignored in compacted clay barriers.

The *latent heat transfer* mode refers to the transfer of latent heat in evaporation of water and condensation of vapour. Evaporation of liquid water provides a positive latent heat flux, whereas condensation provides a negative latent heat flux – hence the ± sign before the last term in equation 6.2. Because of the different stages of evaporation and the highly unsteady nature of vapour diffusion, it is likely that there will be intermediate locations that will exhibit local condensation. This explains why we need to express L_T as a function of spatial distance and time – as shown in the last term in equation 6.2.

Heat capacity

The heat capacity of a clay can be expressed in several different ways, such as specific heat capacity per unit volume, or volumetric heat capacity – depending on the manner in which this property is to be used. The *specific heat capacity* of a clay c_{clay} is defined as the change in heat content of a unit mass of clay per unit change in temperature. Multiplying c_{clay} by the density of the clay ρ will give us $\rho c_{clay} = C$, where C is defined as the *specific heat capacity per unit volume*, or the *volumetric heat capacity* of a clay. As clay is composed of three phases (solid, aqueous and gaseous), and as the heat capacity in each of these phases is different, a weighting procedure is used to obtain the overall heat capacity of a clay. This procedure is not exactly accurate, but will nevertheless provide one with an acceptable estimate of the heat capacity of a clay. To determine the value of C (volumetric heat capacity), for example, one will need to obtain knowledge of the volume fractions of each of the three phases as follows: υ_s for volume fraction of solids, θ for volumetric water content, and υ_g for volume fraction of gas. These are combined with their respective specific heat capacities and densities to obtain C. This gives us:

$$C = \upsilon_s \rho_s c_s + \theta \rho_w c_w + \upsilon_g \rho_g c_g \tag{6.3}$$

where ρ_s, ρ_w and ρ_g refer to the densities of solids, water and gas, respectively, and the subscripts s, w and g, associated with the specific heat capacity c, refer to the specific heat capacities of solids, water and gas respectively. The volumetric heat capacities for most clay minerals (primary and clay minerals) and can be taken to be about 1.88–2.09 J/m³ K, depending on the specific gravity of the mineral under consideration, and the volumetric heat capacities for liquid water and air are 4.184 and 0.0125 J/m³ K respectively.

Thermal conductivity

The *thermal conductivity* k_h of a clay is defined as the quantity of heat passing through a unit area of the clay mass in a unit time under a unit temperature gradient. It is very sensitive to the configuration of the solids in the clay and the relative proportions of solids, water and gas – in addition to the previously stated factors of clay type, water content, density and microstructure. Proper attention needs to be paid to clay microstructure and the distribution of microstructural units. This is because heat transfer by conduction involves heat transfer between molecules, i.e. from one molecule to another. Although conduction occurs in the gaseous and liquid phases, conduction in (and between) the solid fractions is most efficient because of the tight configuration of molecules in the solids. Kasubuchi (1984) has proposed a model that takes these into account in heat conduction in soils, and is applicable to clays and clay barriers. Because of the heterogeneity of microstructural units and their distribution, thermal conductivity is highly variable in a clay mass because of the point- and time-dependencies, or, looking at it from another perspective, because of the spatial and time-variable proportions and distributions of solids, liquid and gaseous phases. Campbell and Norman (1998) have proposed a relationship for k_h as follows:

$$k_h = \frac{\theta \alpha k_w + \upsilon_s \beta k_s + \upsilon_g \phi k_g}{\theta \alpha + \upsilon_s \beta + \upsilon_g \phi}$$

(6.4)

where the subscripts w, s and g associated with the thermal conductivity k refer to the thermal conductivity of water, solids and gas respectively, and α, β and ϕ are weighting factors accounting for heat associated with evaporation and condensation, i.e. latent heat transfer. The thermal conductivities for most clay minerals at a temperature of 20°C are about 2.51–37.66 mJ/cm s K, depending on the specific gravity of the mineral under consideration, and the thermal conductivities for liquid water and air at the same 20°C temperature are 5.77 and 0.251 mJ/cm s K respectively. It is important to note that when the temperature gradient is changed (or fluctuates), the thermal conductivity k_h will change since the changes in other heat transfer factors will impact on the thermal conductivity. Measurement of the thermal conductivity in a clay (or any substance for that matter) requires one to establish a constant temperature gradient across the clay. Experimental procedures for determination of thermal conductivity in clays include (a) the steady-state heat flux method of Manose et al. (2008) and (b) the twin transient-state cylindrical probe method that is based on the heat flow analysis in the unsteady state suggested by Kasubuchi (1977).

Thermal diffusivity

The *thermal diffusivity* D_T, which is a measure of transient heat flow, is defined as the thermal conductivity k_h divided by the heat capacity C. Although the thermal conductivity k_h of a clay is a property that provides a measure of how much heat will flow through the clay, the thermal diffusivity D_T provides one with the information on how rapidly the heat will flow through the clay. Methods of measurement of thermal diffusivity D_T can be difficult when temperature gradients in the clay keep changing, and particularly when we have both time- and point-variable properties as clay characteristics. In short, one needs to have a dynamic measurement of thermal diffusivity. Motosuke et al. (2003) have proposed the use of a forced Rayleigh scattering method (an optical technique) to measure the thermal diffusivity of solids and liquids. They maintain that this provides a thermal diffusivity real-time monitoring system. In another technique for determination of D_T proposed by Osako and Ito (1997), using a transient pulse technique, one has the opportunity to simultaneously determine the thermal conductivity k_h. The interesting aspect of this technique is that it can determine these thermal properties for samples under pressure – a situation that faces the smectite buffer embedding HLW canisters (T and M). Alternatively, one could obtain measurements of temperature response in a clay with respect to thermal pulses and determine D_T using the following relationship:

$$\frac{\partial T}{\partial t} = D_T \frac{\partial^2 T}{\partial x^2}$$

where T represents the temperature, $t =$ time and x is the spatial distance. This relationship assumes that latent heat transfer and sensible heat transfer by convective flow of vapour and/or liquid water are negligible (see equation 6.2).

6.3.3 Interactive phenomena from THMC processes

If we include the chemistry of the groundwater under the general heading of *fluid* in our discussion on liquid transfer processes and interactive phenomena, we can identify the various interactions between heat transfer, fluid transfer, and swelling pressures – shown previously in Figure 6.4. The effect of solutes in the groundwater is seen in the chemical reactions between the solutes and clay particles. The reactions and effects have been detailed in the previous section. To demonstrate the basic elements of the interactions between THMC processes in a swelling clay such as smectite – used as the buffer/barrier material – we consider a simple example that is shown in Figure 6.8. The element shown in the left-hand side of the figure represents a semi-dry volume element of a smectitic clay under physical boundary constraints that limit volume change, as indicated by the laboratory column in the left-middle of the diagram. This volume element is exposed to incoming liquid water, which contains solutes. The various processes represented by the large numerals (1–4) are considered to occur under isothermal conditions, i.e. no heat input. In essence, we are looking at the basic elements of the phenomena depicted in stages 3 and 4 in Figure 6.6 without benefit of heat input.

Without a hydraulic gradient, and under isothermal conditions, there is one main process for liquid water intake into the semi-dry volume element. The second process (condensation), which is relatively minor, is condensation of the vapour produced at the wetting front (water uptake by the semi-dry clay is exothermic). The fluid flow vectors shown in numeral 2, emanating from and into the volume element, are the result of (a) water uptake due to mechanisms associated with clay suction and (b) condensation of the vapour developed ahead of the wetting front. In semi-dry (unsaturated) clays

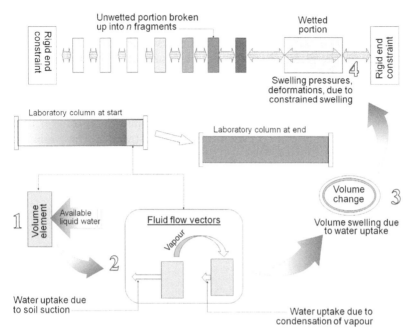

Figure 6.8 Interactive cycle for a constrained laboratory column swelling clay exposed to liquid water. The interactive cycle starts from numeral I and goes through to numeral 4 before returning with a modified or changed volume element – hence affecting ongoing water uptake rate in the next $n-1$ fragment. The sizes of the fluid flow vectors in numeral 2 are meant to portray the relative magnitude (qualitatively) of fluid uptake and intake, and the sizes of the arrows between the n fragments are indicative of the pressures between fragments due to the swelling pressure from the wetted portion.

under isothermal conditions, the amount of vapour developed is a function of unsaturated water content. The sizes of the arrows are relative, i.e. they are meant to portray the relative magnitude (qualitatively) of water uptake/intake. Volume expansion of the element occurs upon water uptake and intake. Constraining the element from swelling produces swelling pressures (or restraining pressures from the physical boundaries). This alters the physical, mechanical and hydraulic properties of the element, and hence will affect the nature of ongoing water uptake and intake rates for the next $n-1$ volume elements. What we have therefore are interactive phenomena – one set of behaviour and values being dependent on applied actions, which, in turn, will impact on the nature of the applied actions. The impact of solutes in the available liquid water is seen as reactions between the solutes and the clay particles. As has been discussed in the previous chapter, these reactions can alter the properties of the clay – especially the swelling behaviour of swelling clays used for buffer/barrier systems.

When a temperature gradient is imposed on the volume element – such as would be obtained in the clay buffer used to embed HLW canisters in repositories – two other water transfer processes are introduced into the total water uptake picture in numeral 2 in Figure 6.8. Vapour transfer now becomes a significant factor. At low moisture contents, vapour movement can be from four to five times larger than would be predicted from diffusion calculations using a Fickian model, because of rapid evaporation and condensation. There are now four processes for liquid water transfer in the volume element (Figure 6.9) in addition to vapour transfer. These are:

1   liquid and vapour uptake due to the matric and osmotic potentials in the semi-dry volume element;
2   liquid water intake due to the hydraulic gradient;
3   condensation of vapour;

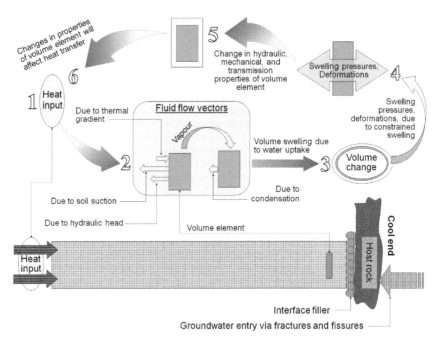

Figure 6.9 Interactions between T, H, M and C processes/reactions in the maturation period. Note that the C (solutes from groundwater) effect is included in the chemical reactions etc. that accompany solute interactions with clay particles. The cause–effect–return cycle starts from numeral I and goes through to numeral 6 before returning with a modified or changed volume element – hence affecting the rate of heat transfer from I to 2, etc. Note, as before, the sizes of the fluid flow vectors in numeral 2 are meant to portray the qualitative relative magnitude of fluid uptake and intake.

4   liquid water depletion (evaporation) and opposing transfer of liquid water at the wetting front due to the thermal gradient.

The cycle of events and processes that follow are similar to those described previously for the isothermal case, with the exception that the impact of the cycle of events is seen in terms of the changes in the rate of ongoing heat transfer. This is due in part to the increased water content, density, permeability, heat capacity and conductivity, etc. The continuous changes in the nature and state of the smectite clay buffer, and the effect on the various properties of the clay, demonstrate the need for a better appreciation of interactive phenomena of T, H and M processes in designing the function and effectiveness of the clay buffer system. The various aspects of these will be discussed in detail in later chapters.

T–H interactive processes

The imposition of a thermal gradient in the clay buffer, as shown in Figure 6.9, results in a combined heat and mass transfer in the clay buffer. Neglecting the swelling phenomenon associated with the type of clay in the buffer, we can examine the nature of the coupled heat–mass transfer phenomenon and arrive at a means to determine the relevant transmission properties. For simplicity in development of the governing relationships – without loss of conceptual problem representation – we can construct an experiment using a partly saturated clay column subject to a heat source at one end with a constant cool temperature at the other end similar in principle to the column shown at the bottom of Figure 6.9.

For the analysis, Yong and Xu (1988) considered the unsaturated column of length L to be fixed at both ends with $x = 0$ at the left and $x = L$ at the right. The temperature at $x = 0$ is T_2 and at $x = L$ it is T_1, and for simplicity in analysis without any loss in conceptualization of coupled heat–mass transfer, we assume both ends to have impermeable boundaries, i.e. closed system for water but not for heat transfer. The initial water content is assumed to be uniformly distributed along the length of the column, $\theta(x,0) = \theta_0$.

From the second postulate of irreversible thermodynamics, the governing sets of relationships pertaining to the heat and mass transfer phenomena represented by the experiment are:

$$\left\{ \begin{array}{c} Q_w \\ Q_T \end{array} \right\} = \left[\begin{array}{cc} L_{ww} & L_{wT} \\ L_{Tw} & L_{TT} \end{array} \right] \left\{ \begin{array}{c} \dfrac{\partial \theta}{\partial x} \\ \dfrac{\partial T}{\partial x} \end{array} \right\} \tag{6.5}$$

where $Q_w = \partial\theta/\partial t$ = fluid flux, i.e. fluid flow through per unit area per unit time; $Q_T = \partial T/\partial t$ = heat flux, i.e. heat flow through per unit area per unit time; $\partial\theta/\partial x$ = thermodynamic force due to water content gradient; $\partial T/\partial x$ = thermodynamic force due to temperature gradient; θ = volumetric water content, $\theta = \theta(x,t)$; L_{ww} = diffusion coefficient for fluid flow due to gradient of θ (note that the total accumulation of moisture due to the combined movements of vapour and liquid is considered as equivalent to liquid water); L_{TT} = thermal conductivity coefficient for heat transfer due to the gradient of T; L_{wT}, L_{Tw} = coupling coefficients; T = temperature, $T = T(x,t)$; and t = time.

From mass and energy conservation considerations, and recalling that C is the specific heat capacity, the following relationships are obtained:

$$\frac{\partial\theta}{\partial t} = \frac{-\partial Q_w}{\partial x}$$
$$C\frac{\partial T}{\partial t} = \frac{-\partial Q_T}{\partial x}$$

(6.6)

From equations 6.5 and 6.6, we now obtain the governing relationships as:

$$\left\{\begin{array}{c} \dfrac{\partial\theta}{\partial t} \\ \dfrac{\partial T}{\partial t} \end{array}\right\} = \left[\begin{array}{cc} 1 & 0 \\ 0 & 1/C \end{array}\right] \dfrac{\partial}{\partial x}\left\{\left[\begin{array}{cc} L_{ww} & L_{wT} \\ L_{Tw} & L_{TT} \end{array}\right]\left\{\begin{array}{c} \dfrac{\partial\theta}{\partial x} \\ \dfrac{\partial T}{\partial x} \end{array}\right\}\right\}$$

(6.7)

The boundary conditions for T are:

$$T(0, t) = T_2 = \text{constant}; \ T(L, t) = T_1 = \text{constant}; \ T_2 > T_1$$

(6.8)

The boundary conditions for θ are:

$$Q_w\bigg|_{x=0} = \left(L_{ww}\frac{\partial\theta}{\partial x} + L_{wT}\frac{\partial T}{\partial x}\right)\bigg|_{x=0} = 0$$

$$Q_w\bigg|_{x=L} = \left(L_{ww}\frac{\partial\theta}{\partial x} + L_{wT}\frac{\partial T}{\partial x}\right)\bigg|_{x=L} = 0$$

(6.9)

The initial conditions for T and θ are given as:

$$T(x,0) = \begin{cases} T_2, x = 0 \\ T_1, 0 < x \le L \end{cases} = T_2 + \left(T_1 - T_2 \right) u(x)$$

$$\theta(x,0) = \theta_0$$

(6.10)

where $u(x)$ is a step function. Yong and Xu (1988) non-dimensionalized the preceding relationships and used trial functions to determine the phenomenological coefficients, which were initially assumed to be certain functions of θ and T. In the non-dimensional form, $\xi = x/L$, where $x \in [0, L]$ and the symbol $\in$ means *belongs to*. For the unsaturated no-volume-change experiment shown at the bottom of Figure 6.9, we now have $\xi \in [0, 1]$. The non-dimensional time τ is given as $\tau = t/t_f$, where t_f is the time taken for completion of the test, i.e. when no further variations in θ and T are observed. $t \in [0, t_f]$ now becomes $\tau \in [0, 1]$. Taking θ' to be the difference between θ and θ_0, i.e. $\theta' = \theta - \theta_0$, the initial condition for θ can be written as $\theta'(\xi, 0) = 0$. Designating T' as the non-dimensional temperature such that $T' = (T - T_1)/(T_2 - T_1)$, the boundary conditions for T for the ends where $x = 0$ and $x = L$ will be $T'(0, \tau) = 1$ and $T'(1, \tau) = 0$ respectively. The phenomenological coefficients are now written as:

$$L_{ww} = \frac{L^2}{t_f} L'_{ww} \qquad\qquad L_{wT} = \frac{L^2}{t_f(T_2 - T_1)} L'_{wT}$$

$$L_{Tw} = \frac{L^2 C}{t_f(T_2 - T_1)} L'_{Tw} \qquad L_{TT} = \frac{L^2 C}{t_f} L'_{TT}$$

(6.11)

The functional forms of the trial functions for the phenomenological coefficients used by Yong and Xu (1988) were determined on the basis of previous experiments as follows:

$$L'_{ww} = \left(a_1 + a_2 T' \right) e^{a_3 \theta'} \qquad L'_{wT} = a_4 + a_5 T' + a_6 \theta'$$

$$L'_{Tw} = \eta L'_{wT} \qquad\qquad L'_{TT} = a_7 + a_8 \theta' + a_9 T'$$

(6.12)

How were these determined? As an example, we will consider the L'_{ww} diffusion coefficient. This coefficient is associated with moisture transport in the unsaturated column in response to volumetric water content gradients, and is similar to $D(\theta)$, the coefficient of diffusivity. Yong and Warkentin (1975) have shown that, for isothermal conditions, $D(\theta)$ varies significantly with θ, and that in general $D(\theta)$ can be written as $D(\theta) = K_1 e^{K_2 \theta}$, where K_1 and K_2 are clay (soil) parameters to be determined from calibration experiments. Taking note of the fact that L'_{ww} varies with T, and assuming a linear dependency between L'_{ww} and T, the relationship for L'_{ww} shown in equation 6.12 is obtained. This example is a demonstration of how the functional forms were obtained for the phenomenological coefficients expressed in equation 6.12.

The constants associated with the functional relationships shown in equation 6.12 were evaluated using the *identification technique*. For this technique, one needs to conduct laboratory tests under different controlled conditions to generate data that would allow one to choose the a_1, a_2, a_3, etc., values that most closely match calculated theoretical values of θ' and T' with measured values. Figure 6.10 shows the theoretically obtained water content–distance (θ–x) relationship for a partly saturated clay column test with a heater temperature of 100°C at one end and a constant temperature of 10°C at the other end. The computed diffusion coefficients used to calculate the θ–x curve are valid only for the chosen temperature gradient and a specified water content. Calculations for other diffusion coefficients will be necessary when other temperature gradients and water contents are considered.

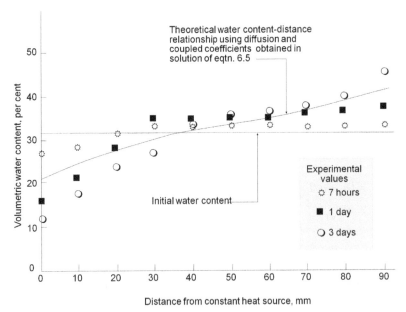

Figure 6.10 Comparison of theoretical water content-distance relationship using phenomenological coefficients (diffusion and coupled coefficients) computed from solution of equation 6.5 with experimental values. The calculated coefficients are for one imposed temperature gradient and one volumetric water content (adapted from results reported by Yong et al., 1992).

Coupling flow system in irreversible thermodynamics

The preceding discussion and analyses confirm that T–H interactions in the heated clay column result in processes that are coupled. In this section, we present a more fundamental treatment of this coupled phenomenon. Water transfer under temperature gradients and heat flow under water potential gradients can be analysed as the simultaneous transfer of heat and water regardless of whether the water is vapour or liquid. This simultaneous transfer is expressed as the coupling flow of heat and mass in irreversible thermodynamics. The fluxes and driving forces of the heat and mass flow can be specified by consideration of the entropy product per unit time in the clay–water system. When both water potential and temperature gradients occur simultaneously in a clay–water system, the entropy production $\Delta S/\Delta t$, that is $\Sigma J_i X_i$ (see equation 4.25), is given as:

$$\frac{\Delta S}{\Delta t} = \sum_{i}^{n} J_i X_i = J_M \, \mathrm{grad}\left(-\frac{\mu}{T}\right) + J_H \, \mathrm{grad}\left(\frac{1}{T}\right)$$ (6.13)

The driving force of heat transfer is given as $X_H = \mathrm{grad}(1/T)$ for the conjugate heat flux J_H, and the driving force of mass transfer is given as $X_M = \mathrm{grad}(-\mu/T)$ for the conjugate mass flux J_M. T represents the temperature and μ is chemical potential of water. If the changes of volume occur, the entropy product will include the driving force of volume flow as $X_V = \mathrm{grad}(P/T)$ for the volume flow. P represents the pressure. Assuming that the linear phenomenological relations (see section 4.3.4) can be given by:

$$J_i = \sum_{k=1}^{n} L_{ik} X_k$$

where $i = 1, 2 \dots n$, J_i is flux of flow of i, X_k is driving force of k ($k = 1, 2 \dots n$) and L_{ik} is the phenomenological coefficient. Recognizing that some swelling occurs in the clays under consideration, but assuming negligible volume expansion, the linear phenomenological equations coupling flow of heat and mass transfer in the compacted clay barriers can be written as:

$$J_M = L_{M1} \, \mathrm{grad}\left(-\frac{\mu}{T}\right) + L_{M2} \, \mathrm{grad}\left(\frac{1}{T}\right)$$

$$J_H = L_{H1} \, \mathrm{grad}\left(-\frac{\mu}{T}\right) + L_{H2} \, \mathrm{grad}\left(\frac{1}{T}\right)$$ (6.14)

where J_M and J_H refer to flux of mass and heat, respectively. The subscript M denotes the flux of vapour or liquid water under consideration. L_{M1}, L_{M2},

L_{H1}, and L_{H2} are the phenomenological coefficients. Onsager's reciprocal theorem gives $L_{M2} = L_{H1}$ (Katchalsky and Curran, 1967).

Introducing the following relationships:

$$\text{grad}\left(\frac{-\mu}{T}\right) = \frac{1}{T}\text{grad}(-\mu) - \mu\,\text{grad}\left(\frac{1}{T}\right) \text{ and } \text{grad}\left(\frac{1}{T}\right) = -\left(\frac{1}{T^2}\right)\text{grad}\,T$$

and introducing new phenomenological coefficients L^*_{M1}, L^*_{M2}, L^*_{H1}, and L^*_{H2}, the relationships in equation 6.14 can now be written as:

$$J_M = -L^*_{M1}\,\text{grad}\,\mu - L^*_{M2}\,\text{grad}\,T$$
$$J_H = -L^*_{H1}\,\text{grad}\,\mu - L^*_{H2}\,\text{grad}\,T$$

(6.15)

These relationships confirm that mass transfer due to temperature gradients occurs under temperature gradients, and heat transfer due to chemical potential gradients occurs when chemical potential gradients exist in a clay–water system. The previous chapters have shown that the chemical potentials of vapour and liquid water are functions of water content. These can be determined through knowledge of the water characteristic curves of clays under consideration. The second relationship in equation 6.15 tells us that heat transfer occurs under water content gradients when the term 'gradμ' is converted into the expression that relates water content to the chemical potential. Nakano and Miyazaki (1979) suggested that because of the several possible choices of driving forces associated with vapour transfer, we will have several different expressions for the relationships shown in equation 6.15. The same holds true for liquid water transfer. Combining the relationships in equation 6.14 with the relevant conservation law, we will obtain the parabolic partial differential equations (Fickian-type second-order partial differential equations) expressing simultaneous transfer of heat and mass in a clay–water system.

Temperature and clay–water potential

The results from an experimental study on the relationship between the clay water potential (determined from psychrometer measurements) and temperature for a 50:50 sand–sodium bentonite mixture are shown in Figure 6.11. The results reported in the figure inform us that the higher the water content of the test sample, the less is the influence of temperature on the clay–water potential. The factors contributing to the dependence on temperature, especially for a swelling clay, include (a) viscosity of porewater η; (b) surface tension of porewater σ; (c) the dielectric constant ε; and (d)

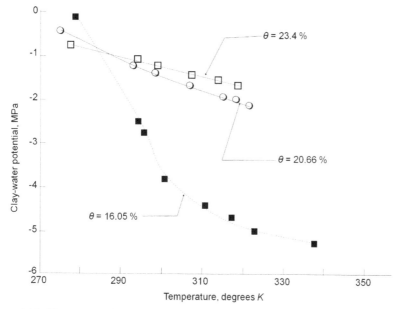

Figure 6.11   Variation   of   clay–water   potential   (obtained   from   psychrometric measurements) with temperature for a 50:50 sodium bentonite–silica sand mixture. The θ values shown in the graph are the volumetric water contents of the samples tested (adapted from data reported by Mohamed *et al.*, 1992).

activation energy of the water molecules and ions. For measurements made with psychrometers, we need to include the effect of temperature on the electromotive force. This increases with temperature because of the increase in evaporation of water between the reference junction and the wet junction of the psychrometer.

The relationship between clay water potential and temperature can be calculated using the diffuse double-layer (DDL) model, using the expression for the midplane potential ψ_c as follows:

$$\psi = \psi_c = \frac{2KT}{Z_i \varepsilon} \ln \left[\frac{\kappa d \sqrt{S}}{\pi} \right] \tag{6.16}$$

where:

$$\kappa = \sqrt{\frac{8\pi Z_i^2 \varepsilon^2 n_o}{eKT}}$$

and K, S, n_o, Z_i, ε, e and d refer to Boltzmann constant, coefficient of effective dialysate concentration, number of ions per unit volume in the bulk solution, valency of the ith species of ion, dielectric constant, electronic charge and half distance between interacting layers respectively. The influence of temperature T on the dielectric constant ε is given as follows: $\varepsilon = [62445/(T + 120)] - 70.91$.

6.4 Post-maturation ageing

The subject of *evolution and ageing* of clay buffers and barriers will be considered in detail in the next chapter when we discuss long-term performance of these buffers and barriers. For this section and this present chapter, we will be concerned only with the THMCB factors that are attributable to ageing processes. For completeness, however, we include the R factor in our depiction of the post-maturation ageing effects in Figure 6.12. In so far as HSW and MSW containment by engineered clay barrier/liner systems is concerned, it is not clear that post-maturation ageing processes strictly apply. This is in part because of the functional requirements of the clay barrier/liner of the multi-barrier system, and in part because of the relatively short service life of the landfill itself – in comparison with HLW containment

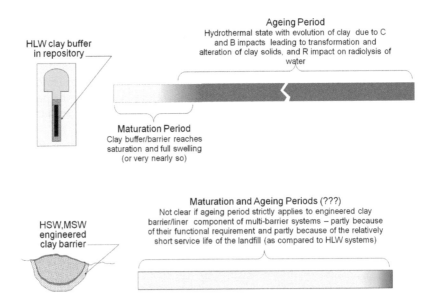

Figure 6.12 Maturation and ageing periods – as they apply to engineered clay buffer/barrier systems for containment of HLW and HSW/MSW. Note that the overlap between the starting point for the ageing period and the ending point for the maturation period for the HLW clay buffer system is deliberate. There is no clear demarcation for these starting/ending points.

and requirements for long-term performance of the associated clay buffer systems (Figure 6.12).

We consider post-maturation ageing for long-term HLW clay buffer/barriers to be a long-term process after clay buffer matures to its full potential. For the HLW repository containment scheme, this is considered to begin when the clay buffer or barrier is near complete water-saturation and full swelling of the clay will be complete, or almost fully complete. This means that we will have also reached maximum swelling (or reaction) pressures within the buffer/barrier. From this point onward, it is important to determine (a) what kinds of processes will directly contribute to, and influence the outcome of the post-maturation ageing of the clay and (b) what kinds of impacts on clay properties and performance are likely as a result of these processes.

6.4.1 Ageing processes and impacts – C factors

The details of the various changes, alterations, transformations, etc. occurring in the clay as a result of reactions and interactions with the processes initiated by long-term THMCB processes are addressed in the next chapter. For now, we address the two questions posed previously – namely, types of processes and their impacts. We have indicated in Figures 6.1 and 6.2 that the C factor is an active agent for ageing processes. There are two aspects of the C factor in the ageing period: (a) contaminant interaction with the clay – with corrosion products and fugitive radionuclides added to the suite of contaminants in the groundwater from the host rock – and (b) alteration of the clay solids such as dissolution of silicic crystalline materials and precipitation of free silica. The chemical reactions that promote dissolution and precipitation, depending on the type of clay mineral, are greatest at higher temperatures and higher moisture contents.

C-factor conversion of smectite

The composition of the reaction fluid has considerable importance in the transformation of smectite to different clay minerals, such as illite, saponite, etc., a process that is also dependent to some extent on the solid–liquid ratio. The following transformation has been widely accepted

smectite + feldspar + mica + others → illite + quartz + chlorite and others

The general chemical reaction of smectite is assumed to be:

$$smectite + K^+ + Al^{3+} + H_2O \rightarrow illite + Si^{4+} + other\ ions \qquad (6.17)$$

$$\text{montmorillonite} + Mg^{2+} + Fe^{2+} + H_2O \rightarrow \text{saponite} + Si^{4+} + \text{other ions} \quad (6.18)$$

Takase and Benbow (Grindrod and Takase 1994) have proposed a model for dissolution of smectite and formation of reaction products that does not preset smectite $\rightarrow$ illite as the basic reaction. The model takes $O_{10}(OH)_2$ as a basic unit and defines the general formulae for smectite and illite as:

$$X_{0.35} \, Mg_{0.33} \, Al_{1.65} \, Si_4O_{10}(OH)_2 \text{ and}$$
$$K_{0.5-0.75} \, Al_{2.5-2.75}Si_{3.25-3.5} \, O_{10}(OH)_2 \quad (6.19)$$

where X is the interlamellar absorbed cation (Na for sodium montmorillonite). The reactions in the illitization process are:

$$Na_{0.33} \, Mg_{0.33} \, Al_{1.67} \, Si_4O_{10}(OH)_2 + 6H^+ =$$
$$0.33Na + 0.33Mg^{2+} + 1.67Al^{3+} \, 4SiO_{2(aq)} + 4H_2O \quad (6.20)$$

causing precipitation of illite and silicious compounds:

$$K \, AlSi_3 \, O_{10} \, (OH)_2 = K^+ + 3Al^{3+} + 3 \, SiO_{2(aq)} + 6H_2 \quad (6.21)$$

Note that log K is a function of temperature.

Quantitatively speaking, the major transformation product is *illite*. Two reactions have been proposed for conversion of smectite to illites: (a) successive intercalation of illite lamellae in smectite stacks forming mixed-layer minerals (Weaver, 1979; Nadeau and Bain, 1986) and (b) neoformation of illite by coordination of Si, Al, K and hydroxyls to an extent controlled by access to any of them.

Whereas this type of process may be valid for the entire smectite family, the activation energy for conversion and dissolution is different amongst the members of the smectite family. Saponite is not affected by calcium and salinity to the same extent as montmorillonite. It is considered to have a higher chemical stability at high temperatures, thereby indicating a higher activation energy than montmorillonite for conversion of saponite to illite. In the case of beidellite, it is possible that potassium uptake can directly lead to collapse of the stacks of lamellae to yield illitic minerals.

Theoretical calculations of the reaction rate can be made assuming that the rate of change of smectite–illite ratio is a function of temperature, potassium content in the porewater and activation energy (Pytte, 1982). The Pytte model is a thermodynamically based Arrhenius-type theory that assumes that the rate of change of the smectite–illite ratio depends on temperature, potassium content in the porewater, and an activation energy that is commonly taken to be 104.6–125.52 kJ/mol. The rate equation is given as:

$$-\frac{dS}{dt} = \left[Ae^{-\frac{U}{RT(t)}} \right]\left[\frac{K^+}{Na^+} - mS^n \right] \tag{6.22}$$

where S represents the mole fraction of smectite in illite to smectite assemblages, R is the universal gas constant, T is the absolute temperature, t represents time and m and n are coefficients. Application of the relationship allows one to determine the rate of illitization from smectite. Figure 6.13 shows the results of calculations using an activation energy of 100.42 kJ/mol. As shown in Figure 6.13, heating to 50°C causes only an insignificant loss of smectite in 10^6 years. On the other hand, about 100 per cent of the original smectite turns into illite in this period of time at 100°C. As the clay buffer/barrier system used to contain HLW is exposed to heat for a period of at least a few thousand years (Figure 6.3), it is important to provide the proper assessment of clay evolution in designing the functional capability of the buffer system. The implications of such a transformation are discussed in the next chapter.

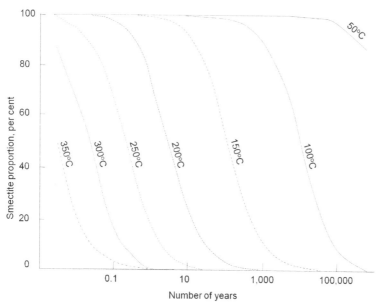

Figure 6.13 Expected conversion of smectite to illite in relation to temperature, for an activation energy of 100.42 kJ/mol, as predicted from Pytte's model. At 50°C, the original 100 per cent smectite drops to 97 per cent in 100,000 years. However, at 100°C, the original 100 per cent smectite drops in proportion to 50 per cent in 1000 years (adapted from Pusch, 1994).

Thermodynamic mineral dissolution kinetics model

In general, dissolution of clay minerals occurs as a result of the thermodynamic reactions at the surface of minerals in clay pores. Assuming that there is an infinitesimal thin stagnant fluid-film at the interface between mineral surface and clay solution in pores, a mineral dissolved at the surface needs to diffuse toward the bulk pore solution that exists outside the thin fluid film. In this model, it is assumed that the rate of the mineral diffused into the bulk solution in pores will give the apparent dissolution rate of clay minerals in engineered clay buffers/barriers, whereas the real thermodynamic dissolution occurs at the surface of clay minerals. Consequently, the apparent dissolution rate of clay minerals in pores is assumed to be given as (Made *et al.*, 1994):

$$R_d = k_d S_{eff} a_{H+} \left(1 - \frac{Q}{K} \right)$$

where R_d is the estimated dissolution rate at a given pH, k_d is thermodynamic dissolution rate at the surface of minerals by chemical reaction, S_{eff} is the reaction surface area, a_{H+} is activity of hydrogen ion, Q is thermodynamic ion activity product of minerals and K is the thermodynamic equilibrium constant. The effect of temperature on the apparent dissolution rate can be evaluated as the real thermodynamic dissolution rate k_d at the surface of minerals – based on the Arrhenius classical theory. This relationship has recently been used by Marty *et al.* (2007) to estimate the long-term transformation of clay buffers/barriers. This will be discussed in the next chapter.

Interaction with groundwater and precipitation in pore spaces

Interaction of HLW containment smectite buffers with groundwater emanating from the host rock will result in:

- dissolution–precipitation of clay minerals as discussed previously, which will result in the formation and precipitation of new chemical compounds because of the release of various ions into the pore spaces from the clay minerals;
- dissolution of calcite, siderite and gypsum, etc. contained in clay buffer/barriers, resulting in their precipitation in porewater according to the following reaction relationships:

$$CaCO_3 + H^+ \leftrightarrow Ca^{2+} + HCO_3^-$$
$$FeCO_3 + H^+ \leftrightarrow Fe^{2+} + HCO_3^-$$
$$CaSO_4 \cdot 2H_2O \leftrightarrow Ca^{2+} + SO_4^{2-} + 2H_2O$$
$$SiO_{2(s)} \leftrightarrow SiO_2$$

- ion-exchange reactions between clay minerals and solutes in the prevailing groundwater, and between clay minerals and metals released into the clay pores from dissolution of calcite, siderite and gypsum contained in clays; the end result is the formation of different clay minerals.

Generally speaking, the precipitation of newly formed chemical compounds in pores will bind smectite particles to each other and clog pores in clays. Binding of smectite particles will increase the mechanical strength of the clay, and clogging of pores will decrease its hydraulic conductivity. These effects are called *cementing effects*. Added to these effects is the loss of swelling characteristics, thereby reducing the sealing characteristics of bentonite – a feature promoted as a desired design property (sealing of the buffer at the interface between buffer and host rock). In the reactions shown in the second bullet point above, the consequent silica release can be detrimental to the ductility of the smectite buffer embedding HLW canisters. Alteration of the smectite minerals and precipitation of the silica will essentially eliminate the swelling capability of the clay and will also reduce its chemical buffering capacity. This makes the buffer incapable of fulfilling its design function as a physical and chemical attenuation barrier.

The thermodynamic precipitation rate of newly formed chemical compounds in the pores can be obtained using the same model used to determine the dissolution rate of clay minerals, that is:

$$R_p = k_p S_{eff} \left(\left(\frac{Q}{K} \right)^p - 1 \right)^q \tag{6.23}$$

where R_p is the estimated precipitation rate, k_p is the thermodynamic precipitation rate at the surface of minerals by chemical reaction, and p and q are experimental constants describing the saturation state dependence of the reaction (Made *et al.*, 1994).

6.4.2   Ageing processes and impacts – B factors

In contrast with C factors, it is not clear when B factors impact and contribute to post-maturation ageing processes. This is in part because of the coupling of B factors with C factors and in part because of the relatively sparse knowledge of the function and impact of microorganisms, especially with respect to smectitic clays in deep geological repository environments with high temperatures. For now, we will discuss some questions relating to their impacts on engineered clay buffer/barrier systems. These include (a) biologically induced oxidation, reduction and dissolution of contaminants, resulting in the changes of pH and Eh in a clay–water system; (b) biologically induced dissolution of clay minerals resulting in formation of new clay minerals and chemicals; and (c) physical clogging of small pores in clays because of the increase of microbe density, leading to a decrease in hydraulic conductivity.

Microbiological activities and impacts

The existence and activities of microorganisms in natural soils and rocks, clay buffers, backfills and host rocks in repository-type environments have been reported by many researchers (e.g. Krumbein, 1978; West *et al.* 1985; Huang and Schnitzer, 1986; Stroes-Gascoyne and West, 1996; Pedersen, 2000; Wang and Francis, 2005; Yong and Mulligan, 2004). The important environmental factors for optimal performance of microorganisms include temperature, oxygen availability, presence of water, nutrients and osmotic pressure.

The three main temperature ranges in which the various classes of microorganisms can grow optimally are as follows: (a) psychrophiles grow at 0–10°C; (b) mesophiles grow from 10°C to 45°C; and (c) thermophiles grow from 45°C to 75°C and above. However, note that reports now show that thermophiles can exist at temperature in the 100°C range (e.g. Pedersen, 2000; Seabaugh *et al.*, 2004). In general, the guidelines are that a 10°C increase in temperature will double the growth rate. Mesophiles are the most common, and as the temperatures at contaminated sites are generally in the same temperature range that favour growth of mesophiles, it will be seen that they are particularly useful for biological treatment processes. Capsule formation enables microorganisms to survive at reduced temperatures and grow as the temperature increases to more favourable conditions (Sims *et al.*, 1990).

Water availability is essential for the survival of microorganisms. This is the principal component of cell protoplasm and is required for nutrient transport into the cell. With respect to engineered clay buffer/barrier systems, and especially for HLW clay buffers that are emplaced in the almost-dry condition, microbial activities will not be a factor at the early maturation stage of the clay buffer. Moisture levels of 50–75 per cent are deemed to be ideal conditions for microorganisms. Fully saturated soils present some problems because of (a) O_2 limitations and transport to bacteria and (b) the low solubility of O_2 in water. Some of the more important chemical factors impacting on microbial activity include pH, toxicity, heavy metals, molecular structure and cometabolism. Although microbes prefer a pH range of between 5.6 and 9.0, there are some that can function in higher- or lower-pH environments. Microbial activities in reduction of electron acceptors will affect pH levels through production of H^+. Dissolution–precipitation and formation of new minerals are some of the direct consequences of the change in the geochemical environment – a process that is sometime identified as a form of biomineralization. *Biomineralization*, which is broadly defined as minerals produced by processes attributable to living organisms, has application in several disciplines. However, the genetic basis for biomineralization is not well understood. Strictly speaking, with respect to soils, we can broadly define *biomineralization* to mean the biologically mediated process by which one obtains amorphous and crystalline materials from aqueous ions. The two general paths are (a) biologically induced mineralization and (b) boundary-organized mineralization.

Heavy metal contaminants, such as mercury, lead, chromium, zinc, cadmium, copper, arsenic and barium, are toxic to microorganisms. Toxicity inhibits microbial growth or substrate utilization. Metal ions can non-specifically inhibit the function of cell membranes by binding to cell surfaces and penetrating cell membranes. Microorganisms can convert toxic compounds to different compounds that are safe or less toxic.

Microbial flora detected in the groundwater include aerobic and micro-aerophilic heterotrophic microorganisms, together with anaerobic iron-reducing and sulphate-reducing bacteria such as *Desulfovibrio* and *Desulfotomaculum*. We should note that chemoautotrophs such as *Thiobacillus thiooxidans* and *Thiobacillus ferrooxidans* could play significant roles in the subsurface environment where organic matter is in short supply – a situation that is particularly true in engineered clay buffer/barriers. This is because chemoautotrophs are able to obtain (a) energy required to sustain life by oxidation of inorganic matter and (b) carbon required for the formation of bodies by decomposing carbon dioxide. Microorganisms react with clay minerals by producing extracellular polysaccharides that coat the microorganisms. An example of how a bacterial cell interacts with clay mineral particles is shown in Figure 6.14. Microorganisms commonly form colonies and biofilms adhering onto the surface of materials. Following adherence of microbial flora on the surface of engineered clay buffer/barriers, or following migration into fissures and cracks in clay buffer/barriers of more than $0.25\,\mu m$ in width (the minimum size of bacteria), they grow and induce biological dissolution/transformation of clays or metals. The microbial dynamics of growth and decay can be expressed using the following generalized logistic equation (see Murray, 1993).

$$\frac{dv}{dt} = \lambda v - f(v) \qquad (6.24)$$

where v is the population of microorganisms, λ is the intrinsic growth rate and $f(v)$ is the decay term that is commonly given by a linear function of v.

B factor and iron-reducing

Microorganisms such as facultative anaerobic bacteria, fungi and even anaerobes can be active in reducing iron. Water in the clay buffer material provides the opportunity for both acid–base and oxidation–reduction reactions, with

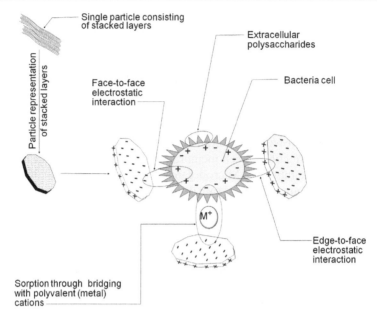

Figure 6.14   Bacterial cell with coating of extracellular polysaccharides in interaction
with charged clay mineral particle, showing the various sorption mechanisms
between the bacterial cell and clay mineral particles. As with the clay mineral
particles, the bacterial cell will generally have a net negative surface charge at
circum neutral pH conditions (adapted from Huang *et al.*, 2005).

the latter being abiotic and/or biotic. The activity of the electron e^- is a
significant factor in oxidation–reduction reactions, as these involve the
transfer of electrons between the reactants. The reduction of the structural
Fe^{3+} in the octahedral and tetrahedral sheets to Fe^{2+} will significantly alter
the short-range forces between the layers in the layer–lattice structure of the
clay minerals and could result in (a) a lower SSA; (b) a high degree of layer
collapse; (c) decreased water-holding capacity; (d) reduced swelling capability; and (e) stacking of the layers. Studies by Stucki and his co-workers over
the past 30 years (e.g. Stucki and Roth, 1976; Stucki *et al.*, 1984; Yan and
Stucki, 2000; Stucki *et al.*, 2002) have helped to explain 'why and how'
reduction of structural iron in smectites reduces swelling performance and
other properties. Reduction of octahedral Fe^{3+} to Fe^{2+} in smectites affects
(a) the interaction of H_2O molecules with the oxygen ions on the basal
surfaces of the mineral particles and (b) the vibrational energies of the Si–O
groups in the tetrahedral sheets. As swelling of smectites involve two types
of interlayers – fully expanding and partially/fully collapsed interlayers (Wu
et al., 1989) – reduction of octahedral Fe^{3+} results in more of the clay layers
collapsing than in the oxidized state. The end result is diminished water
uptake capability. One can conclude also that the water-holding capacity of
the reduced smectite clay will be diminished.

General types of bacteria known to reduce structural iron in clays include *Geobacter, Pseudomonas* and *Bacillus* (Stucki and Getty, 1986). A level of 41 per cent reduction of iron-rich smectite has been reported with metal-reducing bacterium *Shewanella putrefaciens* (Kosika *et al.*, 1996). Microorganisms are significant participants in catalysing redox reactions. Although mineral particles have a high affinity for enzymes, the effectiveness of their catalytic ability is a function of the nature and properties of the sorbing clay minerals (Huang *et al.*, 2005). Bonding between enzymes and clay mineral particles can occur through (a) hydrogen bonding; (b) primary interatomic bonding mechanisms, such as electrostatic, ionic and covalent; and (c) secondary bonding, such as van der Waals.

Biotransformation

Biotransformation of clay minerals occurs as a result of microbial activities under both aerobic and anaerobic conditions. The general three-step process involves dissolution of the original minerals, release of the various structural ions, and formation of new minerals. Figure 6.15 shows the results of interaction of microorganisms with primary and secondary minerals. In the figure, the products obtained are the result of biotransformation

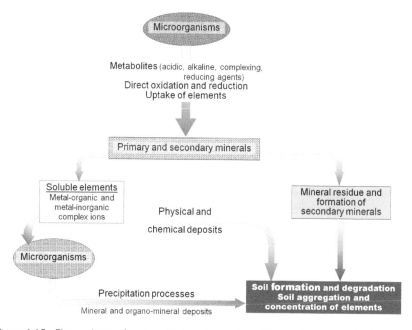

Figure 6.15 Flow chart showing biotransformation (bioweathering and alteration/ degradation) of soil minerals (adapted from Robert and Berthelin, 1986; Huang and Schnitzer, 1986).

processes that include *bioweathering* (biological weathering) and *alteration/degradation* processes. *Bioweathering* is defined as processes associated with biological modification of rates and mechanisms of chemical and physical weathering of minerals, and *alteration/degradation* refers to biologically mediated oxidation–reduction transformation of minerals. As it is difficult to physically determine which of these categories is responsible for the effects of biological transformation of clay minerals, we have included them in the overall category of biotransformation (Figure 6.15). This allows us to define *biotransformation* as the bioweathering and alteration/degradation of clay minerals.

Experiments conducted by Seabaugh *et al.* (2004) provide us with a good example of alteration/degradation of a clay mineral. Specifically, the anaerobic thermophilic bacterium *Bacillus infernus* was isolated from samples obtained from 2700 m below the terrestrial surface. The temperature and pressures existent on the samples before extraction were to some degree akin to those expected in deep geological environments used for containment of HLW canisters. The bacterium, which uses Fe(III) and Mn(IV) as electron acceptors, was used in experiments with a clay that had < 0.5-μm-particle-size fractions of dioctahedral smectite with traces of quartz. By using antraquinone-2,6-disulphonate as an electron shuttle to facilitate electron transfer, the experimental results obtained showed that within 14 h one obtained (a) an 18 per cent reduction of Fe(III) in the dioctahedral smectite; (b) a decrease in the crystalline thickness from 6 to 4 nm; (c) a change in size of crystals from lognormal distribution to asymptotic shape; (d) some bio-genetically produced minerals; and (e) preferential reduction of tetrahedral Fe(III) relative to octahedral Fe(III).

6.5   Mechanical properties of clays – M factor

6.5.1   High-swelling (smectites) and non-swelling clays

The behaviour of natural clays consisting of non-expandable minerals such as illite and chlorite conforms well to the principles articulated in classical soil mechanics, the basic concept of which is that the total stress in a soil element is the sum of the effective (grain or interparticle) pressure and the pore-water pressure. Although grain pressure is easily imagined for such clays, it is difficult to define for smectite clays, as there is normally no direct contact between smectite particles with their hydrate shells. This fact, together with the complex water pressure conditions in the interlamellar space, has raised questions concerning the validity of the effective stress concept (Yong and Warkentin, 1975). Nevertheless, the effective stress concept has found general acceptance and is used widely in the analysis of stability problems and for calculating compressive (consolidation) and expansive strain. Because of the particular characteristics of particle–water interaction in smectitic clay, shearing depends more on elapsed time than in the case of ordinary clays, and creep (defined as shear strain rate) is of fundamental importance because (a) it controls long-term deformation and (b) it may lead to accelerated strain and ultimate failure if the shear stresses are critically high.

In defining and determining creep rates it is practical to express the shear stress as a percentage of the conventionally determined undrained strength (1/4, 1/3, 1/2, etc.) derived from triaxial tests, unconfined compression tests or laboratory vane tests. The determination of porewater pressure is important in conventional soil mechanics theory: it has been proven to be useful in elucidating the effective stress concept and in determination of the strength or shear resistance of non-expansive clays and frictional soils. However, when it comes to high swelling clays, such as smectites, the role of porewater pressure in shearing and shear resistance is not fully understood. To a large extent, this is because the true porewater pressure cannot be accurately measured – especially in high-density smectitic clays. This is because of the impact of interlayer water and forces on the movement and pressure development of porewater in these kinds of clays (Yong and Warkentin, 1975). As to a HLW canister settlement in the borehole of the repository (Figure 2.6), a case that we will consider in detail, it is realized that it may take tens to hundreds of years for the clay buffer to be water saturated. This means that the creep strain of the unsaturated clay under external loading is responsible for the rate of settlement. Consolidation by extrusion of porewater is small during this time period. Instead, settlement will be controlled by shear stresses at the particle contacts and within them. An added factor to consider is the various transient chemical processes that will change the stress–strain properties of the clay buffer. This will make prediction of canister settlement a very uncertain proposition.

6.5.2 Shear resistance

The maximum shear resistance, the shear strength, determines the stability for the following cases (Figure 6.16):

- heavy canister with highly active radioactive waste (spent fuel) resting on smectite-rich 'buffer' clay (left diagram in Figure 6.16);
- clay slope (upper right part of figure);
- clay liner below the erosion-protected landfill (lower right in the figure).

The simplest illustrative case is the liner representing an infinitely long slope that can become unstable by slipping along its lower boundary. In the two-dimensional (2D) case, the stress conditions in 2D are homogeneous if the groundwater surface coincides with the slope surface. The required shear resistance τ_f for eliminating risk of failure is:

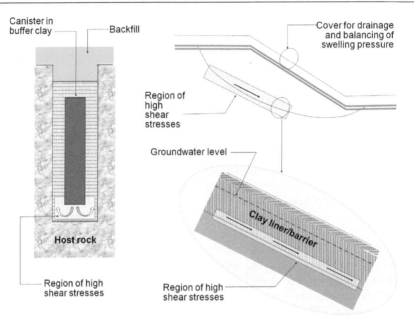

Figure 6.16   Cases of clay affected by shear stresses. Left, heavy canister embedded in clay; upper right, slope in clay fill; lower right, liner below erosion-protected landfill waste.

$$\tau_f = \rho H \left(\frac{n}{1+n^2} \right)$$

where ρ is the density of the clay, H is the vertical thickness of the clay layer, and $1/n$ is the inclination. As an example, for a slope angle $\beta = 17°$ ($n = 0.3$), $H = 1\,\text{m}$ and $\rho = 2000\,\text{kg/m}^3$, the shear strength at the slip condition is 6 kPa. Adding the load of overlying drainage layers and applying a reasonable safety factor, the required shear strength may be more than three times higher. Although this may well be achievable, the real problem is the risk of long-term creep leading to the development of large strains and an associated drop in shear resistance, which will ultimately lead to failure (Pusch *et al.*, 2009).

The slope shown in the upper right part of Figure 6.16 is a somewhat more complex case – especially if the involved soil mass has a short lateral extension ('calotte' slip surface). The stress conditions are inhomogeneous and the central part of the clay at some depth below the slope slip surface is exposed to the highest shear stresses. These can cause failure by sliding along a continuous slip surface. Long-term creep strain may reduce the shear strength where the shear stress is highest and mobilize more shear resistance in adjacent parts of the clay mass, resulting in further loss in shear resistance and causing failure after a sufficiently long time.

The most complex case is the heavy canister resting in a smectite clay buffer (left-hand part of Figure 6.16). Consolidation will occur if the clay is completely water saturated when the canister load is inserted, i.e. compression of the clay by expulsion of porewater. Following placement of the canister, the entire load will be carried by the porewater and the canister will settle until the porewater overpressure is dissipated – going by classical soil mechanics and geotechnical engineering theories. Under constant volume conditions, it is possible, from a theoretical point of view, for the canister to continue to sink further, as the shear stresses in the clay will result in generation of creep strain. Theoretically, the canister can even reach the bottom of the hole, and although this would mean that the isolating capacity of the buffer is largely lost, there is very little risk of sudden failure of the canister. As for the two other cases shown on the right-hand side of Figure 6.16, long-term creep strain may reduce the shear strength and cause large strain, leading to possible failure.

6.5.3 Stability

For the three basic cases shown in Figure 6.16, one must be able to determine the stability of the system, i.e. the safety factor with respect to development of large and uncontrolled slip. For the simple cases with uniformly distributed shear stresses, this factor is simply expressed as the ratio of the shear strength and the shear stress. For other more complex cases, such as the embedded canister and landfill slope cases shown in Figure 6.16, the distribution of the shear resistance and the shear resistance at large strain ('residual' strength), will determine their stability. As stated earlier (Chapter 3), as the simple Mohr–Coulomb failure model is not exactly appropriate for use in these situations, an extended version of the Drucker–Prager plasticity model is more commonly used. This model takes into account the influence of the intermediate principal stress and dilation (see section 3.10). However, this failure concept still relies on the validity of the effective stress concept, and some significant doubt remains in so far as its applicability to types of swelling clays used for clay buffer/barriers, so there remains some considerable uncertainty in stability evaluations and calculations. As the strain is intimately associated with shear resistance, the development of creep in the form of time-dependent shearing is of fundamental importance in the stability of the typical cases shown in Figure 6.16.

6.5.4 Creep strain mechanisms

General

The density, composition and microstructure of clays determine their shear resistance. As described in preceding chapters the microstructure of natural clays is characterized by aggregation, meaning that it consists of more or less dense peds (microstructural units) that are mutually coupled. For artificially prepared clay-engineered barriers in repositories housing radioactive or toxic chemical wastes, the aggregated nature of the smectites is clearly evident and important, as it determines the spectrum of microstructural energy barriers that have to be surmounted in the creep process – according to rate process theory.

If the energy required to activate slip units is constant, implying that all interparticle bonds are the same and equally spaced, and all slip units are activated, the strain rate is constant. The accumulated strain will then ultimately be so large that the coherence of the material is lost, thereby leading to failure. If the number of slip units or bonds that can be activated increases with the macroscopic strain, or if the strength of the bonds increases with the strain, the strain rate will drop. If the criterion of preserved coherence is not fulfilled on account of large strains, the strain rate will increase and ultimately cause failure, except if there is a geometrical restraint. Such a restraint is demonstrated in the case of a canister resting on the rock at the base of a deposition hole.

Definition of slip and creep

If slip along a plane through a cubic element with edge length L in Figure 6.17 takes place by an amount b, the resulting shear is b/L. If a slip unit, which can consist of a patch of hydrogen bonds or a physical contact between adjacent particles, does not traverse the whole element but produces displacement

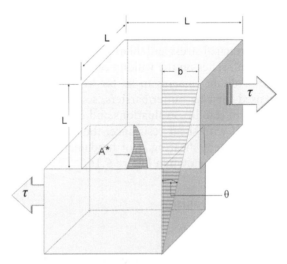

Figure 6.17 Soil element with internal displacement produced by a slip unit jump. τ, shear stress.

over a limited area A^*, it contributes to the bulk creep strain an amount A^*/L^2 times smaller than b/L (Pusch and Feltham, 1981), i.e.:

$$\delta_\gamma = \left(\frac{b}{L}\right)\left(\frac{A^*}{L^2}\right)$$

The key mechanism that produces bulk shear deformation is as follows: at a given point j in the material, slip is held up at an energy barrier u_j, which is determined by the intrinsic nature of the obstacle u_{j0} as well as by the local deviatoric stress σ_j acting on it at time t (Feltham, 1979). One can therefore represent the stress and time dependence of u_j by:

$$u_j = u_j\,(u_{j0},\,\sigma_j);\; \sigma_j = \sigma_j(t) \tag{6.25}$$

As not all u_{j0} are of the same type, equation 6.25 implies that the energy barriers form a spectrum that may vary with t, i.e. time from the instant of the application of a shear stress to the material. For cases where the number of slip units or bonds that can be activated increases with the macroscopic strain, or where the strength of the bonds increases with the strain, the frequency of jumps is assumed to be given by Arrhenius's rate equation: $t_A^{-1} = v\,e^{-u/kT}$, where u is the barrier height and v a lattice frequency in the interval $10^{10}–10^{11}$/s. The term t_A is *dwell time*.

The energy spectrum

The spectrum of interparticle bond strengths represents a first-order variation in barrier heights, a second one being represented by the interaction of differently sized aggregates of discrete particles. The peds behave as strong units that remain intact for small strains but yield at large strains, and contribute to the bulk strength by generating dilatancy. This means that the energy spectrum is not a material constant but changes with strain, and hence with time. The practical importance of this is that the microstructural network may stay relatively unchanged for low and moderate bulk shear stresses, whereas higher shear stresses can cause irreparable changes of the network leading to bulk failure at a certain critical strain.

In principle, the response of the structure to the shear stress is that the overall deformation of the entire network of particles is accompanied by translations and rotations of some of the microstructural units and larger non-clay particles, such as silt grains. In turn, this entails a breakdown of dense microstructural units and transforms the clay to a laminated structure of flaky particles (Figure 6.18).

Figure 6.18   Example of a sheared clay rich in montmorillonite (MX-80 clay). The shear direction is indicated by arrows. This is a micrograph of an ultramicrotome-sectioned acrylated-embedded sample (300 Å), using a transmission electron microscope (magnification x10,000). The dry density of the clay is about 1300 kg/m³).

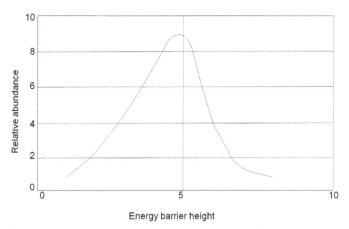

Figure 6.19   Energy barrier spectrum (adapted from Feltham, 1979).

Shearing takes place through thermally aided activation of barriers producing slip on the microstructural scale. The barriers form a spectrum of the type shown in Figure 6.19.

Creep model

At a given temperature the u-spectrum in Figure 6.19 is characterized by a distribution of barrier heights within a defined interval: $u_1 < u < u_2$. Following Feltham (1979), the number $n(u,t)$ of potential slip units per unit volume that are held up at time t at barriers of height u forms the energy spectrum that can be subdivided into equal intervals δu. In terms of slip units, we will have $\delta n(u,t) = n(u,t)\delta u$, where δu is an increment per unit energy.

By considering consecutive activation energy intervals the thermally aided creep induced by the shear stress can be explained in terms of in- and out-flow of slip units. For a case that is considered to be stable, slip is held up at successively higher barriers, meaning that the creep is successively retarded. One can show that it is proportional to $1/(t + t_0)$.

The slip units may alter their geometrical configuration as a result of displacement, and they may also find themselves in modified internal stress fields. One can identify the following slip units and activation energies (Pusch and Yong, 2006):

- Domains, i.e. microstructural elements consisting of expanded stacks of particles (lamellae in smectites). The activation energy for slip consists of weak hydrogen bonds (0.1–0.5 eV). The strain required to mobilize these bonds is very small, i.e. a few angstroms.
- Domains consisting of dense stacks of particles. The activation energy for slip consists of strong hydrogen bonds (0.5–0.9 eV). The strain required to mobilize these bonds is probably less than for weaker H-bonds.
- Contacts between mechanically stable particle aggregates and non-clay particles. The activation energy for slip consists of primary valence bonds (2–4 eV). The lower value may be valid for soft clays (smectite with bulk dry densities lower than 1700 kg/m³), whereas the higher value may be valid for clays with densities exceeding 1700 kg/m³ including dilatancy effects. The activation energy is assumed to be an exponential function of the density. The strain required to mobilize these bonds is relatively large, i.e. a few hundredths to a few tenths of a micrometre.

Considering the case of a clay element subjected to a constant deviator stress, one can assume that the number of energy barriers of height u is $n(u,t)$ δu where δu is the energy interval between successive jumps of a unit, and t is the time. The entire process is stochastic. The change in activation energy in the course of evolution of strain means that the number of slip units is determined by the 'outflux' from any u-level into the adjacent, higher energy interval and by a simultaneous inflow into the interval from u-δu (Pusch and Adey, 1986).

Each element of clay contains a certain number of slip units in a given interval of the activation energy range. Displacement of such a unit occurs as the shifting of a patch of atoms or molecules along a geometrical slip plane. In the course of the creep process, the low energy barriers are triggered early and new slip units are triggered at the lower energy end of the spectrum in Figure 6.19. This end represents a *generating barrier*, whereas the high *u*-end is an *absorbing barrier*. A changed deviator stress affects the rate of shift of the energy spectrum only to higher *u* values provided that the shearing process does not significantly reduce the number of slip units. This is the case if the bulk shear stress does not exceed a certain critical value – of the order of one-third the conventionally determined bulk strength. It implies that the microstructural constitution remains unchanged and that bulk strain corresponds to the very small integrated slips along interparticle contacts. In principle, this can be termed *primary creep*.

For low shear stresses, allowing for 'uphill' rather than 'downhill' jumps, the rate of change of *n(u,t)* with time is obtained as:

$$\frac{\delta n(u,t)}{\delta t} = v\left[-n(u+\delta u)e^{\frac{-u+\delta u}{\kappa T}} + n(u,t)e^{\frac{-u}{\kappa T}}\right] \tag{6.26}$$

where δu is the width of an energy spectrum interval, v is the vibrational frequency (about 10^{11}/s), t is time, κ is Boltzmann's constant, and T is the absolute temperature. Coupling this relationship with Feltham's transition probability parameter to describe the time-dependent energy shifts, and assuming that each transition of a slip unit between consecutive barriers gives the same contribution to the bulk strain, one obtains, for a boundary condition of $t < t_0$, the bulk shear strain rate as:

$$\frac{d\gamma}{dt} = B\left(1 - \frac{t}{t_0}\right)$$

As the constant B and the value of t_0 depend on the deviator stress, temperature and structural details of the slip process, the creep can be expressed as $\gamma = \alpha t - \beta t^2$ for $t < a/2\beta$. This means that creep initially varies linearly with time before dissipating to zero. An example of this is given in the creep diagram (Figure 6.20) where a rapid drop in strain rate occurs after a couple of months.

For higher bulk loads, microstructural strain causes some irreversible changes associated with local breakdown and reorganization of structural units. However, repair by inflow of new low-energy barriers will cause strain retardation – due also to confrontation with higher energy barriers. This type of creep can go on forever without approaching failure. Following Feltham (1979), the process of simultaneous generation of new barriers and

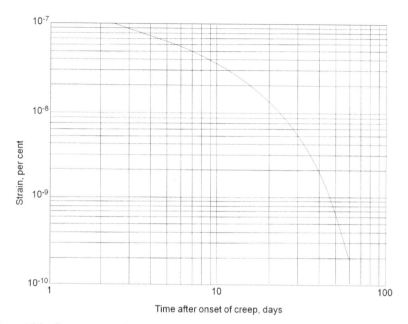

Figure 6.20 Creep strain of smectite-rich clay (MX-80) with a density at saturation with distilled water of 1500 kg/m³. Average shear stress = 6 kPa.

migration within the transient energy spectrum lead to the expression for the creep shear rate as:

$$\frac{d\gamma}{dt} = A^* \int_{u1}^{u2} n(u)v(u,t)\,du = A^* \int_{u1}^{u2} n(u,t)e^{\frac{-u}{\kappa T}}\,du \qquad (6.27)$$

where A^* is the contribution of each jump to the shear strain of the specimen. This relationship implies that the lower end of the energy spectrum refers mainly to breakage of weak bonds and establishment of new bonds (where stress relaxation has taken place owing to stress transfer from overloaded parts of the microstructural networks to stronger parts), whereas the higher barriers are located in more rigid components of the structure.

The model implies that for thermodynamically appropriately defined limits of the u-spectrum, the strain rate relating to logarithmic creep is expressed as:

$$\frac{d\gamma}{dt} = BT\left(1 - \frac{t}{t_0}\right)$$

where B is a function of the shear stress t, and t_0 is a constant of integration. This leads to a creep relationship that closely represents the commonly observed logarithmic type of creep, i.e. creep strain is proportional to $\log(t + t_0)$. The significance of t_0 is realized by considering that in the course of applying a deviatoric stress at the onset of the creep test, the deviator stress rises from zero to its nominal final value. A u-distribution exists at $t = 0$, i.e. immediately after full load is reached. This may be regarded as equivalent to one which would have evolved in the material initially free from slip units, had creep taken place for a time t_0 before loading. Thus, t_0, which can be evaluated from creep experiments (Figure 6.21), is characteristic of the structure of the prestrained material. This model implies that, for moderate deviator stresses that allow for microstructural recovery, there is successive retardation of the creep rate according to the logarithmic time law without any risk of failure even after a very long period of time.

A further increase in the deviator stress leads to what is conventionally termed as *secondary creep* – a condition where the creep rate tends to be constant. This condition results in a creep strain that is proportional to elapsed time. In contrast to what is typical for lower stresses, one can imagine that critically high rate creep makes microstructural self-repair impossible. Comprehensive slip changes the structure without allowing reorganization to occur, thus leading to ultimate failure.

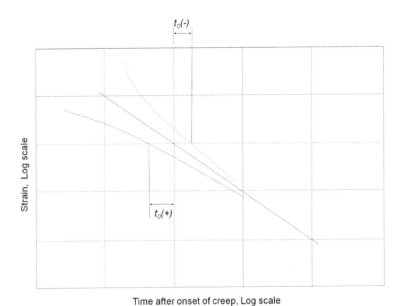

Figure 6.21   Generalization of creep curves of log-time type.

6.5.5   Experimental

A first experimental example involves creep tests of a smectitic clay of Triassic age from northwestern Lithuania. This is a primary candidate clay for a planned Lithuanian repository for low-level and short-lived intermediate-level radioactive wastes. The smectitic clay has an average liquid limit of 55 per cent and a plastic limit of 22 per cent. The total 30 per cent clay content ($< 2 \mu m$) of the dispersed clay consists of 20 per cent smectite content, with illite and chlorite making up about 10 per cent. The remaining 70 per cent of the clay consists of quartz, feldspars and mica grains. To determine the shear strength and creep properties of the clay, a double-ring shear box was used. This was designed to accommodate a large sample. It consisted of two stationary shear box halves separated by a ring that was loaded as indicated in Figure 6.22.

The general shear stress conditions in the sheared sample at failure can be estimated as the load F divided by the total shear plane area, i.e. $2\pi D^2/4$, where D is the diameter of the shear box. However, the shear stress distribution at lower stress levels must be obtained from numerical techniques. In utilizing the ABAQUS code (see Chapter 9) for the calculations, the following properties were used: shear modulus $G = 1.5$ MPa, Young's modulus $E = 4.44$ MPa, Poisson's ratio $\nu = 0.48$, the yield shear stress $\tau_f = 15$ kPa, the angle of internal friction φ and the friction between clay and ring were 0,

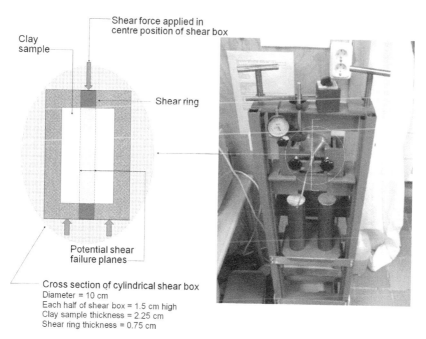

Figure 6.22   Ring shear box with dead load and no friction in the load arrangement.

and the dilatation angle was taken to be 0. In the current model, the yield concept was incorporated by using a Drucker–Prager model in the material plastic property definition process. The formula:

$$d = \frac{\sqrt{3}}{2}\tau\left(1+\frac{1}{k}\right)$$

where d = displacement, τ = shear stress, and k = constant, is used to calculate cohesion with regard to shear strength and flow stress ratio. For a 1-mm ring displacement (1 per cent) the theoretical contact force between ring and clay would be 50 N. For 3- and 5-mm displacements (3 and 5 per cent) it would be 53 N and 52 N, respectively. Representative average shear stresses can be derived from the expression: $F/2 = [A_p\tau_{crit} + (A_{tot}-A_p)\tau_{av})]$, where F is the ring load, A_p is the area of the plasticized part of the whole cross-section area A_{tot} and τ_{av} is the average shear stress of the non-plasticized cross-section area. Figure 6.23 shows the calculated shear stress distribution for 5 per cent uniaxial strain.

The calculations show that the average shear stress in the larger part of the clay sample is not significantly affected by the shear strain. This suggests

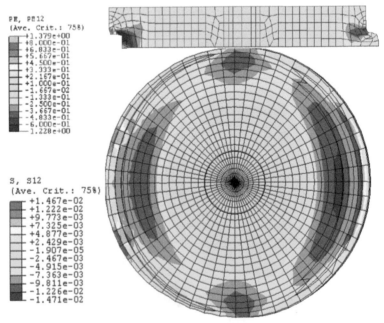

Figure 6.23 Theoretical shear stress distribution at 5 per cent uniaxial strain (movement left to right). Maximum shear stress is the critical value for plasticization i.e. 0.01471 MPa, minimum is 0.0048 MPa. Upper image, cross-section; lower image, normal view.

that the clay behaves more as a viscous medium than as an elastoplastic substance, in agreement with the fact that this clay, as well as all other artificially prepared clays, has no preconsolidation pressure and no definite shear strength – only a shear resistance.

Actual ring shear testing was performed with clay with a density of $1800\,kg/m^3$ at complete water saturation. The laboratory-prepared clay was made with clay powder and compacted in the ring shear box with filters at the bottom of each box half. Two 200-cm^3 samples were consolidated to a density of $1750\,kg/m^3$ (dry density $1375\,kg/m^3$) and $1940\,kg/m^3$ (dry density $1526\,kg/m^3$). These represent the density span of practical interest in the design of clay barrier caps for waste piles.

For the softest sample, the shearing load steps corresponded to average shear stresses of 7, 11 and 14 kPa. For the densest sample ($1940\,kg/m^3$ at water saturation), the average shear stresses were increased in 3- to 4-kPa steps from 7 to 33 kPa, and subsequently to 39 kPa and 65 kPa. The shear strain was recorded with an accuracy of 0.001 mm. Step loading was increased when the creep rate had dropped to less than 8 per cent per second. Figure 6.24 shows the recorded strain of the softest sample at the 14-kPa average shear stress. It shows a strain rate approaching log-time creep. By increasing the shear stress to 17 kPa the shear strain rate tended to become constant. This would indicate a risk of ultimate failure.

Figure 6.25 shows the creep curves of the denser sample ($1940\,kg/m^3$). The graphical results show that for the two lowest stress levels, the curves approach the log-time condition after 20,000–80,000 s, indicating that the

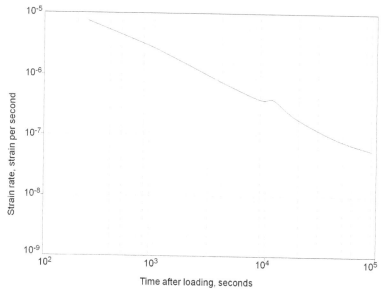

Figure 6.24 Creep behaviour of the softest sample at the average shear stress of 14 kPa.

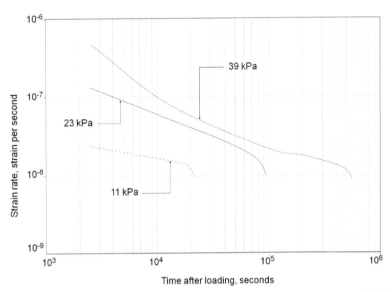

Figure 6.25 Creep curves of the denser sample for the average shear stresses of 11, 23 and 39 kPa.

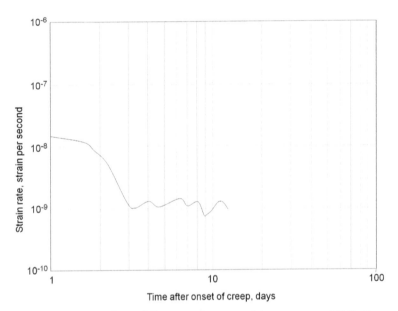

Figure 6.26 Creep of soft clay following an increase of shear stress to 23 kPa. The creep rate attenuated and became a constant, relatively low strain rate without resulting in failure in the 2-week-long test performed at room temperature.

strain will approach zero after a few weeks. For the highest stress level the same tendency appears. However, the results show that a longer testing time of at least a few months is needed before it is possible to determine whether the creep state is a primary or secondary creep state.

The evolution of strain rate evolution shown in Figure 6.26, i.e. a nearly constant strain rate, was recorded after the average shear stress was increased to a level that was expected to yield instant failure of the soft clay. The exhibited phenomenon is believed to be a strain-hardening effect caused by successive dispersion of the initially dense particle aggregates. These formed an anisotropic matrix of aligned and partly disintegrated stacks of lamellae oriented in the direction of the maximum shear stress. The result is an increasing specific surface area (SSA) with most of the shear resistance provided by a steadily increasing, but ultimately constant, number of quickly formed and broken hydrogen bonds. This suggests that smectite clays may not undergo failure like illitic clays, i.e. by slip along discrete failure planes, but undergo zone failure. Accordingly, one would expect such clays to exhibit properties similar to those of high-viscous fluids.

The experimental results show that the stress conditions in a ring shear box with large diameter are uniform in a large part of the sample, and are in agreement with those predicted by numerical calculations. The results also validated the applied creep theory, implying that the strain rate drops quickly at low shear stresses and follows the log time strain rate rule at intermediate shear stresses. The tendency of the strain rate to become constant at higher stresses shows that artificially prepared smectite clay behaves as a viscous substance. It provides shear resistance and has no preconsolidation pressure.

6.5.6 Calculation of creep strain and rate

In practice one commonly uses empirical expressions of the creep rate such as equation 6.28 (Börgesson, 1990):

$$\dot{\gamma} = \dot{\gamma}_0 \cdot e^{\alpha \frac{(\sigma_1-\sigma_3)}{(\sigma_1-\sigma_3)_f}} \cdot e^{-\alpha \frac{(\sigma_1-\sigma_3)_0}{(\sigma_1-\sigma_3)_f}} \left(\frac{t}{t_r}\right)^{-n} \tag{6.28}$$

where t is the time after stress change, t_r is the reference time (10^5 s), $(\sigma_1-\sigma_3)_0$ is the reference deviator stress [$0.5(\sigma_1-\sigma_3)$], $(\sigma_1-\sigma_3)_f$ is the deviator stress at failure, $\dot{\gamma}$ is the creep rate, $\dot{\gamma}_0$ is the creep rate at time t_r, and n and α are parameters derived from laboratory tests.

The shear strength q can be expressed as a function of the mean effective stress p as $q = ap^b$, where $a = q$ for $p = 1$ kPa and b is the inclination of curve in log$p$/log$q$ diagrams (see section 3.10). Limited experimental triaxial tests have obtained a-values of between 2.8 and 5.5 depending on the porewater chemistry and type of adsorbed cation. The shear strength relationships are empirical and are not based on any microstructural concept associated with a constant b value ($b = 0.77$). To derive reliable theoretical expressions of creep and creep rate as functions of stress and clay density, it is necessary to identify the real mechanisms involved in the creep process. This will be highlighted in the next section when we examine certain applications.

6.5.7   Applications

Canister settlement

The results from an application of the boundary element code BEASY to predict the settlement of HLW canisters with 25 T weight in smectite-rich buffer clay (MX-80) have given a log-time evolution leading to a settlement of about 10 mm in one million years (Pusch and Adey, 1986).

Design of a surface repository for hazardous waste

For this practical example of the mechanical performance of smectites, we will examine the long-term creep of smectitic clays used to contain short-lived intermediate-level radioactive waste (Figure 6.27). The clay plays the same role as the clay core in an earth dam, i.e. to minimize water flow through the construction, which is a repository for low-level and short-lived intermediate-level radioactive waste. The repository consists of concrete

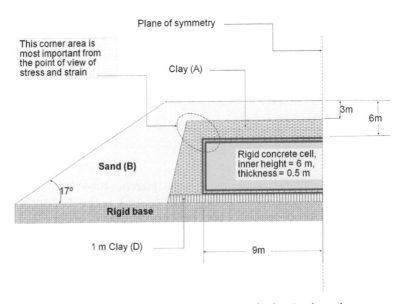

Figure 6.27 Concept of repository with concrete vaults hosting hazardous waste and surrounded by smectitic clay for retarding water inflow.

Table 6.1 Barrier components from top

Material	Layer thickness (m)	Hydraulic conductivity (m/s)
Boulder/pebble (50–500 mm) over sandy gravel and silty sand	3	$>10^{-3}$
Clay in top liner	≤ 1.5	10^{-10}
Moraine (filter function)	≥ 0.2	10^{-8} to 10^{-6}
Sand (gas pressure distributor)	0.2	10^{-5}
Concrete vaults	Wall and roof thickness ≥ 0.5 m	10^{-10}
Clay (10–15%) bottom bed	1.5	10^{-10}

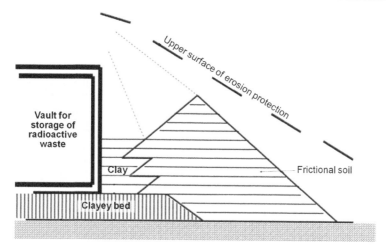

Figure 6.28   Diagram showing how parallel placement and compaction of clay close to the concrete vaults and supporting frictional material can be made.

vaults resting on a low-permeable foundation bed and surrounded by on-site compacted smectitic clay. The layers above 'clay A' in the figure serve to protect it from erosion and freezing. The composition of the various barriers is listed in Table 6.1. The construction of the clay embedment of the concrete vaults is of the type used in earth dam construction, i.e. parallel placement and compaction of the clayey 'core' and the outer, supporting and erosion-protecting filter (Figure 6.28).

Geotechnical performance

The geotechnical performance of the engineered multi-barrier systems is of significance because they have to remain sufficiently homogeneous and coherent to be able to maintain their isolating capacity for a long time – for as much as 300 years and more, depending on the type of waste to be contained. The most important criteria are (a) no critical strain (shear failure or fracturing) and (b) no significant drop in density of the clay component.

The stress/strain situation of the clay embankment is similar to that of the clay core of an earth dam, i.e. the clay located adjacent to the concrete vaults is supported and maintained in position by the mass termed 'sand B'. This exerts an earth pressure that ranges from K_o pressure ('earth pressure at rest') to passive earth pressure, which is equal to the swelling pressure of the smectitic clay. For a swelling pressure of 100–200 kPa the whole system is estimated to be stable and in equilibrium. It is in fact self-adjusting, as any tendency of the clay to displace the surrounding mass causes a drop in clay density and thereby in swelling pressure, and vice versa.

Calculation using numerical methods

The stress/strain performance of the system has been analysed by elastic finite element method (FEM) analyses using the code ANSYS, ascribing the properties in Table 6.2 to the simplified system in Figure 6.27. The system was considered to be linearly elastic with the concrete vaults and clays A and D on site before sand B was placed – resulting in generating the stress and strain conditions to be determined. Sand B load is represented by a uniformly distributed vertical load of 100 kPa in the calculations. The dotted area, the 'knee' in Figure 6.27, is most important, as this is where stress concentrations and horizontal tension are expected.

Figure 6.29 shows the outcome of FEM calculations of the horizontal strain in the system. The strain in the upper part of the clay, near the upper edge of the vaults, is most important as it can lead to loss of coherence in the clay if this leads to significant tension. The bottom clay layers will be exposed to a vertical pressure of about 200 kPa, which is balanced by the swelling pressure of the clay material for a dry density of around 1700 kg/m³. This means that the settlement of the vaults by compression of these layers

Table 6.2 Material data

Component	Thickness (m)	E-modulus (kPa)	Poisson's ratio	Density (kg/ m³)
Clay (A)	1.5	10,000	0.3	2000
Sand (B)	2.8	Uniformly distributed load (20,000)	0	2000
Sand (C)	0.2	50,000	0.2	2000
Clay (D)	1.0	30,000	0.3	2500
The walls of the concrete vaults (6 m high) are 0.5 m thick; the roof is 0.5 m thick and 9 m wide		200,000	0.3	2500

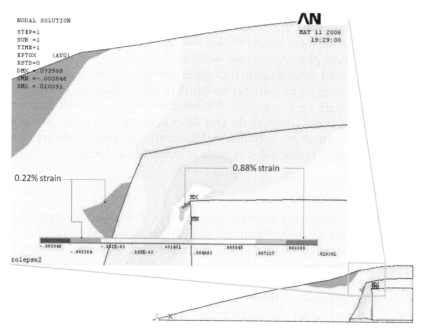

Figure 6.29   Horizontal strain ranges from compression up to about 0.3 per cent (left-end dark shading in the shading bar) to tension by up to 0.8 per cent (right-end shading). The soil supporting clay A embedding the vaults undergoes very little strain. The top 'blow-up' shows horizontal strain ranging from compression by about 0.2 per cent to tension by 0.8 per cent. The clay A adjacent to the concrete vaults undergoes some slight lateral expansion (around 0.2 per cent), except for the part at the upper edge of the vault where the horizontal strain is near 0.8 per cent. At this point, there is a slight drop in density.

will be insignificant. A number of degrading processes can occur in top liners that are related to the microstructural constitution of the liners: freezing, desiccation, erosion, chemical impact and liquefaction.

Freezing

Freezing of porewater in ordinary clays takes place at around 0°C or somewhat lower depending on the thermal properties of the clay–water system, and on the chemistry of the porewater. Frost heaving and formation of ice lenses occur in soils if they are frost susceptible and if water for ice lens growth is available. The frost heaving pressures created at the ice lens interface with the overlying soil layer can be theoretically calculated (Pusch and Yong, 2006). Experimental frost heaving tests conducted for fine inorganic silts show that the predictions accord well with measured values of heaving pressures. A certain amount of local consolidation or compression of the unfrozen material ahead of the freezing front occurs, the extent of which is a function of the water available for ice lens growth and the degree of heave experienced in the layer behind the ice lens. A most important consequence of the frost heave phenomenon for swelling clays is that these clays will not retain their initial microstructural constitution after thawing; they will be heterogeneous and much more permeable than before freezing. To avoid this situation, the clays need to be buried below the frost line. In most instances, in temperate climates, locating the clay layer in the top cover of the repository at about a 4 m depth is sufficient. However, one needs to always check temperature and freezing index records at potential site locations.

Desiccation

Desiccation of a clay top liner can cause cracks if the water content drops below the shrinkage limit (Pusch and Kihl, 2004). It is the breakpoint of the relationship between water content and volume of a drying clay sample, representing the water content below which no further change in volume takes place, i.e. when capillary forces are at maximum. Desiccation cracks can be filled by pervious soil material falling from the overburden, primarily from the adjacent filter that contains silt in its lowest part. These will serve as conduits to the top liner. If the clay layer is effectively compacted at a low water contents and located at about 4 m depth below the ground surface, where the temperature is relatively low and the clay moisture largely preserved, no continuous desiccation cracks are expected.

Piping and erosion

Physical disturbance by piping and erosion may cause transport of fine particles within and emanating from clay-based top liners, resulting thereby in a more permeable liner (Pusch and Kihl, 2004). The risk of such degradation, which depends on the flow rate and hence on the hydraulic gradient, is high for liners with low clay contents such as the Triassic clay. On the other hand, the risk is small for soils with moraine-type gradation, which is valid in principle for the type of clays used for buffer/barriers. Hence, erosion by percolating water is considered not to be a problem. The risk of loss of particles downwards requires that the clay be placed on a bed of well compacted silt with a void size smaller than about 100 μm.

Chemical impact

Top liners will normally be saturated with low-electrolyte rain water, and are not expected to undergo significant mineralogical changes for many hundreds or thousands of years (Pusch and Kihl, 2004). The exception would be if the liner and covering layers contain sulphide minerals and if precipitation is in the form of acid rain. Both would cause a drop in pH of the porewater. However, the chemical buffering capacity of the thick overburden is estimated to be sufficient to maintain a pH value in the range of 6–8.

A most important issue is the chemical interaction of clay and concrete in the vaults. Ordinary Portland cement possesses a very high pH. This is known to degrade smectite clay when it comes into contact with the cement. Theoretical considerations and studies using batch tests with KOH/NaOH/ $Ca(OH)_2$ solutions at 90°C have indicated that the reaction produces zeolites, such as phillipsite and analcime, and that they would be formed in a few months. Interstratified smectite–illite (S/I) with up to 15–20 per cent illite was found when the solution contained much K^+. Uptake of Mg^{2+} in the montmorillonite crystal lattice resulted in the smectite species saponite. These reactions are believed to occur also at room temperature but at a very low rate and they are believed to have the preceding impacts (Pusch and Yong, 2006).

Liquefaction

Liquefaction may be a serious problem for any water-saturated landfill clay of low density. Seismically induced shearing will cause contraction and development of a porewater overpressure that can reduce the effective pressure to a critically low level, particularly in slopes. For densities exceeding about $1800\,kg/m^3$, this is not a problem as long as the expected seismic events represent values on the Richter scale lower than 6 (Pusch, 2000).

Constructability

The density of a clay determines its hydraulic conductivity and swelling pressure. Pilot field tests must be conducted to determine the compactability of a clay material and to ascertain that the desired density can be reached. Samples taken from a compacted clay must be investigated to determine the important physical properties as functions of density. Constructability should be determined in relation to the undrained shear strength of the freshly placed clay material with its natural water content. These tests must conform to national regulations. Test equipment and protocols should be designed to accommodate volumes of clay samples that are not too small. An example of equipment used is the shear box apparatus type PPG-PSH of Russian origin; it has a volume of $140\,cm^3$ and a diameter of 7.14 cm and is 3 cm in height. Undrained quick shear tests can be performed by using shear boxes and the German standard DIN 18137 or similar standards. The normal pressures applied are 100, 200 and 300 kPa. Shear stresses are applied in 5-kPa steps. The shearing continues until the total shear strain was 5 mm, the duration of each shear test being 5 min.

6.6   Stress relaxation and strength regain

Tectonically induced shearing along a fracture in the rock can occur resulting in induced shear in the buffer clay surrounding an embedded HLW canister. A question of great practical importance is whether sheared 'buffer' clay surrounding the HLW canister will self-heal and regain its original strength and stiffness after termination of shearing. Assuming that the shearing terminates when the host rock strain has stopped, the strain-induced shear stresses in the clay will drop with time in accord with a time law that is related to creep. Theoretically, one would expect a log-time increase of the value resulting from the shearing. For ordinary natural clays with a sensitivity (ratio of undrained shear strength of undisturbed clay to the undrained shear strength after remolding) of 2–5, the re-established strength would be 20–50 per cent of the original undisturbed shear strength. Consolidation under prevailing external pressure could increase this strength.

A collapsible non-smectitic clay, such as a 'quick-clay', will generally have a very high sensitivity – usually more than 50. This means that it turns liquid on shearing. On resting under constant volume conditions the shear strength increases with time. This thixotropic hardening is largely due to microstructural reorganization as concluded from shear strength measurements of a quick illitic clay from the Swedish west coast. Figure 6.30 shows the strength regain as measured by use of laboratory vane boring for a marine clay with

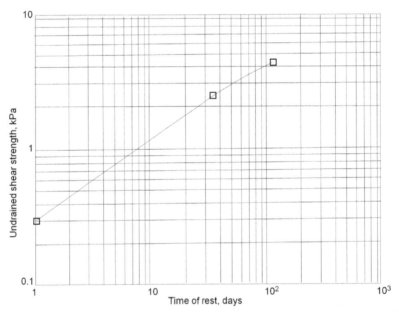

Figure 6.30   Strength regain of a quick clay. The undisturbed undrained shear strength, which was determined by laboratory vane testing, was 24 kPa (data from Pusch, 1983).

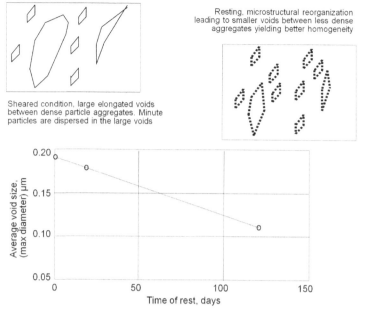

Figure 6.31   Microstructural reorganization of sheared quick clay. The upper images show the general principle of microstructural transformation of dense aggregates with sharp boundaries to more homogenized state with expanded aggregates and smaller voids, whereas the lower image shows the time-dependent drop in void size deduced from transmission electron micrographs (data from Pusch, 1983).

a density of $1700\,kg/m^3$, a water content of 58 per cent and a liquid limit of 46 per cent. This clay was percolated by fresh water for thousands of years after deposition. Figure 6.31 shows the change in void size.

With respect to changes in the macrostructure and microstructure of the clay, it is not difficult to imagine that large shear strains will cause breakage of numerous interparticle bonds. The result of bond breakage will be a dispersion and release of particles from the initially dense aggregate groups to form gels. As a result, there will be a reduction in the size of the initially large voids as illustrated in Figure 6.31. The evolution of the macrostructure and microstructure of the clay has been validated by the microstructural analysis – as shown in the bottom part of the figure.

The denser the clay particle assemblies (microstructural units), the less mobile are the water molecules in the interstitial space. It takes about 2.3 s for free water, after momentary reorientation of the molecules by exposure to a very strong magnetic field, to return to the original water matrix (Pusch and Yong, 2006). For water molecules interacting with adjacent clay lattices the return to the original state is much quicker. In the study of the thixotropic strength regain of the quick clay by Jacobsson and Pusch

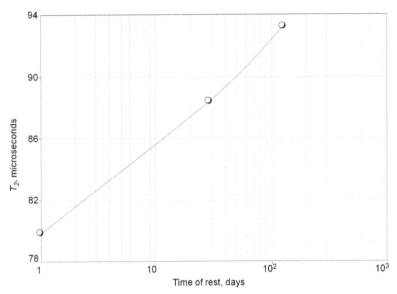

Figure 6.32 Spin–spin coherence time T_2 of the sheared quick clay at room temperature, indicating successive loss in organization of water molecules because of particle dispersion (data from Jacobsson and Pusch, 1972).

(1972), impulse spectrometrometry was used to obtain measurements of the water molecule mobility by determining the spin–spin coherence time T_2. Although the content of paramagnetic elements, such as iron, can affect conclusions concerning proton relaxation, comparison of T_2 of the same material at different water contents is still relevant. The increase in T_2 with time according to Figure 6.32 indicates that the average state of ordering of the water lattices decreased with time. This is compatible with the idea of softened aggregates.

Smectite-rich clay, which has a large part of the porewater located in the interlamellar space at high densities and a SSA that can be 10 times higher than for illitic clay, would be expected to show significantly lower T_2 values for similar densities. This is also the case as indicated by Figure 6.33, which shows that for a water content of 58 per cent, i.e. that of the illitic quick clay, smectite-rich clay has a T_2 value that is about 10 per cent lower than the value for the illitic clay. The 10 per cent lower T_2 value for 10°C in comparison with the 25°C T_2 value demonstrates that the rigidity of the water lattices is higher for the lower temperature. However, considering the large difference in SSAs between illitic and smectitic clays, the difference between the T_2 values for these clays is rather small. The probable explanation for this could be that the fraction of interlamellar water begins to drop when the water content exceeds about 42 per cent (Table 6.3). Figure 6.34 illustrates the microstructural constitution of artificially prepared smectite-rich clay

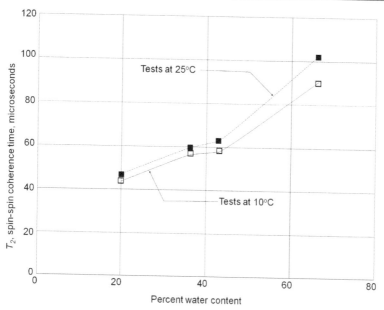

Figure 6.33 Measured spin–spin coherence time T_2 of undisturbed smectite-rich clay (MX-80) (data from Pusch, 1983).

Table 6.3 Fraction of interlamellar, organized water of the total porewater content[a]

Density at water saturation (kg/m³)	Water content (%)	Fraction of interlamellar water in sodium montmorillonite	Fraction of interlamellar water in calcium montmorillonite
2130/1800	18	93	93
1850/1350	37	88	62
1795/1265	42	85	53
1700/1110	53	72	40
1570/900	74	48	27

Note

a The maximum number of interlamellar hydrates is taken as three in sodium montmorillonite and two in calcium montmorillonite (Pusch and Yong, 2006).

(MX-80), characterized by considerable homogeneity but still with voids of various sizes.

The importance of the increased spacing of adjacent stacks of smectite lamellae that is assumed to take place in the homogenization process, has been documented by Ichikawa *et al.* (1999), as seen in Figure 6.35. This figure shows that the viscosity, which is a measure of the strength of the water lattices, drops from 0.01 to 0.003 Pa/s when the distance from the montmorillonite surface increases from 3 to 10Å, representing one and three water molecule diameters respectively.

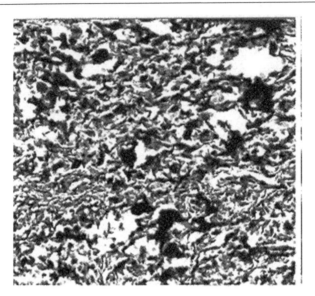

Figure 6.34 Example of digitized micrograph of Wyoming bentonite (MX-80) with a bulk density at saturation of 1800 kg/m³ (dry density 1270 kg/m³). The darkest parts are the densest parts of the clay matrix. For the various shades of grey, the density variation of the clay matrix is in the same order as the variation of grey colours, ranging from darkest grey for the denser portion to lightest grey the least dense clay. The white portion represents open parts of the matrix. Edge length is 3 μm.

6.6.1 *Rheological properties of sheared buffer clay*

The sheared buffer clay zone in the case of tectonically induced shearing along a fracture in the rock will undergo thixotropic stiffening through a reorganization of particles consisting of released stacks of smectite lamellae similar to what takes place in a *quick* clay. In addition, the much larger SSA of the smectite-rich buffer clay is expected to have a strengthening effect due to the establishment of new integrated water lattices. The combined processes would significantly increase the bulk shear strength and reduce the strain rate under constant shear stresses. The bulk shear strength increase is demonstrated in the initial shear period in Figure 6.36 for a clay with about 25 per cent smectite (montmorillonite). This figure shows the creep rate at repeated application of the same shear stress 19 kPa after 3, 5, 22 and 65 days after shear failure. The strain rate dropped significantly after 5 days of rest under no external pressure. However, a third loading after 22 days of rest gave nearly the same strain behaviour as the original shearing (0 days) and a fourth loading after another 65 days of rest gave instant failure in the form of quick shearing at a constant strain rate.

The reason for the stronger retardation of creep after 22 days of rest than after the initial loading (0 days) can be explained by more comprehensive

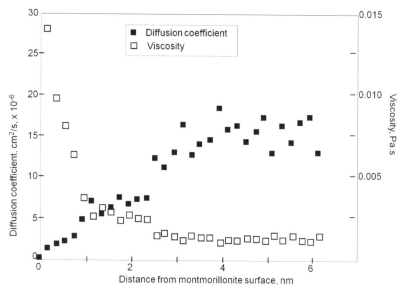

Figure 6.35 Diffusion coefficient and viscosity of montmorillonite calculated using molecular dynamics (data from Ichikawa *et al.*, 1999).

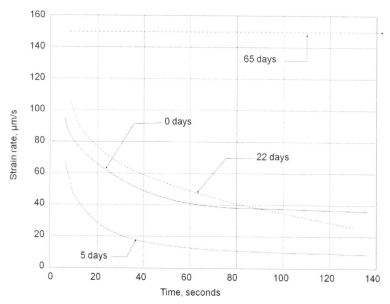

Figure 6.36 Strain rate at shearing of smectite clay with a density at saturation of 1700 kg/m³ at different times after shear failure. Repeated shearing after 5, 22 and 65 days of rest gave initial self-repair but accumulated strain finally led to failure.

structural reorganization as suggested by the large total strain in the direction of shear, 8750 µm, of the 100-mm-diameter sample. The much larger strain rate after 22 days in comparison with that obtained after 5 days indicates that a critical state was approached. This was also validated by the quick shear failure followed by strain that is typical of dense fluids – as in the earlier described creep test (Figure 6.20). It is believed that the microstructural reorganization initiated at the first loading and further developed at the subsequent ones gave a high degree of alignment of the stacks of smectite lamellae, the bonds between which could have been dominated by hydrogen bonds in the continuous films or organized water molecules in the clay matrix.

6.7   Concluding remarks

The THMCB processes initiated in the clays used to construct engineered clay buffers and barriers are the result of waste containment constraints and site specificities. The provoking forces or activities have been discussed in this chapter, with a view to highlight the issues that need attention. The main concern is with respect to how well one (a) understands how, or if, the clay in the engineered clay buffer/barrier will change; (b) can predict or have knowledge of what these changes will be; (c) can anticipate and plan for these changes; and (d) could model for prediction and assessment of long-term performance of the clay buffer/barrier system.

Although the various processes have been identified and discussed, attention should be directed to the fact that some of these are independent (uncoupled), whereas others may be interdependent (coupled). For example, the *T factor* that generates various actions and reactions can be considered as independent and the processes that it generates are to a large extent uncoupled. However, when the independent *H factor* is introduced into the system, one needs to consider coupling between *T* and *H* processes – the extent of which is not always clearly evident. The strength and nature of the couplings between processes such as heat, mass, and pressure have yet to be fully resolved. When the *C* and *B factors* are introduced into the system, the problem becomes much murkier. Many of these issues will be discussed in the next few chapters – on long-term performance – and also in the chapter dealing with conceptualization for modelling/prediction.

Chapter 7

Clay evolution and long-term buffer/barrier performance

7.1 Introduction

7.1.1 Geochemical/biogeochemical evolution – clay ageing

Geochemical/biogeochemical evolution of natural soils occurs as a normal time-related ageing process, the results of which are manifested in terms of changes in the nature of the soil and its characteristics and properties. This is why soils are often referred to as *dynamic living systems*. When soils, such as clays, are used as construction or liner/buffer/barrier materials, they are said to undergo maturation and ageing from the day they are emplaced. The same time-related ageing processes apply to clays used as clay buffers and barriers – either as stand-alone systems or as clay buffer/barrier components in engineered multi-barrier systems for containment of high-level radioactive waste (HLW) and hazardous solid waste (HSW)/municipal solid waste (MSW) (see Figures 2.6–2.8 and 2.10–2.12). The term *engineered clay buffer/barrier* has been used in the literature to mean both a stand-alone barrier and a part of a multi-component barrier (multi-barrier) system. In this book, an *engineered clay buffer or barrier* is a stand-alone clay buffer or barrier that is designed (engineered) to function as a containment system. When the clay is used as a barrier or liner component in an engineered multi-barrier system (EBS) (as shown in Figures 2.10–2.12) we will refer to this as a *clay barrier component*. As in the previous chapters, we will use the term *clay buffer/barrier* to mean both the stand-alone system and the clay barrier component in the EBS. It is implicitly understood that the clay has been *engineered* to provide the properties deemed desirable as a clay buffer and/or clay barrier for the EBS (see section 2.5.2).

For simplicity in discussions, we will refer to geochemical/biogeochemical evolution of clays as *clay evolution,* and the processes involved in clay evolution as *ageing processes*. This by no means ignores the possibility that time-related physical processes such as compaction, consolidation, piping and erosion will alter the properties of a clay. These processes can and should be anticipated in design considerations, as they are part of containment design analysis. An example of this (design consideration) is shown in Figure 7.1 in a permeation test conducted with a mixture of MX-80 smectite and ballast material. By increasing the fines content of the ballast A material (creating ballast material B), piping and erosion of particles shown in the smectite-A mixture have been essentially eliminated in the smectite-B mixture. The table shown in the top left-hand portion of the graph gives the details of permeation tests conducted on the two samples. The percentage of particles with less than two microns (<2 µm) dimensions in the three sections of the sample are shown – in contrast with the initial percentage (10.0±0.5). Sample B shows no change in the proportion less than 2 µm, whereas sample A shows migration of the less than 2 µm from the upper third to the bottom third of the sample.

Although the effects from time-related physical processes are relatively minor when compared with the effects of geochemical/biogeochemical evolution, they need to be part of the considerations in design of the clay buffer/barriers components in engineered multi-barrier systems. The results of ageing processes on the properties of clays can prove disastrous to their capability to function well as clay buffers and clay barrier components. Proper accounting for the effects of time-related ageing processes on clay properties and performance in designing the service function of buffers and

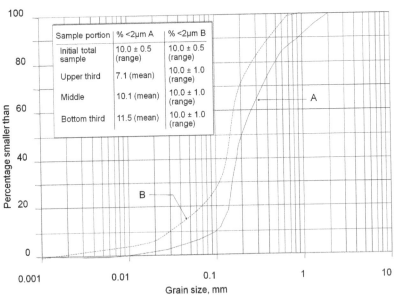

Figure 7.1 Change in content of < 2 µm particles of a mixture of 10 per cent smectitic clay (MX-80) and 90 per cent of type A or type B ballast material (total density of 1600 kg/m²) after 1 week of permeation under an average hydraulic gradient of 10 m/m. The grain size distribution for type B ballast shows a higher proportion of finer material than type A ballast.

barrier components is needed. The focus of the material developed in this chapter will be on the biogeochemical evolution of clays, particularly smectitic clays, and its impact on long-term performance in buffer and barrier components of EBS.

Design long-term performance, to be distinguished from actual long-term performance, is determined on the basis of (a) the nature of the wastes being contained; (b) the processes generated or initiated by the factors described previously as THMCB factors in Chapter 6; (c) the requirements for disposal/containment of the waste as mandated by the respective regulatory and oversight agencies; and (d) safety assessment protocols. Although frequently overlooked, it is important to note that items (c) and (d) do have significant impacts on design requirements. Specification of clay buffer/barrier performance provides one with a basis for selection of the type of clay to be used for the buffer/barrier component of EBS – a procedure that most generally requires determination of clay properties and performance (clay attributes) under specific conditions. It is relevant to note that clay evolution – and hence possible degradation and deterioration of clay properties/performance – is not always considered. If the impact of clay evolution or clay ageing is felt in terms of deterioration of clay attributes, the consequences on design performance expectations can be disastrous.

7.1.2   Long-term service life

The answer to the question of length of service life, with respect to performance requirements of clay buffer/barriers, cannot be given without reference to the type of waste being contained. In general, one can state that the design service life specified in regulatory requirements will define *long term*. The commonly accepted regulatory requirements for design service life of HLW and HSW/MSW containment facilities are 100,000 years and 35 years respectively. There are some who have argued that the 100,000-year design service life for HLW clay buffer containments systems is a reluctant compromise to a former one million-year design service life requirement – 'reluctant' because the half-lives of some activation and fission products will not be reached in the 100,000-year period (see Table 2.1 in Chapter 2). For example, the half-lives of zirconium-93, caesium-135 and iodine-129 are 1.5×10^6, 2.3×10^6 and 1.59×10^7 years respectively.

There appears to be no rational reason for specification of a 35-year design service life for HSW/MSW landfill containment system – except for an expectation that dissolution of the wastepile would probably be complete, or almost so, by this time period. Those that have argued for a much longer time requirement have been concerned with a lack of clear evidence of wastepile toxicity and hazard reduction in a 35-year time period. That being said, the engineered clay barrier that underlies the geomembrane–leachate collection system and other protective layers should function as an attenuation barrier for some considerable period of time after the 35-year design service life. For how long? The trite answer to this question is 'for as long as it takes for leachates generated from the wastepile to become harmless to the environment and public health'. The processes and their potential impact on the serviceability of the clay buffer/barrier system are shown in Figure 7.2. Much study and investigation are required to assess their impacts on the service life of the clays used for the buffer/barrier systems. Bearing in mind the highly varied nature of materials and chemicals disposed and generated as dissolution products in the HSW/MSW landfills, specification of a service life-time period and especially for the life of the clay barrier for such landfills requires proper assessment of risks posed to the environment and public health at the end of service life.

7.2   Impact of clay evolution

7.2.1   The negative impact factor

We consider the *negative impact* of clay evolution on long-term buffer/barrier performance to mean failure of design performance standards of any of the clay buffer/barrier attributes. These attributes include, for example, hydraulic conductance, swelling potential, rheological properties and

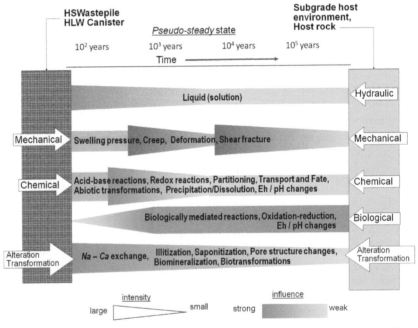

Figure 7.2 Source, intensity and influence of various processes in long-term service life of clay buffer/barrier components in engineered barrier systems.

thermal properties. Positive impacts of clay evolution (if any) are expected to add value to the performance of the attributes. Knowledge of the negative impacts of clay evolution is necessary if proper long-term performance of clay buffer/barriers is to be obtained. The various processes driven by the THMCB and containment factors have been described in the previous chapter. The B (biology) factor comes directly from the clay itself and the host environment, i.e. the microorganisms that participate directly in biogeochemical evolution of the clay are either indigenous to the clay and the host environment, or extraneous microorganisms transported by flow of ground water into clays. All of the other factors, i.e. THMC, add their own overlays to the natural biogeochemical processes of clay evolution. By this we mean, for example, that (a) high temperatures (T factor) will have decided impacts on the rates of geochemical reactions and (b) canister corrosion products and fugitive radionuclides, and contaminants in leachate streams (C factor), are added chemical components to the natural *in situ* spectrum of biogeochemical reactions. How the natural biogeochemical processes are affected by the THMC overlays, and how they individually and collectively impact on the characteristics and properties of the clay used for buffers and barriers over long time periods have yet to be fully determined. This is considered to be the ageing process.

With respect to the performance of clay buffer/barriers in the long term, at this stage (a) all the processes and their coupled phenomena have yet to be studied; (b) the kinetics of abiotic and biotic reactions have yet to be fully determined – especially the reaction rates and interdependencies; and (c) experiments have yet to be conducted that will satisfactorily compress and mimic 10^5 years of physico-chemical, chemical and biological reactions and interactions into some manageable laboratory time frame – under specific temperature and pressure conditions. Prediction of long-term behaviour of the buffer/barrier clays must rely in part on theoretical and basic knowledge of biogeochemical reactions, and in part on a study of natural analogues. A practical approach for assessment of the negative impact of clay evolution is to look at (a) what constitutes failure of the design function of the clay buffer/barrier system; (b) the processes/factors and conditions that are active in ageing of the clays; and (c) which of those processes/condition contribute to 'what constitutes failure', and how they contribute to service failure of the clay buffer/barrier.

7.2.2   Design function failure

Failure of a clay buffer/barrier essentially means that the fundamental purpose of buffer/barrier systems in waste containment facilities in land environments has not been met, i.e. meaning that the clay buffer/barrier cannot deny exposure of the waste substances and their reaction products to humans, other biotic receptors and the environment. We define *clay design function failure* to mean (a) failure of the clay buffer/barrier in repository containment of HLW to seal the canister and isolate it from its surroundings and (b) failure of the clay barrier component in EBSs to meet specified/ required performance – for example, failure to provide for contaminant attenuation and partitioning during transport of contaminants through a clay buffer or barrier in HLW and HSW containment (Figure 7.3). In general, this signals degradation or deterioration of individual or multiple attributes (permeability, strength, swelling, buffering and accumulative capacities, etc.) to the point of failure (yet to be defined).

For clay buffer/barriers used in HLW repository containment in subsurface geological environments, failure of the design function for the clay means failure to provide a competent cushion seal for the canister, and failure to seal the canister from the surrounding environment. This occurs when (a) corrosion products and fugitive radionuclides are not attenuated/ mitigated and are allowed to enter into the geosphere and onto the land surface environment; (b) solutes in the groundwater in the host rock are permitted unimpeded entry into the clay buffer to interact with the clay; and/or (c) the clay buffer is not capable of supporting the canister load and overburden loads. These could occur from deterioration and/or degradation

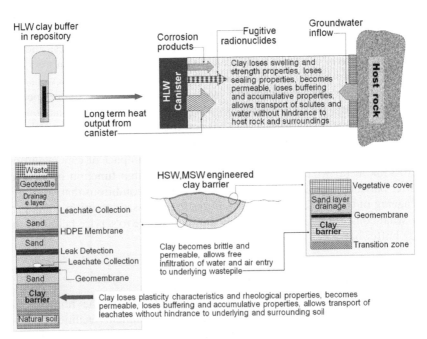

Figure 7.3   Failure indicators for clay properties as a result of ageing processes for clay barrier component in engineered multi-barrier systems for containment of HLW and HSW/MSW.

of any or many of the attributes, such as permeability, accumulative and buffering capacities, swelling pressure, rheological properties, as a result of ageing processes. The properties or characteristics of the clay in the buffer that are responsible for design function failure include, amongst the many attributes, (a) loss of swelling and strength (rheological) properties; (b) loss of permeability constraint (i.e. increase in permeability); (c) loss of chemical buffering and accumulative properties; (d) unhindered transport of ground-water solutes; (e) unimpeded transport of corrosion products and fugitive radionuclides through the clay buffer into the surrounding host rock and geosphere; and (e) facile transport of heat and water through the clay buffer. They are some of the principal indicators of failure (Figure 7.3). There are no definitive trigger limits that indicate failure in any one of the required design functions, for example swelling pressure or permeability control (Figure 7.4). This is because onset of failure of any one of the attributes is difficult to define definitively. The degradation or deterioration of any specific attribute is a gradual phenomenon, and when this occurs in com-bination with degradation of other attributes it is virtually impossible to identify a tipping point – the onset of failure. This is illustrated in Figure 7.4.

In the case of surface covers and side/bottom barriers for HSW/MSW containment, failure of design function of the clay barrier component occurs when (a) contaminants that have breached the overlying multi-layers are allowed to pass freely through the clay barrier component into the

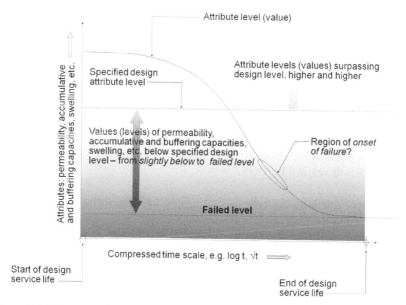

Figure 7.4 Variation of attribute level or value in relation to time. It is assumed that the deterioration in attribute level is due to negative impacts from ageing processes.

surrounding subsurface environment in the side/bottom barrier, and/or (b) water and air are allowed free access through the clay barrier component of the multi-layer cover to the underlying waste pile. We should be careful to point out that ageing processes may not always be the cause of failures in the service life of the clay buffer/barrier components for HLW and HSW containment. Any of these failures included in the list of failure indicators can occur at any time in the service life of the HLW repository and/or landfill owing to unforeseen events, construction and placement deficiencies, or design errors. However, the concern in this chapter is directed towards identifying the effects of long-term (ageing) processes impacting on the principal attributes of the clays used in EBS. The use of failure indicators provides one with a picture of a 'failed' clay – the type of clay that must not result from ageing processes.

7.2.3   Clay buffer/barrier attributes

The attributes that make a swelling clay such as a smectite attractive for use as a clay buffer/barrier in engineered multi-barrier systems are:

*   *very low hydraulic, gaseous and heat transmission properties*, i.e. very poor hydraulic conductance, poor gaseous transmission characteristics and good thermal properties, attributes that are important in attenuating flow of heat and restricting movement of fluids and gases;
*   *high swelling potential* – this is important for sealing purposes when used in confined situations such as in HLW containment repositories and engineered multi-barrier systems for HSW/MSW containment;
*   *high contaminant accumulation and attenuation capability*, i.e. good physical and chemical buffering potential/capability, which is essential for reduction and/or control of transport and fate of contaminants to the surrounding environment;
*   good *rheological properties* – essential for support of HLW canisters, HSW/MSW wastepile loads, overburden pressures, etc. and for resistance against shear forces exerted by canister and wastepile loads, especially on side and vertical barriers.

The above attribute list tells us that ageing processes must be prevented from degrading the clay used as clay buffer/barrier components to the point where any of the listed attributes are altered or changed for the worse, i.e. compromised. This means that one does not want good hydraulic, gas and heat transmission, poor or no swelling potential, poor or no physical/chemical buffering capability, and poor rheological properties. What kind of a clay would this be? At the extreme, the antithesis of a swelling clay is a non-cohesive (granular-type) clay. An example of this is a soil made up of clay-sized quartz particles. This is a granular-type clay and is non-cohesive.

This type of clay would have good hydraulic and heat transmission properties, no swelling capability, no contaminant accumulation and attenuation capability, but good rheological properties.

As noted in Chapter 3, the montmorillonite minerals in smectites consist of stacks of 10-Å lamellae separated by up to three water layers. According to Pusch and Yong (2006), for densities at water saturation of the smectite higher than 2500 kg/m³ (water content < 7 per cent), there is virtually no water layer present in interlamellar (interlayer) space. For densities between 2350 and 2500 kg/m³ (water content 7–10 per cent), one would expect one layer to be present, and for densities between 2100 and 2350 kg/m³ (water content 10–20 per cent) there would be between one and two layers of water. There would be two water layers for densities between 2000 and 2100 kg/m³ (water content 20–25 per cent), two to three water layers for densities between 1875 and 2000 kg/m³ (water content 25–35 per cent), and three water layers for densities lower than 1875 kg/m³ (water content > 35 per cent). The stacks of lamellae remain coherent and behave as colloidal gels for densities higher than about 1050 kg/m³ for situations in which the electrolyte concentration in the porewater is very low. The same coherence and gel-like behaviour would be true for densities higher than 1600 kg/m³ if the porewater contains calcium as the dominant cation.

As with 'regular' clays, consolidation of the smectite results in an increase from a low density to a higher density, whereas expansion of the consolidated sample occurs upon unloading. Unlike most other clay types (i.e. 'regular' clays) these processes are nearly reversible for smectites, although unloading is delayed by internal friction. Voids in the buffer, caused by imposed strain or by heat-induced desiccation early after installation of the canisters, have to be sealed by expansion of the clay. The involved processes are redistribution of porewater and clay minerals, and expansion of dense stacks of lamellae to form soft clay gels that are successively consolidated under the swelling pressure exerted by adjacent dense clay. If expansion is not allowed to take place spontaneously, self-sealing will be deficient. Two mechanisms can cause lack of expansion: extreme drought, i.e. inadequate supply of water, and precipitation of cementing agents. A combination of the two can result in a significant loss of the capability of the smectite to expand.

In practice, the smectitic clay buffer is manufactured by compaction of clay granules under 100–200 MPa of pressure. This produces a mass with only small voids between the clay granules. The voids have the form of channels of varying apertures. The tightest portions of the apertures will be completely filled with capillary water, whereas the wider ones will contain air. We illustrate this by showing the 'isothermal' state in the upper portion of Figure 7.5. The changes resulting from the application of a thermal gradient are shown in the lower portion of Figure 7.5. The heat given off from the canister is transported through the buffer and further out into the surrounding rock. In the initial phase of this occurrence, when no water has

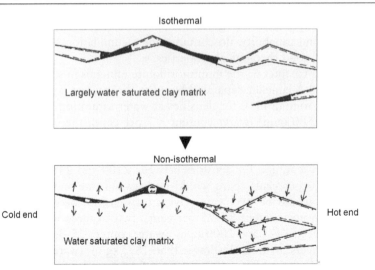

Figure 7.5 Microstructural changes in the initial process of redistribution of water and change in density of the clay matrix under a thermal gradient (adapted from Pusch *et al.*, 2009).

yet been taken up by the buffer from the rock, desiccation of the hot part of the buffer results, causing migration of water in liquid and vapour forms towards the rock. In the hottest part of the clay buffer the stacks of lamellae will contract, thereby causing widening of the channels voids and formation of steep, radial fractures to several centimetres' distance from the hot canister. Once this part has ultimately become water saturated, the fractures and widened channels will be closed by the swelling pressure that is exerted by the colder part of the buffer and self-sealing will result, provided that the heated clay has not lost its capability to fully expand.

As clay minerals constitute the underpinning that provides the clay with the attributes necessary to serve as buffer/barrier material, the set of problems reduce to determination of (a) what ageing processes will result in alteration/transformation of clay minerals; (b) how these alterations/transformation manifest themselves, for example loss of swelling potential, rearrangement or restructuring of microstructural units and macrostructure, collapse of clay layer; and (c) what this means for long-term performance of HLW clay buffer containment and clay barriers in HSW/MSW landfill engineered multi-barrier systems.

7.3 Clay evolution

Geochemical and biogeochemical processes occurring in clays will affect (a) the status of both the organic and inorganic clay fractions and (b) the properties and behaviour of the clays. It is not easy to distinguish or differentiate between the results obtained from geochemical reactions and those obtained from biogeochemical processes. From a theoretical standpoint, we can consider geochemical processes and reactions separately. However, in field situations with naturally present microorganisms, the activities of these microorganisms can affect for example the pH and Eh of the clay–water system, resulting thereby in influencing or affecting geochemical reactions. A sort of cycle of activity and interaction of microorganisms–geochemistry-microorganisms–geochemistry, etc., occurs in a clay. Substrate availability for the microorganism, to some extent, depends on the geochemistry of the clay (chemistry of porewater and surface chemical characteristics at particle surfaces). In turn, the activities of the microorganisms will affect the chemistry of the porewater. And so the cycle goes on (Figure 7.6). It can be argued that rather than seek to separate geochemical processes/reactions from biogeochemical processes/reactions, it is more expedient to consider all of these processes as geochemical processes with a microbiological overlay.

The record shows that a clear deficiency exists in total understanding of the inter-relationships and interactions between microorganisms and inorganic/organic fractions in the soil, and their influence on the various transformations and alterations of clay constituents. Accordingly, it is simpler and more expedient to consider reactions in the clay–water system as geochemical

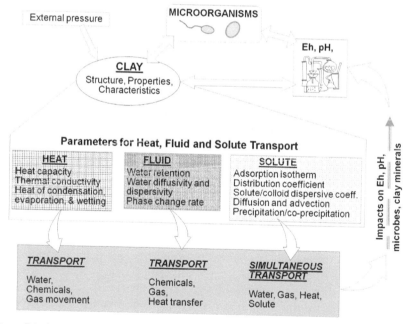

Figure 7.6   Long-term interactions and their effects on clay structure, properties and characteristics. The cycle between reactions and clay status is a continuous cyclical process over the service life of the clay buffer/barrier component of an engineered barrier system.

reactions, with the clear understanding that microbiological activities may have set the stage for these reactions or that the reactions will be intertwined with the activities of microorganisms. Clay evolution is a complicated process that involves these two categories of processes and their reactions. The problem becomes more complicated in clay buffer/barriers because of the temperature, pressure, hydraulic, and chemical (contaminants, leaching and corrosion products, etc.) factors, as shown in Figure 7.6.

Geochemical reactions such as dissolution–precipitation of carbonates, sulphates and silica, oxidation–reduction reactions and protonation–deprotonation are some of the more significant reactions that could change the nature of a clay – to such an extent that it would lose its design functional role as a clay buffer/barrier. Although abiotic and biotically mediated transformation of clay minerals can occur, Moodie and Ingledew (1990) show that some transformations are preferentially biotically mediated and others are abiotically mediated. They give the examples of oxidation of S^0 into SO_4^{2-} with reduction of Fe^{3+} into Fe^{2+} for the former and oxidation of S^{2-} into S^0 with reduction of Fe^{3+} into Fe^{2+} for the latter. In the context of HLW and HSW/MSW containment by clay buffer/barriers as stand-alone and in multi-barrier systems, the results of these reactions include (a) clay mineral transformation/alteration and clay compositional changes; (b) alteration of the transport and fate of contaminants through the clay buffer/barrier; (c) degradation of chemical and physical buffering potential; and (d) changes in the hydraulic, mechanical, physical and chemical conditions and properties of the clay and of the host site and environment.

7.3.1 Clay alteration and transformation

It is not uncommon to find the terms *alteration* and *transformation* being used in the literature in relation to clays to mean physical and chemical changes in the nature of the clay. That being said, it is not always clear from the same literature what the differences are between *clay alteration* and *clay transformation*. To avoid confusion in usage of terms, for the purpose of discussions in this book, the term *clay alteration* is used to refer to (a) physical changes in a clay, such as changes in density, porosity, restructuring of microstructural units (Figure 7.7), and (b) changes in properties and characteristics of the clay, such as hydraulic conductance, strength characteristics, swelling characteristics. Although the changes in properties and characteristics are a direct outcome of physical changes, the causes of physical changes are time-related physical and geochenmical/biogeochemical processes.

The term *clay transformation* refers to the changed nature of a clay resulting from physical and chemical transformation of the various constituents (fractions) in a clay, such as clay minerals, organics and amorphous materials. Pusch *et al.* (forthcoming) have shown that by subjecting MX-80 smectite to hydrothermal treatment, substitution of Al^{3+} by Fe^{3+} in the octahedral

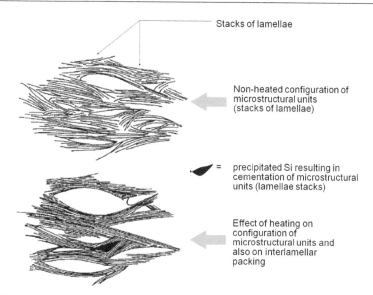

Stacks of lamellae

Non-heated configuration of microstructural units (stacks of lamellae)

= precipitated Si resulting in cementation of microstructural units (lamellae stacks)

Effect of heating on configuration of microstructural units and also on interlamellar packing

Figure 7.7   Changes in microstructural units and interlamellar packing due to hydrothermal effects on MX-80 smectite (adapted from Pusch *et al.*, 2008).

layer occurs in the montmorillonite fraction, thereby causing higher lattice stresses because of the larger ion radius of Fe^{3+} in comparison to Al^{3+}. This reduces the resistance of the montmorillonite in the smectite to dissolution – a long-term potential problem that can lead to severe changes in the nature of the MX-80 clay. Dissolution of the other clay fractions (accessory minerals such as calcite, quartz, and feldspars) in addition to the montmorillonite in the MX-80 in hydrothermal treatment released Si and Fe, which were precipitated in the clay as cementing agents (see Figure 7.7) when cooling of the clay occurred. Section 6.4.1 has provided a discussion of the conversion of smectite to illite – with a demonstration of the expected conversion of smectite to illite using calculations based on the Pytte model (Figure 6.13).

Chemical transformations include both abiotically and biologically mediated transformation reactions. Abiotic transformations occur without mediation of microorganisms and include chemical reactions such as hydrolysis, oxidation–reduction and dissolution. It is not uncommon for all three kinds of mechanisms to participate in the transformation of clay minerals, either in combined action or in sequential form. The transformation products are generally other kinds of minerals, organic compounds, carbonates, etc. Iron, manganese, aluminium, trace metals in clay minerals together with adsorbed oxygen can catalyse chemical oxidation of many organic chemical at ambient pressure and temperature, and many substituted chemical organic chemicals can undergo free radical (homolytic) oxidation, as opposed to heterolytic oxidation (polar reaction/oxidation).

Biologically mediated (microbially mediated) transformation reactions, also known as biotic transformation processes, include associated chemical reactions such as oxidation–reduction, cation exchange reactions, and dissolution–precipitation of silica and carbonates. Microbial activity can contribute to these reactions as catalytic agents or through precursor processes, especially with respect to the spatial structuring of redox dynamics in clays, resulting in transport of several ions and colloids released (Murphy and Ginn, 2000; Regnier et al., 2002; Thullner et al., 2005) – governed by solute transport equations, as shown in Chapter 5. According to Kurek (2002), microbial activities can increase weathering of minerals by a million times. The accounts given by Kurek (2002), Ehrlich (2002), and Huang et al. (2005), regarding the modes of microbial attack leading to dissolution of minerals by microbes under both aerobic and anaerobic conditions, are repeated as follows:

- *Direct attack*: (a) direct enzymatic oxidation of a reduced mineral component and (b) direct enzymatic reduction of an oxidized mineral component.
- *Indirect attack*: (a) with a metabolically produced redox agent or inorganic and organic acids; (b) with a metabolically produced alkali; (c) with a metabolically produced ligand, forming a highly soluble product with a mineral component; and (d) by biopolymers.

As clay minerals have the ability to affect the growth or metabolism of microbial habitats because of their geochemical properties and characteristics, the rates and extent of microbially mediated transformations will be conditioned by types of clay minerals and microorganisms. The reaction of microorganisms with clay particles resulting in the production of extracellular polysaccharides emanating from their surfaces can alter the arrangement and packing of particles, i.e. change the nature of the microstructural units and, in turn, change the micro-/macrostructure of the clay (see Figure 6.14). Similar effects are obtained when adsorption of enzymes by clay particles occurs. The adsorption mechanisms include electrostatic, covalent, ionic, hydrogen bonding and van der Waals forces.

Further to the discussion on biomineralization and biotransformation in section 6.4.1, it is useful to note that organic macromolecules are important agents in the fostering of biomineralization. Growth and arrangement of the inorganic crystals are conditioned by assemblies of these macromolecules (Mann et al., 1993). The ability of bacteria to bioconcentrate metals in solution is a feature that makes biomineralization an important factor in the evolution of clay buffer/barriers in engineered barrier systems.

7.3.2   Clay mineral transformation

Weathering of clay minerals

Transformation of clay minerals occurs naturally in soils as a result of weathering processes, which include chemical and biologically mediated reactions. Figure 7.8 shows the roles of magnesium (Mg) and potassium (K) in weathered products of basic and acid igneous rocks. The weathering of basic igneous rocks containing significant amounts of Mg can lead to the formation of montmorillonite or kaolinite, depending on the Mg content of the weathered product. According to Grim (1953), montmorillonite is the alteration product if Mg is allow to remain in the weathered material upon breakdown release from the parent mineral in basic igneous rocks with high Mg content. If, on the other hand, Mg is immediately (very quickly) transported away from the weathered material, the alteration product will be kaolinite. The same result is obtained if Mg is rapidly leached from the alteration product montmorillonite, i.e. kaolinite will be obtained from montmorillonite with rapid leaching of Mg. In acid leaching environments, the weathering processes involved in the removal of Mg will cause breakdown of the minerals, leading to the release and transport of Al and Fe and the retention of Si in the weathered product. In contrast, if the leaching environment were neutral or mildly alkaline in those same weathering

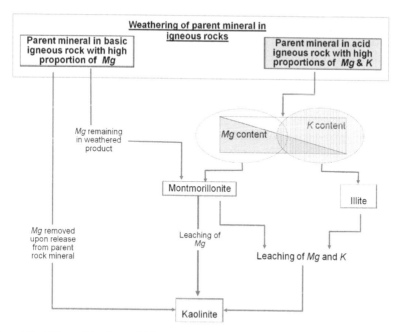

Figure 7.8   Effect of leaching of Mg and K in parent minerals of basic and acid igneous rocks.

processes, Fe and Al would remain with the weathered product, while the released Si would be transported away. The pH of the leaching environment is of significant importance in as much as (a) acid leaching environments will most likely result in the removal of Fe and Al and concentration of Si in the weathered product and (b) neutral or mildly alkaline leaching environments will most likely result in the concentration of Al and Fe in the weathered product.

In the case of acid igneous rocks containing significant proportions of Mg and K, weathering breakdown of the parent mineral will produce alteration mineral products that could be montmorillonite or illite, depending on the proportion of Mg or K present in the alteration product. Where Mg is the highly dominant proportion, montmorillonite will be obtained. If on the other hand K is the highly dominant proportion, illite will be obtained. In between, one expects that a mixed-layer mineral will be obtained. Subsequent leaching of Mg and K will result in the formation of kaolinite as the alteration product.

The dissolution of some minerals and the re-precipitation of others contribute to the variety of transformed minerals. Temperature and moisture are key elements in the weathering process. The intensity of weathering increases with increasing temperature and the degree of weathering is greater in most soils. Eberl and Hower (1976) showed from their experiments that transformation of smectite to a mixed-layer clay at temperatures of 60°C would take at least 10^6 years in comparison with a time requirement of 10^8 years at temperatures of 20°C. Different minerals have different degrees of robustness, i.e. more resistant to weathering. Figure 7.9 shows a weathering scheme that was developed by Jackson (1964), with different minerals obtained as weathering products.

Structural iron reduction

Dissimilatory reduction of structural iron Fe(III) to Fe(II), i.e. use of Fe(III) as an electron acceptor in metabolism, in the octahedral sheet of montmorillonite minerals in smectitic clays can lead to severe loss in those attributes that provide the mineral with its swelling capability. This, in turn, will impact adversely on the design hydraulic conductivity, ion exchange capacity and accumulative potential of the clays.

7.4   Microbial impact

7.4.1   Ecology of bacteria

Microorganisms are classified into six categories, i.e. protozoa, microalgae, fungi, Eubacteria, Archaea and virus. Eubacteria and Archaea are collectively called 'bacteria'. It has long been considered that bacteria could have an

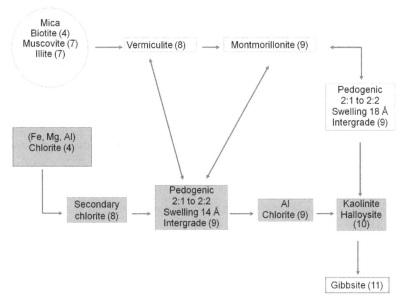

Figure 7.9 Transformation of various clay minerals from weathering processes. Numbers in parentheses are weathering indices with more stable minerals having higher numbers (adapted from Jackson, 1964).

impact on clay minerals, with respect to their transformation, through mechanisms of weathering and dissolution over long time periods in the natural environment. There exist various species of bacteria in nature, i.e. alkalophilic and asidophilic bacteria in the ground as well as barophilic bacteria in sediments residing in the seabed. At present, there are (a) no assured methods to indicate or identify what species of bacteria will survive after a long time period and (b) no reliable means to determine what characteristics and functions the various species of bacteria will possess. In this section, we present an overview of the world of bacteria – as a means to provide a view of the general and common features of bacteria that are commonly associated with engineered clay buffer/barriers.

The living bodies of bacteria are composed of various elements, such as C, H, O, N, P, S and trace elements. On average, almost all the bacteria can be chemically expressed as $[A]CH_{1.7}O_{0.4}N_{0.2}$, where A designates P, S, K, Mg, Mn, Ca, Fe, Co, Cu, Zn, Ni, Mo, etc. Bacteria take these elements *in vivo* as nutrients and these elements are used to make (a) the body of bacteria and (b) such metabolites as gas and extracellular polysaccharides, as well as ATP (adenosine triphosphate), which supplies the energy required for their growth and life maintenance. Bacteria grow and live in environments where there are water and organic and inorganic matters that can supply these elements. Bacteria that utilize organic compounds as a carbon and energy source are called *heterotrophic bacteria*. Bacteria that utilize inorganic compounds such as H_2, NH^{4+}, NO^{2-}, S, H_2S and Fe^{2+} as energy sources and utilize CO_2 as a carbon source (not organic matter) are called *autotrophic bacteria*. Aerobic bacteria grow in environments where there is oxygen, whereas anaerobic bacteria grow in environments devoid of oxygen, and facultative bacteria grow in both environments with and without oxygen. They live in a wide range of temperatures, i.e. 0–120°C. The optimum temperature range is 25–50°C for a great number of bacteria. Sulphate-reducing bacteria survive in temperature regimes that are higher than 100°C (see section 6.4.2). Depending on the species and circumstances, the lifespan of bacteria is different, i.e. from a few days to about 300 days at the maximum. The alternation of generational species or ecological succession will be continued for a long time.

The mass of bacteria is assumed to be about 10^{-12} g/cell, and their size is around 0.25–2.0 μm. The surface of bacteria has a positive charge at an extremely low pH and a negative charge at a pH higher than 4 (Marshall, 1976). In general, bacteria live on the surface of solid substrates, with a maximum density, for example 4–6×10^{13} cells/kg mineral for *T. ferrooxidans* and 2–4×10^{13} cells/kg mineral for *T. thiooxidans* (Konishi *et al.*, 1990, 1995). They form microcolonies and produce biofilms that are composed of the stacks of microcolonies on the surface of compacted clay as shown in Figure 7.10. The mass of bacteria in microcolonies is assumed to be about 90 mg/cm³ colony. Their size is about 10–20 μm in diameter and 5–10 μm in thickness (Molz *et al.*, 1986). The maximum thickness of biofilms is assumed to be 10–50 μm.

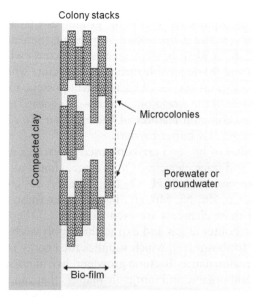

Figure 7.10 Ecology of bacteria. Bacteria form microcolonies, and the stacks of microcolonies form a biofilm on the surfaces of compacted clay.

7.4.2 Growth dynamics of bacteria

Indigenous bacteria in clays and the extraneous bacteria that are transported into clays by groundwater will grow on the wall of clay in large pores or fractures with widths that are larger than 5 μm, provided that water and nutrients are available for growth in the microenvironment. Otherwise, bacteria will grow on the surface of clay buffer/barriers at the interfaces between clays and rocks or between clays and canisters.

The term *growth dynamics of bacteria* is used in this section to mean both (a) growth and decay of bacteria and (b) changes of the population or density of bacteria with time. In general, the growth dynamics of bacteria is given by a generalized logistic curve expressed by the relationship shown in section 6.4.2 (see equation 6.24) (Hofbauer and Sigmund, 1988; Murray, 1993), i.e.:

$$\frac{d\rho_m}{dt} = \mu\rho_m - f(\rho_m)$$

where ρ_m is the density of bacteria, t is time and μ is the intrinsic growth rate coefficient. The term $\mu\rho_m$ on the right-hand side of the relationship indicates the net growth rate, whereas the second term $f(\rho_m)$ expresses the net decay rate. If we use the intrinsic decay rate coefficient κ, we can replace $f(\rho_m)$ with the term $\kappa\rho_m$. By doing so, we can now express the growth rate of living bacteria $d\rho_m/dt$ as (Monod, 1949):

$$\frac{d\rho_m}{dt} = (\mu - \kappa)\rho_m$$

Nakano and Kawamura (2008) have suggested that because of the ageing and lifespan of bacteria, the intrinsic growth and decay rate coefficient can be considered to be a function of time as follows:

$$\mu = \mu_{max}\left(1 - a_\mu \cdot \exp(-b_\mu t^2)\right) \tag{7.1}$$

$$\kappa = \kappa_{max}\left(1 - a_\kappa \cdot \exp(-b_\kappa t^2)\right) \tag{7.2}$$

where μ_{max} and κ_{max} are the maximum values of μ and κ respectively, t is time and a and b are constants. When the biofilms are formed, the population of living bacteria is expressed as follows:

$$\rho_m = \rho_0 \cdot \exp\left[\int_0^t \left\{(\mu - \kappa) - \frac{\partial}{\partial t}(\ln\chi)\right\}dt\right] \tag{7.3}$$

where ρ_m refers to the mean density of bacteria in biofilms, ρ_0 is the initial density of bacteria, and χ is the thickness of biofilms. Figure 7.11 shows several growth curves of living bacteria calculated from equation 7.3.

As bacteria grow in their habitat, the population of living bacteria gradually increases during a certain characteristic time in the initial stage, and thereafter reaches the saturation state at the final stage as shown in Figure 7.11. At this time, the thickness of biofilms would reach a certain maximum thickness, and can be calculated as follows:

$$\chi = \chi_{max}\left(1.1 - a_\chi \cdot \exp\left(-b_\chi t^2\right)\right) \tag{7.4}$$

where χ_{max} is the maximum value of χ and a and b are constants.

7.4.3   Microbial dissolution of clays

Because of the various minerals nutrients contained in clay minerals, they constitute a nutrient source for bacteria. Accordingly, it is natural for bacteria to interact with clays. The attack of bacteria on clays results in dissolution of clay minerals. Amongst the various possible microbial impacts on clays and their performance as engineered clay buffers and barriers, this is considered to be the most significant. The dissolution of clay minerals releases the bonding energy established between atoms that make up the clay minerals, resulting in the release of these atoms. Estimation of the rate and extent of dissolution

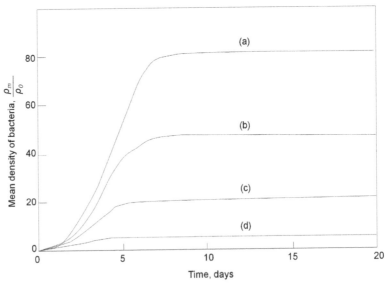

Figure 7.11   Logistic curves with different biological parameters μ, κ and χ [$b_\kappa = b_\chi = 0.06$ (a), 0.07 (b), 0.09 (c) and 0.15 (d)] when $\mu_{max} = \kappa_{max} = 2.0$, $a_\mu = 0.5$, $a_\kappa = a_\chi = 1.0$, and $b_\mu = 4.0$, in equations 7.1, 7.2 and 7.4 (see Nakano and Kawamura, 2008).

can be obtained by taking into account (a) mass conservation and (b) energy balance or conservation law. In cases when the estimation or measurement of mass can be properly determined for dissolved atoms that are recomposed into various phases (as determined for example from batch experiments), the mass conservation law can be usefully applied to determine the rate and extent of dissolution. In nature, however, it is impossible to precisely track the atoms released from clay minerals, as some ionized atoms are discharged into the environment. Under these circumstances, the alternative method is to use the energy balance or conservation law for estimation of dissolution of clay minerals by bacteria. In essence, this requires determination of the balance between energy consumed by bacteria and energy released from clay minerals.

The energy within the clay minerals, i.e. the bonding energy between atoms, can be utilized for growth and life sustenance for bacteria, and can be chemically and physically transformed into the bonding energy for resultant precipitation, metabolites and gas *in vivo* and *in vitro*. A part of the bonding energy will be dissipated as heat. Bacteria would not utilize the whole energy released by dissolution of clay minerals. Clay minerals exposed on the wall of pores or on the surface of compacted clays would not be thoroughly dissolved; a part of them would remain undissolved. Taking all of these into account, the energy balance established between clay minerals and bacteria can be described as:

$$\varepsilon_{cm} \frac{\partial(\xi\rho_{cm}L)}{\partial t} = E_{\mu} \frac{\partial}{\partial t}\left(\int_0^{\chi} \rho_{\mu}\, dx\right) + E_l \int_0^{\chi} \rho_m\, dx +$$
$$E_{met} \frac{d}{dt}\left(\int_0^{\chi} \rho_{met}\, dx\right) + (1-\varsigma)\varepsilon_{cm} \frac{\partial(\xi\rho_{cm}L)}{\partial t} \tag{7.5}$$

where L is the corrosion depth of clays, ρ_{cm} is the density of clay minerals, ρ_{μ}, ρ_m and ρ_{met} are the density of bacteria increasing by the net growth, the density of living bacteria, and the density of metabolites respectively. ε_{cm} is Gibbs free energy of clay minerals, and E_{μ}, E_l and E_{met} are Gibbs free energy required for the net growth, living maintenance and the production of metabolites respectively. ξ is a corrosion factor $(0 < \xi \leq 1)$ by which the extent of corrosion is defined when clays of L in thickness are partly dissolved by bacteria, ς is the ratio of the energy used per unit time by bacteria to the Gibbs free energy released from clays of thickness $L (0 < \varsigma \leq 1)$, and it denotes the energy consumption efficiency of microorganisms. χ is the thickness of biofilms, x is a distance from the surface of clays, and t is time. The term on the left-hand side expresses the energy released from clays of thickness L. The first, second and third terms on the right-hand side are the energy used per unit time for the net growth, life maintenance and metabolism respectively.

Rearranging equation 7.5, we can obtain the following equation expressing the surface corrosion depth L of engineered clay buffer/barriers as:

$$L = \frac{\rho_0}{\xi \rho_{cm}} \cdot \frac{\varepsilon_\mu}{\varsigma \varepsilon_{cm}} \int_0^t \left\{ \left(\frac{\varepsilon_l}{\varepsilon_\mu} + \mu \right) \chi \left(\frac{\rho_m}{\rho_0} \right) \right\} dt \tag{7.6}$$

As $\mu \to \mu_{max}$, $\kappa \to \kappa_{max}$, $\chi \to \chi_{max}$ and $\rho_m/\rho_0 \to (\rho_m/\rho_0)_{max}$ when t becomes extremely large, i.e. larger than about 20 days, we can conveniently approximate equation 7.6 with the following simple expression, although the equation will slightly overestimate the corrosion depth:

$$L = \frac{\rho_0}{\xi \rho_{cm}} \cdot \frac{\varepsilon_\mu}{\varsigma \varepsilon_{cm}} \cdot \left\{ \frac{\varepsilon_l}{\varepsilon_\mu} + \mu_{max} \right\} \cdot \chi_{max} \cdot \left(\frac{\rho_m}{\rho_0} \right)_{max} \cdot t \tag{7.7}$$

where t is time, ε_μ is the apparent energy required for the growth that would be approximately equivalent to the energy measured in batch experiments, and ε_l is the apparent energy required for life maintenance. As it is difficult to choose the appropriate values for ε_l, one could hypothesize that $\varepsilon_l = \alpha \varepsilon_\mu$,

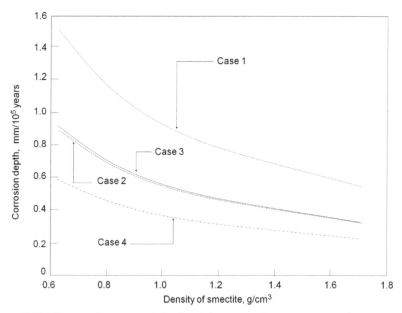

Figure 7.12 Estimated corrosion depth of bentonite for various possible microbial parameters. Case 1 represents corrosion for the case with a density that will be obtained in slightly polluted water as the initial density of bacteria. Case 4 shows the corrosion obtained for a density of the indigenous bacteria in bentonite as the initial density.

where the coefficient α is assumed to be $0.1 < \alpha < 2$. The subscript 'max' expresses the maximum of each microbial parameter in the stable state.

The corrosion depth of bentonite that includes beidellite is illustrated in Figure 7.12 for several cases, with possible microbial parameters for a long term of 10^5 years. The curves of cases 1–4 have been obtained for the initial bacteria density of 4×10^4, 4×10^3, 4×10^2 and 4×10^1 cells/cm³, respectively, The corrosion depth of case 3 is nearly equal to that of case 2, although slightly above, even if the initial bacteria density is smaller than in case 2, as the intrinsic growth coefficient is larger in case 3 than in case 2. This situation suggests that in addition to the initial bacteria density adhering on the surface of compacted clays, microbial ability for growth would have a considerable effect on the extent of corrosion. Nakano and Kawamura (2008), after a detailed examination of the extent of influence by microbial activity for growth, have suggested that the possible corrosion depth of compacted engineered buffer/barriers would be in the range of 0.2–$5.3\,mm/10^5$ years – provided that the bentonite has been compacted to a maximum density of $2.1\,Mg/m^3$ with smectite density of about $1.6\,Mg/m^3$. It has been observed that a smaller density of bentonite or smectite would yield a larger corrosion depth.

7.5   Concluding remarks

Strictly from a theoretical standpoint, and intuitively, we know that chemical reactions, geochemical and biogeochemical processes occurring in clays will, in time, affect both the nature of clays and their properties/behaviour. This is especially true when we take into account the very long time period (thousands and thousands of years) for HLW repository containment by clay buffers, and to a lesser extent the shorter time period associated with engineered clay barriers and liners. The reactions and processes can and do occur simultaneously and sequentially. It is not easy to distinguish between the results of any of these actions, nor is it feasible or even necessary to do so if one accepts that there will be alterations and transformations of the clay minerals and other clay fractions – to the extent that these will impact on the design performance of clay buffers and barriers.

We lack the necessary research and information on the various reactions and processes, and especially on the results of these actions. Batch experiments from laboratory research studies can only give us some idea on rate constants and kinetics of reactions. However, given that it is not possible to compress time scales in the order of thousands of years for application to studies involving geochemical and biogeochemical processes, we can only project from theoretical expectations and from natural analogues. The study of natural analogues gives us an opportunity to extend our laboratory and theoretical knowledge of clay transformations. Much more needs to done to allow us to be more knowledgeable about the evolution of clays.

Chapter 8

Field and mock-up experiments

8.1   Introduction

Time requirements and economic considerations do not allow for full-scale experiments to be conducted for clay-engineered barriers intended for use as engineered multi-barrier systems or as clay buffers in repositories housing canisters containing radioactive waste. In particular, if we are to provide reliable performance information on clay-engineered barriers over their service life, full-scale experiments would need to be conducted over a period of at least 35 years for engineered multi-barrier systems and 100,000 years for clay buffers in repositories. In attempts to shorten the time requirement in long-term experiments, laboratory scale-up and full-scale experiments, methods, devices and techniques have been devised to (a) accelerate physical and chemical reactions in the clay; (b) heighten and increase biological activities; and (c) shorten time-dependent processes and increasing rate-dependent processes etc. without compromising their effect. These, however, have not proven to be successful or reliable in producing the spectrum of information needed to allow decision-makers to render their judgements on anticipated performance of the candidate clay to be used in multi-barrier systems in the long term. For these, and many other reasons, organizations responsible for safe disposal of high-level radioactive wastes (HLWs) and hazardous solid wastes (HSWs), and especially the former, have tended to rely on conceptual and theoretical system-performance models for prediction of long-term performance.

The need to provide proper validation for calculations and predictions made by analytical computer models is obvious. In part, this need can be met with controlled laboratory scale-up experiments for short-term performances. However, for longer terms, this becomes a problem. Techniques have to be devised that would allow one to reconcile observations of short-term clay-engineered barrier performance with actual long-term performance records. Problems associated with scaling, time, time-dependence, rate-dependence and 'accelerated reactions/activities' are at the root of most of the difficulties.

The foregoing notwithstanding, it is nevertheless important and necessary to conduct, where possible, full-scale experiments. This will at least allow one to account for the effects of size, heterogeneity, bulk phenomena, site specificities, etc. as controls in system performance. The question of short-term extrapolation to long term can be theoretically examined through *trend analysis*, assuming that one has the proper handle on the various extrapolation and reaction kinetics models.

In this chapter, we will pay attention to the HLW deep geological containment scenario discussed in earlier chapters – and in particular, we will have the opportunity of comparing model-derived predictions with recordings of investigations of the various time-dependent processes that are assumed to take place in a repository located in the various geological host media, such as crystalline, argillaceous and salt rocks. A few of them can be studied individually, assuming stationary boundary conditions in the laboratory as described in preceding chapters. However, the large majority are coupled experiments that take place under transient boundary conditions. These are hard to simulate in ordinary experiments. Hydraulic boundaries need to be stationary to avoid highly complex conditions and to allow for more representative interpretation of 'readings'. Mock-up experiments conducted on the ground surface are a sample of these types of experiments. In the case of HLW deep geological containment, realistic conditions can only be provided by underground research laboratories (URLs). We will focus our attention on a few large-scale experiments of fundamental importance designed to increase our understanding of the evolution of the clay buffer under conditions controlled by the host rocks (crystalline and argillaceous rocks) in this chapter. Supporting laboratory experiments and some relevant theoretical models will also be discussed. The next chapter will consider the problems of modelling for performance prediction of these large-scale experiments.

8.2   Buffer evolution

8.2.1   *General*

A number of predictions and recordings of the actual evolution of buffer clay under assumed repository conditions have been made in the comprehensive international study on the performance of clay-engineered barriers in underground laboratories (Pusch, 2008). These have been concerned with the so-called near-field that consists of HLW canisters, clay buffer, tunnel backfill, and surrounding rock to a depth of a few metres from the buffer/rock interface. In this discussion, we will provide a few examples of predictions and recordings and indicate the shortcomings and possible problems related to modelling and practical application of common design principles.

8.2.2   Temperature evolution

Temperature and temperature gradients drive all the major physical and chemical processes in the near-field. They have been studied extensively in the last decades, with the intent of obtaining a basis for the design-engineered barriers and for the selection of suitable sites for repositories. A number of predictions of the thermal evolution of the clay buffer have been made and compared with recordings in URLs. An example of prediction of the temperature distribution in the early phase of hydration of the buffer (10 years) is illustrated by Figure 8.1. The data show that in the first decade or so, the temperature at the canister surface is typically about 100°C, and that it is about 70°C at the buffer/rock contact and about 50°C at a 5 m distance from this interface. The average radial temperature gradient is about 1°C per centimetre in the buffer after 10 years and this remains sufficiently high in order to control the chemical and hence also the physical evolution of the buffer for centuries.

8.2.3   Maturation of buffer

Redistribution of porewater

The thermal gradient impact on the initial water content of the buffer is demonstrated in terms of its redistribution, as illustrated in Figure 8.2. With

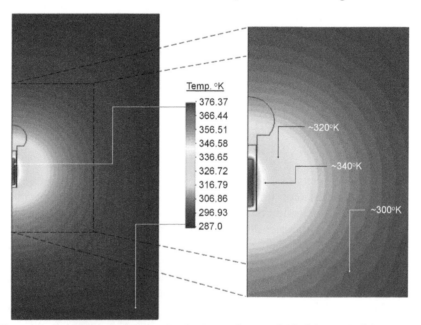

Figure 8.1   Predicted temperature distribution in the near-field of the vertical deposition hole (adapted from Svemar, 2005).

time, the water content will increase through uptake of water from the rock and the buffer. Eventually, the buffer will be fully water saturated – with the rate of saturation being dependent on the water pressure and hydraulic conductivity of the rock. The initial drying of the hottest part of the buffer close to the canister can result in contraction of the smectite particles, so much so that spontaneous expansion in the subsequent wetting phase may not occur.

The rate of water saturation of the buffer depends on the rate with which the rock provides water to the buffer. The controlling factors include the rock structure, its hydraulic conductivity and piezometric conditions. They are fundamentally different for crystalline and argillaceous host rocks: the former is normally fractured and has a hydraulic conductivity that is a few orders of magnitude higher than that of very tight buffer clay, whereas the latter can be as tight as the clay buffer. The consequence of this is that it may take hundreds or thousands of years before the clay buffer is provided with water. The end result of this is that the buffer next to the canisters will undergo substantial drying, as indicated by Figure 8.2.

Salt precipitation in maturing buffer

The absolute temperature in the buffer soon after placement of a HLW canister is high enough to initiate dissolution of the smectite minerals and

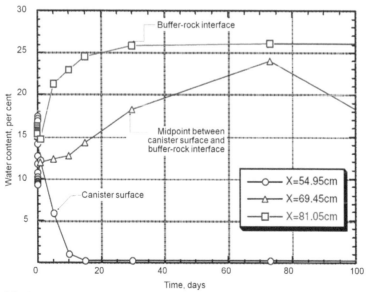

Figure 8.2   Example of predicted change of the water content as a function of time (0–100 days). x distance is measured from centre of canister (adapted from Svemar, 2005).

accessory minerals: primarily feldspars, carbonates and sulphur-bearing minerals. The direction of the temperature gradient means that minerals with 'reversed' solubility, i.e. carbonates, chlorides and some sulphur-bearing minerals, will precipitate in the hottest part of the buffer. These will generate a sink that causes additional ionic migration and precipitation of new minerals, eventually leading to formation of a 'salt crust' at the canister surface, involving Na^+, Ca^{2+}, Mg^{2+}, Cl^- and SO_4^{2-} ions present in the groundwater of the surrounding rock or released from dissolved accessory minerals in the buffer.

The salt accumulation process can be studied using accelerated laboratory experiments. Figure 8.3 shows the results of such tests conducted on a sodium smectite-rich clay with a dry density of 1270 kg/m³ and an initial 24 per cent degree of water saturation that was exposed to a temperature gradient of 14°C/cm. The cold end, at 30°C, was exposed to a 3.5 per cent NaCl solution (Pusch, 1994, 2008). The results shown in the figure portray the redistribution of salt and porewater in the sample after a 1-month period of testing.

Under repository conditions with lower temperature gradients than in the laboratory experiment, the salt accumulation process will be slower. However, this accumulation process will proceed for at least as long as the clay buffer takes up water from the rock. For very low hydraulic conductivities in

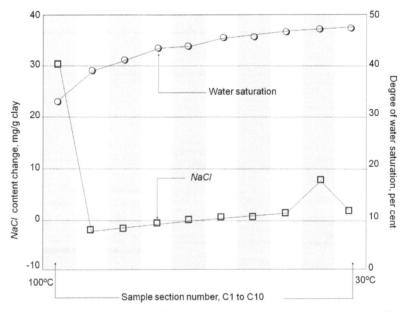

Figure 8.3 Change in water and NaCl content from the initial concentrations at uniform temperature. The successive wetting of the clay and the enrichment of NaCl are obvious (adapted from Pusch, 1994).

the host rock, this process may continue for hundreds of years. During the hydration period, water migrating from the rock through the buffer gives off salt, as seen in Figure 8.3. After full water saturation, salt accumulation proceeds by diffusion of ions. These will promote precipitation of salts with reversed solubility, such as gypsum and chlorides, so long as the temperature gradient prevails. Salt accumulation is expected to be most important issue in the water saturation period of the clay buffer. Accordingly, it is necessary to determine what the rate of wetting of the buffer will be in rocks with different water-bearing capacities. If a rock is sufficiently permeable, water saturation of the buffer will be controlled entirely by the suction forces in the buffer. If the rock is not permeable, the time taken for the buffer to suck water from the rock will be substantial. This will obviously impact on the time taken to achieve complete water saturation of the buffer, thereby increasing the risk of salt accumulation near the canisters.

The Stripa Underground Laboratory Study, Sweden

In the period 1981–85, a large-scale experiment, named BMT, was conducted at a depth of 360 m in SKB's (Swedish Nuclear Fuel and Waste Management Company) underground laboratory in granitic rock, the purpose of which was to investigate the performance of smectite-rich, dense bentonite clay under repository-like conditions (Gray, 1993). The test drift, which is shown in Figure 8.4, had six core-drilled 'canister' deposition holes, 0.76 m

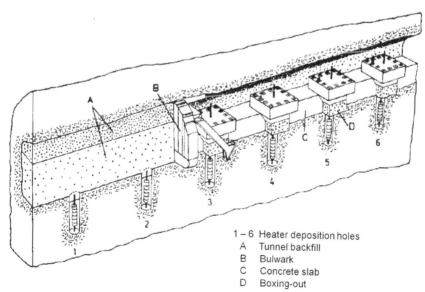

Figure 8.4 Perspective view of the BMT test drift at the Stripa underground laboratory.

1 – 6 Heater deposition holes
A   Tunnel backfill
B   Bulwark
C   Concrete slab
D   Boxing-out

in diameter and about 3 m in depth. Clay buffer with a dry density of about 1650 kg/m³ and 10 per cent water content by weight was used to surround electrical heaters with 600 to 1200 W power simulating the heat production of HLW canisters. The power was up to 1800 W in certain tests, resulting in a canister surface temperature of up to 150°C.

The thermal, hydraulic and mechanical (THM) evolutions of the clay buffer were recorded by various pieces of instrumentation for a period of 3 years. The evolution of the buffer under different wetting conditions was determined through examination of the distribution of the water content in the buffer at the termination of the tests, and the swelling pressures were recorded by pressure cells. The degree of water saturation at the termination of the tests was determined by quick sampling, and the swelling pressure was accurately measured using Glötzl pressure cells cemented in recesses that had been cut in the rock at three levels.

The tests indicated that water uptake from the rock was clearly related to the rock structure and its ability to supply the clay buffer with water. A fracture zone was located where holes 1 and 2 had been bored. This provided an unlimited supply of water for maturation of the buffer. Holes 3 and 4 were intermediate with respect to water inflow, whereas hole 5 was intersected by a discrete, steep fracture that supplied this hole with water to nearly the same extent as the fracture zone. Hole 6 was not intersected by any significant water-bearing fracture. The rock in this region represented poorly water-bearing 'dry' rock. Information on water inflow in the holes before placement of canisters and blocks of buffer clay can be seen in Table 8.1. The table also shows heating data for the electric heaters that simulated the HLW canisters.

The tests conducted in holes 1 and 2 lasted for 3.3 years and led to practically complete water saturation and water distribution in the buffer, and are therefore not included here. For the other holes, the conditions at mid-height of the heaters – the hottest parts – are shown in Table 8.2. The evolution of the buffer in hole 4 was similar to that in hole 3 and is not described here. The apparent concentric pattern of the 'iso-moisture' curves obtained at all the levels examined in the holes demonstrated that the water sorbed by the buffer clay had been radially distributed in a very uniform fashion, despite the fact that water access from the rock was in most cases from one or two fractures. The results shown in Figure 8.5 for three different hole-wetting conditions (intermediate wet, wet and dry) are average values for the corresponding levels (heights) as a function of the distance from the heaters at the termination of the tests. The different accesses to water in holes 3, 5 and 6 (compare with Table 8.1) gave very different water content distributions in the buffer. The curves representing holes 3 and 6 are concave upwards, i.e. of diffusion type, whereas the shape of the curve for hole 5 is upward convex, suggesting that pressure-driven flow contributed to the wetting of the buffer.

Table 8.1 Heating data (Pusch et al., 1985)

Hole number	Inflow of water in hole before test start (L/day)	Electric power (W)	Time (months)	Max temperature at heater/buffer contact (°C)	Max temperature at rock/buffer contact (°C)
1	6.2	600	40	68	36
2	5.6	600	40	66	38
3 and 4	3.0	600	15	81	35
5	1.9	600	27	64	34
6	0.2	600	25	81	35

Table 8.2 Parts from which data were derived

Hole number	Electric power (W)	Initial water content (%)	Time (months)	Buffer at mid-height; from the floor of the drift (m)
3	600	10	15	1.50–1.80[a] 1.80–2.05
5	600	13	27	1.50–1.80[a] 1.80–2.05
6	600	10	25	1.50–2.05

Note
a Two intervals were investigated because of obvious differences in water content.

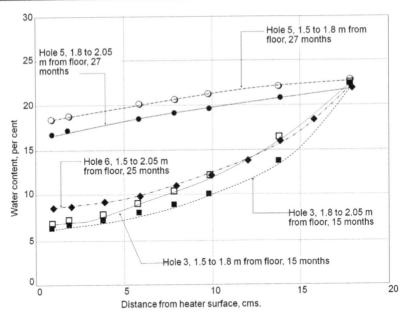

Figure 8.5 Water content distribution in (a) 'intermediate-wet' hole 3 after 15 months; (b) 'wet' hole 5 at termination after 27 months; and (c) 'dry' hole 6 at termination after 25 months. Note that the heaters for all three holes were 600-W heaters, and that for holes 3 and 5, there is a fracture-rich interval at 1.50–1.8 m from the floor.

Swelling pressure development for holes 1, 2 and 5 can be seen in Figure 8.6 and for holes 3 and 6 in Figure 8.7. The swelling pressure was highest in the wetter holes 1 and 2 (4–8 MPa after 20 months) and somewhat less in hole 5, which was less wet (Table 8.1), resulting in a swelling pressure of 3 MPa after 20 months. Significantly lower swelling pressures were obtained for the other holes, which were even drier.

The swelling pressures obtained in the various holes varied much more than the water content distribution. Thus, although the water content was 17–23 per cent at a distance of 15–18 cm from the heater surface in holes 3 and 6, and 20–23 per cent at the same distance from the heater surface in hole 5, the swelling pressure after about 20 months was around 3 MPa in the last-mentioned hole but only 0.3–0.5 MPa in holes 3 and 6. In the interval that was 10–15 cm in distance from the heater, the water content was 10–17 per cent for holes 3 and 6 and 19–23 per cent for hole 5. This shows that the buffer in hole 5 had undergone much more wetting closer to the heater than in holes 3 and 6 in which the pressure exerted by the largely water-saturated peripheral part of the buffer was not effectively transferred to the less saturated part. In hole 5, as well as in holes 1 and 2, the largely saturated clay in the vicinity of the heater provided sufficient support to the peripheral part to make the swelling pressure rise to a high value.

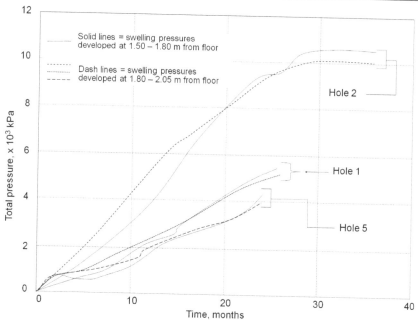

Figure 8.6   Evolution of swelling pressure in holes 1, 2 and 5. Complete water saturation with build-up of full swelling pressure occurred in hole 2.

The tests at Stripa led to the following major conclusions:

- The ability of the rock surrounding canister deposition holes to supply the clay buffer with water is of fundamental importance for its maturation rate. Where the average hydraulic conductivity of the rock is less than 10^{-11} m/s, complete water saturation of the buffer will take tens to hundreds of years. When the hydraulic conductivity ranges between 10^{-11} and 10^{-10} m/s, water saturation of the buffer clay will take from several years to decades. For hydraulic conductivity values higher than 10^{-10} m/s, full water saturation will be attained in a few years.
- Irrespective of whether the surrounding rock provides water through a single fracture or several fractures, the water content in the buffer grows uniformly in a tangential direction, and hence also so does the density of the buffer. More fractures result in more rapid wetting, implying that the discharge capacity of the rock initially controls the time for the clay buffer to reach water saturation.
- The driving force for attaining homogeneity is the creep potential. This is related to the water content. Where there are gradients in water content, there will be a tendency to seek minimization of the differences in water contents and densities. If the density is uniform in the buffer after a long time, the classical concept of internal friction in smectite

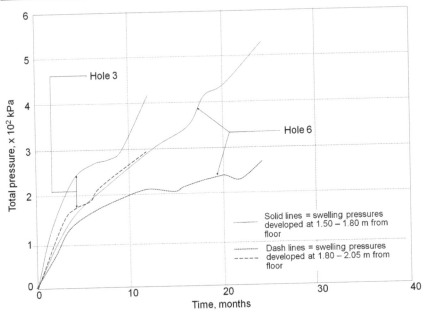

Figure 8.7 Development of swelling pressures in holes 3 and 6 with time.

clay would not be valid. As discussed in Chapter 6, creep may lead to such a condition.

Febex study, Grimsel, Switzerland

The Febex study has been used for investigating and developing techniques for disposal of HLW in crystalline rock. Febex (Full-scale Engineered Barrier Experiment) was an international project operated at Grimsel, Switzerland, undertaken by France, Germany, Spain, Japan and Switzerland. The objective of the project was to test a concept of HLW containment that included two horizontal canisters surrounded by smectite-rich, dense buffer in horizontal 2.4-m-diameter tunnels in granite. It was co-financed by the EC under contract FI4WCT950006 (Svemar, 2005).

The three main objectives of the Febex proejct were defined as follows:

- demonstrating the procedures and equipment required for a specific HLW disposal concept, with particular emphasis on the production, handling, and emplacement of highly compacted bentonite blocks on a semi-industrial scale;
- evaluating, improving and validating numerical models of coupled thermohydromechanical evolution of the EBS; the natural piezometric pressure in the test area was 0.85 MPa and the heat production of the HLW was simulated by electrical heaters;

- assessing geochemical processes within the EBS, including canister corrosion, gas generation/transport and solute migration.

In general, all of the observations confirmed the predictions of the pre-operational THM modelling, which was made using CODE_BRIGHT (Svemar, 2005). Figure 8.8 shows the predicted temperature evolution, which was close to the recorded values. The heater power was adjusted to maintain a temperature of 100°C at the canister/buffer contact (i.e. inter-face). The maximum temperature at the rock/buffer contact was about 40°C.

The buffer material used for the Febex project consisted of sector-shaped blocks of highly compacted montmorillonite-rich Spanish bentonite of Ca type that had been converted to Na form. They were placed around heaters that simulated canisters with reprocessed HLW. The clay was prepared by compressing clay powder to a maximum dry density of 1600 kg/m³ in order to limit the swelling pressure to 5 MPa. One important outcome of the test is illustrated by the graphs shown in Figures 8.9 and 8.10. The evolution of the maturation in terms of swelling pressure was naturally of utmost importance.

Figure 8.10 shows similar changes in water content to the Stripa project, i.e. drying at the heater content ('inner ring') and nearly complete satura-tion from start at the rock contact ('outer ring'). At mid-distance from the heater the water content increased in the first 2 months and then dropped

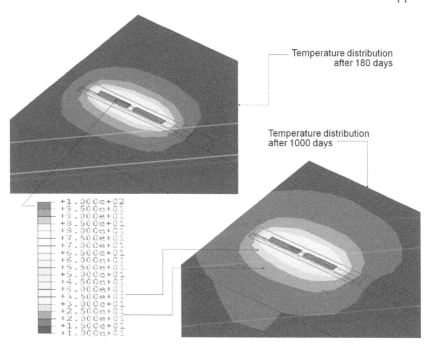

Figure 8.8   Temperature distribution (°C) in the near field rock and Febex tunnel after 180 days and 1000 days (adapted from Svemar, 2005).

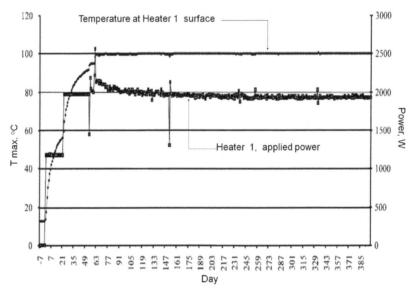

Figure 8.9 Power and temperature recordings for one of the heaters (adapted from Svemar, 2005).

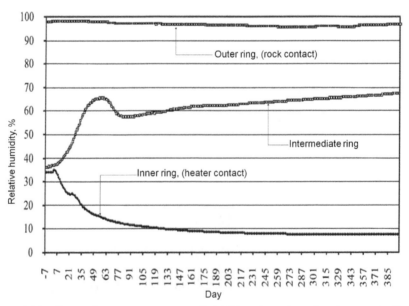

Figure 8.10 Changes in water content as indicated by relative humidity changes signalled by moisture sensors (adapted from Svemar, 2005).

somewhat in the subsequent few weeks, after which it slowly grew almost linearly with time. The peak after 400 days (not shown in Figure 8.10) was most probably caused by leakage of water along the instrument cables, followed by self-sealing of the flow path by expansion of the surrounding clay as indicated by the subsequent temporary pressure drop. This phenomenon, which is common in field and mock-up experiments, can cause serious misinterpretation of the rates of hydration and associated growth of swelling pressure. The impact of improper instrumentation on results is an important consideration that cannot be overlooked.

The question of the accuracy of the predicted maturation of the buffer is illustrated in Figure 8.11. This shows the development of the total pressure with time. A most important observation is that the predicted pressure growth near the heater was underestimated, whereas it was highly overestimated for the buffer close to the rock in the first 2 years. The same is true for the buffer at mid-distance from the heater in the first 3 years and presumably much longer than that. The models used for predicting the evolution of the buffer are responsible for the frequently observed discrepancies.

The Febex project included measurement of gas pressure evolution. This was achieved through the use of filter pipes installed in the clay buffer around one of the heaters. It was found that there was no pressure in the first months of heating. This was explained by the assumption that released gas escaped through the small gap between buffer and the bulkheads that

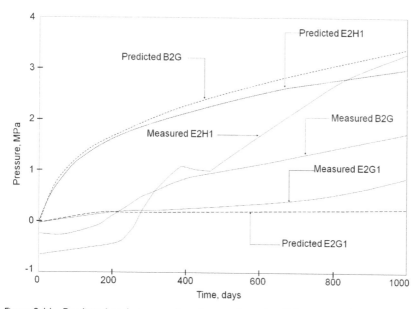

Figure 8.11   Predicted and measured evolution of total radial stress with time at point E2G1 (radius 1.2 m), at point E2H1 (radius 0.48 m) and total axial stress at point B2G (radius 0.80 m in section B2) (adapted from Svemar, 2005).

confined the tested part of the tunnel. However, gas pressure developed early in the filter pipes near the rock wall. Gas conductivity of the buffer was determined by injecting nitrogen into the filter pipes and recording the pressure developed during injection and in the subsequent shut-in phase. As long as the joints between the blocks were not closed, the permeability of the backfill was too high to be recorded. At a later phase, when self-healing occurred through wetting of the buffer, recording of gas permeability could be implemented. This was determined to be $10^{-18}\,m^2/s$.

The study, which included determination of the chemical composition of the gas, provided the following conclusions:

- Carbon dioxide was generated in significant amounts at temperatures exceeding 50°C, i.e. up to $0.1\,m^3$ per 1000 kg dry buffer, and up to $0.35\,m^3$ per 1000 kg of wet buffer.
- There was enough oxygen adsorbed by the buffer for oxidation of the organic components and the hydrocarbons.

Prague study

A three-year mock-up test simulating SKB's concept KBS-3V on half-scale was undertaken at the Technical University in Prague during the earlier part of this century, giving detailed information on the performance of the Czech clay buffer candidate 'RMN' under repository-like conditions (Pacovsky *et al.*, 2005). The test is reported here because it indicates how smectite clay that is rich in montmorillonite, with more than 5 per cent iron, performs under such conditions. The test arrangement is illustrated in Figure 8.12.

In the mock-up experiment, the host rock was represented by an 80-cm-diameter insulated steel tube with a filter consisting of a clay pellet system attached inside for uniform wetting of the buffer (Figure 8.12). The buffer consisted of highly compacted blocks of clay with 85 per cent calcium montmorillonite, 10 per cent finely ground quartz and 5 per cent graphite powder (Figure 8.13). The dry density of the clay blocks was $1800\,kg/m^3$. The radial thickness was 18 cm and the water content was 7 per cent. The 50-mm gap between the steel tube and the heater was filled with 'RMN' granules. Samples could be taken by moving 3.5-cm pipes through the steel tube, all the way to the heater. The thin-walled pipes had cutting edges that could penetrate the buffer blocks by rotation and pushing. Sample holes were sealed with equally dense clay plugs. The desired constant 95°C temperature at the heater surface in the larger part of the experiment gave a temperature of 45–48°C at the outer boundary of the clay buffer at mid-height (temperature gradient 2°C/cm). The filter was saturated with water that was under 60 kPa of pressure. Nearly complete saturation was reached after 2 years.

The objective of the study was to determine what major mineralogical changes occurred in this type of buffer clay under repository-like conditions.

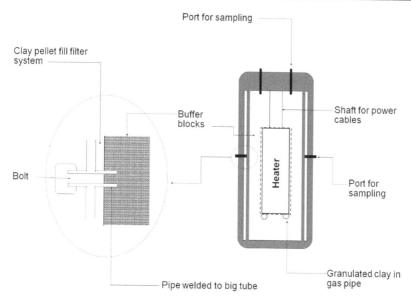

Figure 8.12 Czech mock-up test. The detail shows the arrangement for sampling in the course of the experiment. Circles in the bolt-head represent O-ring seals.

Figure 8.13 Sector-shaped block of 85 per cent smectite-rich, iron-holding montmorillonite clay, with some vermiculite, mixed with 10 per cent quartz powder and 5 per cent graphite to enhance heat conductivity. The compaction pressure used was 100 MPa (adapted from Pusch and Yong, 2006).

The water supplied for artificial wetting was natural granitic groundwater with 1000 ppm dissolved salt – with Na as the dominant cation and with negligible K and Ca content. The major anion was Cl. No theoretical predictions of the THMC performance were made. The first phase of the experiment involved heating the 1-kW heater under closed conditions, i.e. with no uptake of water of the buffer and no possibility for escape of vaporized porewater. This condition, which led to the development of a temperature of about 82°C at the buffer/heater contact, represented the environment of the buffer clay in a tunnel excavated in low-permeable rock. The graphical record of temperature and water content changes with time is shown in Figure 8.14.

The first dry phase was followed by allowing water access to the circumferential filter. The temperature of the heater surface was increased to, and maintained at, 95°C. Heating, wetting and development of swelling pressure were recorded through a series of instrumentation. Samples were extracted two to three times per year to check the wetting rate. Samples taken at the termination were used for mineralogical analyses and determination of pertinent geotechnical properties (Table 8.3). The results show that a large part of the clay buffer had undergone significant changes. The ratio of the swelling pressures of the M2 and M4 samples, which had practically the same density, was $650:310 = 2.09$. The coldest zone had the same hydraulic

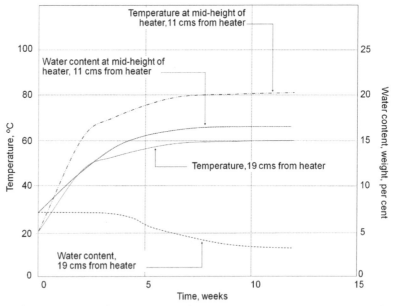

Figure 8.14 Evolution of temperature and water content at mid-height heater of the buffer under closed conditions. The 19-cm location from the heater is at the outer boundary of the clay blocks, just inside the pellet fill (adapted from Pacovsky et al., 2005).

Table 8.3 Geotechnical data of samples M1–M4 in oedometers

Sample	Distance from heater (cm)	T (°C)	ρ_{sat} (kg/m³)	p_s (kPa)	k (m/s)
M1	16–18	45–47	1800	430	1.6×10^{-11}
M2	12–14	54–56	1945	650	1.9×10^{-11}
M3	6–8	67–69	1910	355	2.3×10^{-10}
M4	0–2	85–90	1925	310	2.1×10^{-10}

Note
The data p_s = swelling pressure, ρ_{sat} = density at water saturation, and hydraulic conductivity = k, represent equilibrium reached after about 40 days (Pusch et al., 2005).

conductivity as virgin clay with the same density. On the other hand, for samples extracted at a distance of 10–18 cm from the heater, the hydraulic conductivity was about 10 times higher than that of an equally dense unheated RMN clay. The clay closer than 10 cm from the heater was up to 100 times more permeable than unheated clay with the same density. The swelling pressure showed the corresponding pattern, i.e. a marked drop for the samples taken close to the heater.

Mineralogical characterization of the original clay was made using transmission electron microscopy (TEM) with energy dispersive X-ray (EDEX) and coherent scattering domain data (CSD) to determine the thickness of particles or stacks of lamellae. The Koester diagram in Figure 8.15 shows the charge distribution in the various layers in untreated RMN clay. The information in this diagram can be used to compare with the distribution of charge in the various layers after termination of the experiment (Figure 8.16).

The changes in charge distribution occurring in M4 at the end of the experiment are obvious in Figure 8.16. The results tell us that significant dissolution of the montmorillonite and complete disappearance of the intergrowth of illite and kaolinite had occurred as a result of migration and precipitation of released elements at different distances from the heater. TEM-EDX photos of M4 indicated that desiccation fissures had been formed before resaturation. These fissures remained after water saturation in the oedometers, demonstrating that the clay was not sufficiently expandable to self seal. The X-ray diffraction (XRD) analyses of samples M3 and M4 showed a clear reduction in montmorillonite content and an obvious increase in illite particle thickness compared with M2 (compare with Table 8.4).

The main results from the Prague study were as follows:

- The swelling pressure had dropped by more than 50 per cent in the larger part of the buffer. The hydraulic conductivity had increased by 10–100 times.

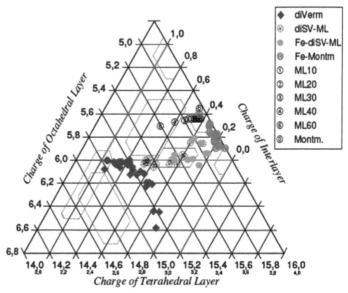

Figure 8.15 Koester diagram, showing distribution of charge in tetrahedral, octahedral and interlayers in untreated RMN buffer clay. ML, mixed-layer minerals. Note that the numbers shown are percentages (adapted from Pusch *et al.*, 2005).

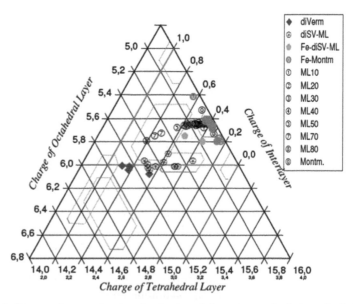

Figure 8.16 Koester diagram showing distribution of charge in tetrahedral, octahedral and interlayers in sample M4 (adapted from Pusch *et al.*, 2005).

Table 8.4 Mineralogical changes in the mock-up test (adapted from Pusch et al., 2005)

Mineral	Sample M2	Sample M3	Sample M4
Montmorillonite	Slight dissolution	Moderate dissolution	Significant dissolution, low crystallinity
Illite	High disorder, minor illite	Significant illite formed or less disordered	
Talc and kaolinite	Low disorder, particle growth	Increased disorder or neoformation	

- Intergrowth of illite and kaolinite was obvious in the dioctahedral vermiculite in untreated RMN clay. In these aggregates, illite dominates over kaolinite but in sample M4 most of the intergrowths of illite/kaolinite had dissolved.
- Fe set free by the dissolution of the iron montmorillonite could have formed iron complexes resulting in cementation in the entire buffer mass, especially in the most heated part (M4).
- Replacement of octahedral Al by Fe could have caused a drop in coherence of the montmorillonite crystals, thereby promoting easier dissolution.
- Formation of illite in the hottest part of the buffer (sample M4) may have been associated with uptake of K from dissolved vermiculite.

SKB/Andra (Agence Nationale pour la gestion des déchets radioactifs) study

Laboratory-scaled hydrothermal tests were performed on montmorillonite-rich clay (MX-80) under open conditions under a temperature gradient of about 6°C/cm for 1 year. The samples had been saturated with slightly brackish sodium-dominated and practically potassium-free water before placing them in 7-cm-long hydrothermal cells of iron. The confined sample had been saturated with weakly brackish Na as the dominant cation before the start of the experiment. The temperature of one of the ends, equipped with a filter, was held at 90°C, while the opposite end was heated to 130°C and exposed to strong γ radiation in one of the two otherwise identical experiments (Pusch et al., 1993). Evaluation of the impact from 1-year-long experiments with and without strong γ radiation gave the results shown by the XRD spectrum in Figure 8.17 and in Table 8.5.

The conclusion from these tests was that there were no significant changes of the clay. This conclusion is supported by the cation-exchange capacity (CEC) data showing that untreated MX-80 had a CEC of 99 mEq/100 g in comparison with the most harshly treated clay, which had a CEC of 93 mEq/100 g. Accessory minerals changed substantially, however. Potassium and sodium feldspars were partly lost at 95°C and completely dissolved at 130°C. At this temperature, strong precipitation of gypsum and anhydrite

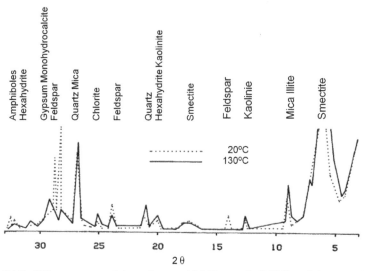

Figure 8.17   Diffractograms of the reference MX-80 sample (20°C) and the most heated part of the hydrothermally tested sample (130°C). Feldspars, amphibole, some of the quartz and smectite disappeared in the hot part.

Table 8.5 Summary of major changes in the hydrothermally treated montmorillonite-rich clay (MX-80)

Temperature (°C)	Loss	Increase
130–125	Feldspars***, quartz*, amphibole**, smectite*	Sulphate*, chlorides*, 10Å minerals*
125–120	Feldspars***, quartz*, smectite*	Chlorite*, sulphate*
120–115	Feldspars**	Chlorite*
115–110	Feldspars**	Anhydrite*, gypsum*
110–105	Feldspars*	Anhydrate*, quartz*
105–100	Feldspars*	Anhydrate**, quartz*
100–95	Feldspars*	Anhydrate**, quartz***

Impact: *insignificant, **significant, ***strong.

occurred at the hot end. It was estimated that the formation of chlorite was generated by iron that was released from the iron plate at the hottest end of the cell. The same mineralogical changes were identified in the tests with and without γ radiation, but radiation gave more obvious changes and migration of Fe to a larger distance from the hot end.

SKB studies in 'prototype repository'

A full-scale 5-year field experiment at SKB's underground laboratory at Äspö with very dense MX-80 clay buffer surrounding a full-size copper-lined

KBS-3 canister in an 8-m-deep, 1.75-m-diameter deposition hole in granite
was evaluated in 2007 with respect to the performance of the clay at the
end of the experiment. A filter had been placed at the walls of the hole
to provide the clay with access to groundwater. The groundwater had a
total salt content of about 6000 ppm with a Na/Ca ratio of about unity. The
canister had electrical elements simulating the heat produced by the highly
radioactive content of real-life HLW canisters. The surface temperature of
the canisters was maintained at 85°C for a couple of years and at successively
lower temperatures later in the experiment. The average radial thermal gradient was 1°C/cm. The evolution of temperature and swelling pressure at the
buffer/canister contact at mid-height canister, previously shown as Figure
6.7, is repeated as Figure 8.18.

A representative sample numbered 'R8:225: Canister' was taken from the
proximity of the heater and subjected to oedometer testing to determine if
the physical properties deviated from those of equally dense virgin MX-80
clay. In addition, studies were undertaken to determine what possible
chemical and mineralogical changes could have occurred. Virgin MX-80
clay, saturated and percolated with distilled water and with the same density
as a sample R8:225 taken from the vicinity of the canister, had an average hydraulic conductivity that was less than one-hundredth of that of this

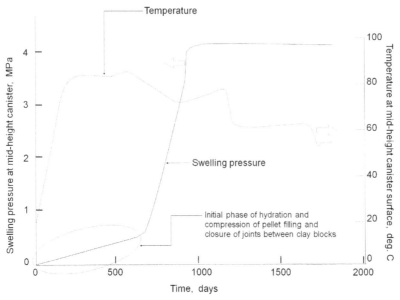

Figure 8.18 Evolution of temperature and swelling pressure at canister mid-height in
full-scale 5-year experiment in SKB underground laboratory using MX-80
bentonite as the buffer. Step-wise reduction in temperature is the result of
power failure of heater elements. The temperature ordinate scale is shown
on the right and the swelling pressure ordinate is on the left (adapted from
Pusch et al., 2007).

sample. In contrast, there was no difference between the swelling pressure of this sample and that of virgin MX-80. This tells us that although the swelling pressure of the clay adjacent to the canister did not indicate any physico-chemical changes, the 100-fold increase in hydraulic conductivity and the reduced dispersibility of this clay still showed that significant changes had taken place. Table 8.6 summarizes the data from tests of samples from the field experiment. The data were derived from comprehensive (> 100 speci-mens) particle-wise TEM-EDX measurements.

The most important findings were as follows:

- MX-80 clay contains two general types of montmorillonite: (a) montmorillonite with a normal charge as end member of IS–ml series and (b) low-charge montmorillonite as end member of diVS–ml series. The ratio IS/diVS varies from 1:20 to 2:3.
- The hydrothermally treated clay had undergone substitution of Al^{3+} by Fe^{3+} in the octahedral layer, causing higher lattice stresses because of the larger ion radius of Fe^{3+} than that of Al^{3+}, leading to a reduction in the resistance of the montmorillonite to dissolution. The change in chemical composition of the montmorillonite is illustrated by the derived formulae:

  - montmorillonite as end member of IS–ml series in virgin MX-80:

  $$Ca_{0.04}Mg_{0.09}K_{<0.01}Al_{1.64}Fe^{3+}_{0.13}Mg_{0.23}Ti_{<0.01}(OH)_2Si_{3.96}Al_{0.04}O_{10}$$

  - montmorillonite as end member of IS–ml series in hydrothermally treated clay:

  $$Ca_{0.07}Mg_{0.02}Na_{0.04}K_{0.07}Al_{1.46}Fe^{3+}_{0.23}Mg_{0.25}Ti_{0.03}(OH)_2Si_{3.98}Al_{0.02}O_{10}$$

- The hydrothermally treated clay had undergone step-wise alteration from a normally charged montmorillonite to a low-charge montmorillonite through replacement of original octahedral Mg^{3+} by Al^{3+}. The composition of the montmorillonite as end member of diVS–ml series in the hydrothermally treated clay had the following form:

  $$Ca_{0.02}Mg_{0.04}Na_{0.01}K_{0.03}Al_{1.62}Fe^{3+}_{0.16}Mg_{0.17}Ti_{0.04}(OH)_2Si_{3.99}Al_{0.01}O_{10}.$$

- Montmorillonite is the dominant mineral phase in both virgin and hydrothermally treated clays. However, mineralogical changes induced by the hydrothermal conditions are obvious, as indicated by an increase in smectite layer frequency in the IS–ml phases of the heated clay (95 per cent) compared with the typical number for virgin MX-of 80 (85 per cent), and by an associated drop in smectite layer frequency in the

Table 8.6 Summary of physical data of the sample adjacent to the canister at mid-height, and of virgin MX-80 clay with the same density, i.e. 1970 kg/m³ (1540 kg/m³ dry density) saturated with distilled water or artificially prepared ocean water (3.5% salt content) (adapted from Pusch, 2008)

Sample	R8:225: Canister, parallel to canister	R8:225: Canister, perpendicular to canister	Virgin MX-80, distilled water	Virgin MX-80, ocean water
Hydraulic conductivity (m/s)	2×10^{-11}	2×10^{-11}	8×10^{-14}	2×10^{-13}
Swelling pressure (MPa)	4.2	4.3	4.0	3.5–4.0

diVS–ml phases from 90 per cent of the virgin MX-80 to 80 per cent in the heated clay (Table 8.7).

- The higher interlayer charge of R8:225 in comparison with the virgin MX-80 would imply a lower swelling pressure according to classical double-layer theory. However, laboratory investigations show the opposite, as shown by the results portrayed in Figure 8.19, and by the fact that the recorded swelling pressure of R8:225 was higher than that of equally dense virgin MX-80 clay (Table 8.6).

Hydrothermal treatment under water-saturated conditions alters the microstructural constitution as illustrated by Figure 8.20, the main impact being contraction of stacks of lamellae provided that the clay is confined. It results in development of larger and more interacting voids and denser particle network, which increases the hydraulic conductivity and the shear strength (Pusch and Yong, 2006). If no chemical changes take place then the microstructural change would be reversible, presumably with some hysteresis.

The mineralogical investigation of sample R8:225 showed altered particle morphology. The typical appearance of interwoven thin stacks of montmorillonite lamellae has been locally changed to a mass of kaolinite-like, discrete particles (Figure 8.21). This change, which is assumed to have taken place

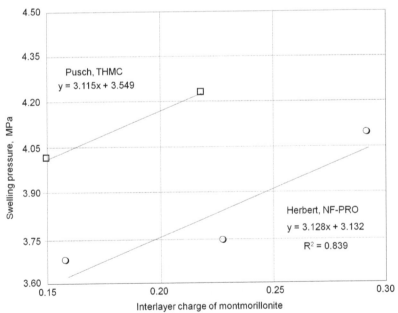

Figure 8.19 Comparison of swelling pressure versus charge for MX-80 clay, showing the trend of increased pressure for increased charge in different experiments using different techniques.

Table 8.7 Parameters of mineral formula per $[(OH)_2O_{10}]$ from TEM-EDX analyses (after Kasbohm and Thao) (adapted from Pusch, 2008)

Parameter	Virgin MX-80	R8:225: Canister
Phases	Montmorillonite (low-charge montmorillonite), diVS–ml	Montmorillonite (low-charge montmorillonite), diVS–ml
Traces	IS–ml	IS–ml
	PSV–ml	KSV–ml
		'Albite'
Smectite layer-frequency (S%)		
In IS–ml phases	85	95
In diVS–ml phases	90	80
Interlayer charge (average of all)	0.222	0.305

Note

diVS–ml, dioctahedral vermiculite–smectite mixed-layer phases; IS–ml, illite–smectite mixed-layer phases; KSV–ml – kaolinite–smectite–dioctahedral vermiculite mixed-layer phases; PSV–ml, pyrophyllite–smectite–dioctahedral vermiculite mixed-layer phases.

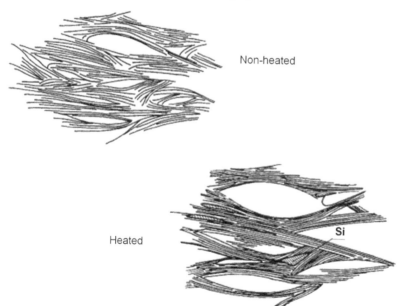

Non-heated

Heated

Si

Figure 8.20 Microstructural changes in hydrothermally treated MX-80 clay (adapted from Pusch et al., 1991).

in the softest parts of the heterogeneous microstructural network, may have resulted from significant crystal reorganization or neoformation contributing to the very significant increase in hydraulic conductivity found in the oedometer testing (Pusch and Yong, 2006).

Figure 8.21 Partially hypidiomorphic, pseudohexagonal particles in sample R8:225.

8.2.4 Long-term mineralogical changes of the buffer

Mechanisms involved in conversion of smectite to non-expandable minerals

The described physical and mineralogical changes appeared early in hydro-thermal experiments and prolonged exposure to repository conditions is expected to cause further changes of the same type. The most important long-term process is considered to be conversion of smectite to illite and precipitation of silica according to equation 8.1, defined earlier in the book:

$$\text{Smectite} + K^+ + Al^{3+} \rightarrow \text{illite} + Si^{4+} \qquad (8.1)$$

Although the reaction shown in equation 8.1 is valid for the entire smectite family, the activation energy for conversion and dissolution varies. Montmorillonite is the most common smectite species in nature, and is commonly proposed for use as a buffer clay as it has the best gelation properties. However, extensive experience in deep drilling shows that the physical properties of trioctahedral smectite saponite is less strongly affected by calcium and salinity. Furthermore, it is judged to have a higher chemical stability than montmorillonite at high temperatures and saline conditions, implying thereby a higher activation energy for conversion of saponite to illite than that of montmorillonite (Gueven *et al.*, 1987).

Although the exact mechanism involved in the conversion of smectite to illite is not known, one of two possible reactions has been proposed: (a) successive intercalation of illite (I) lamellae in smectite (S) stacks forming mixed-layer (IS) minerals, or (b) neoformation of illite by coordination of Si, Al, K and hydroxyls to an extent controlled by the access to either of them. If one assumes that illite is formed through mixed-layer phases, the release of silica can be explained by heat-induced transformation of montmorillonite of *trans-* to *cis-*form. The two forms are illustrated in Figure 8.22, both being proposed by German mineralogists some 70 years ago. The Hofmann, Endell and Wilm (HEW) model has been interpreted to represent low-temperature conditions whereas the Edelmann and Favajee (EF) model, i.e. the commonly assumed crystal structure, has been claimed to represent the high-temperature version. According to Forslind and Jacobsson (1975), theoretical crystal modelling and nuclear magnetic resonance experiments show that heating and saturation with calcium instead of sodium or lithium cause inversion of the HEW lattice constitution to the EF stage. Water is thereby released as a consequence of the collapse of the interlamellar water structure and silicons from tetrahedral sites are set free, thereby facilitating replacement of this ion by octahedral aluminium. In turn, the aluminium can be replaced by magnesium or iron available in the porewater.

Further investigation of the assumed transformation of montmorillonite to illite has been made with the use of the 'magic angle spin/nuclear magnetic resonance' (MAS/NMR) technique (Pusch *et al.*, 1991). Recent experiments

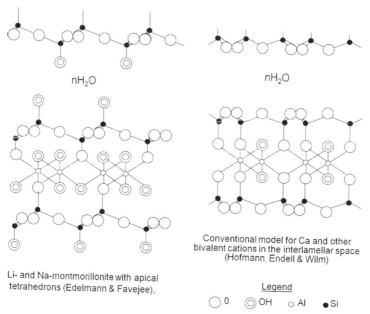

nH_2O nH_2O

Li- and Na-montmorillonite with apical tetrahedrons (Edelmann & Favejee).

Conventional model for Ca and other bivalent cations in the interlamellar space (Hofmann, Endell & Wilm)

Legend
○ 0　◎ OH　o Al　● Si

Figure 8.22 Proposed temperature-dependent montmorillonite crystal models. Left, lithium and sodium montmorillonite with apical tetrahedrons (Edelmann and Favejee); right, conventional model for Ca and other bivalent cations in the interlamellar space (Pusch and Yong, 2006).

using the same technique have shown similar results (Mantovani *et al.*, 2008). The ^{29}Si tests in the aforementioned study showed that step-wise heating to 200°C of iron-poor montmorillonite indicated a successively lower intensity of the main 93 ppm peak. This was interpreted to be the result of transition of EF to HEW crystal structure, as association of silicons with OH-groups as in the EF model would have yielded a stronger signal at cross polarization than when no protons are close to the silicons. The effect was shown to be much weaker in the heating of Ca-montmorillonite, suggesting that the HEW structure prevailed both before and after the heat treatment when Ca was in interlamellar positions.

The change in Si and Al correlation by heating of montmorillonite in distilled water using MAS/NMR testing is further supported by the diagrams in Figures 8.23 and 8.24. The first one shows the obvious temperature-related change of the 'second' Si peak (–107 to –118 ppm) at cross-polarization. The change in shape may be interpreted as crystallization of cristobalite of amorphous Si, but since the content of such matter is almost none it is much more likely associated with permanent transition from *trans-* to *cis*-coordination of silica (Si-O).

The same, very iron-poor montmorillonite used in the aforementioned Si studies was saturated with Na and Ca in different ^{27}Al magic angle spin (MAS) experiments of unheated and samples hydrothermally treated to 90–130°C (Pusch *et al.*, 1991). The study showed an absence of tetrahedral Al in the virgin material. However, some Al had entered tetrahedral positions in the hydrothermally treated clay – the small extent of which is ascribed to the processes associated with the very small amount of cations that could replace released octahedral Al, such as magnesium, iron or external aluminium in the porewater.

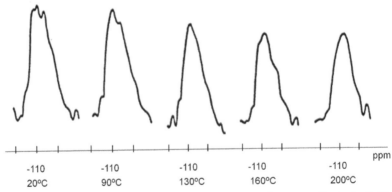

| -110 | -110 | -110 | -110 | -110 | ppm |
| 20°C | 90°C | 130°C | 160°C | 200°C | |

Figure 8.23   Change in shape of the ^{29}Si spectrum by heating of montmorillonite in distilled water using MAS/NMR. The diagram shows the obvious temperature-related change of the 'second' Si peak (–107 to –118 ppm) at cross polarization for the temperature interval 20–200°C.

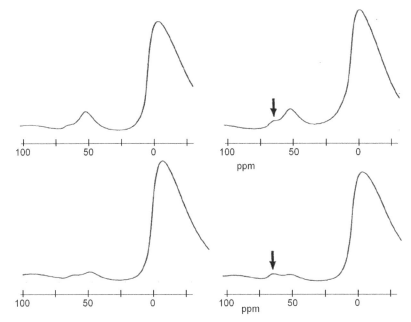

Figure 8.24   Change in shape of the ^{27}Al spectrum by heating of Na-saturated (upper) and Ca-saturated (lower) montmorillonite using MAS/NMR. Left row shows spectra of unheated material and right row shows spectra of hydrothermally treated montmorillonite (200°C). Arrows indicate characteristic peaks of tetrahedral aluminium. The treatment gave some small changes.

Recent ^{29}Si and ^{27}Al MAS/NMR investigations of montmorillonite and kaolinite associated with XRD and SEM microscopy of hydrothermally treated samples at temperatures up to 300°C (Mantovani *et al.*, 2008) show that in high-pH environments, akin to those found in concrete, trioctahedral smectites, such as hectorite and saponite, are much more stable than dioctahedral smectites (montmorillonite, beidellite and nontronite). Summarizing these findings, it can be stated that although the detailed processes in hydrothermally induced changes in smectite buffer under repository conditions are not very well understood, the major mechanisms in conversion are known. Thus, it is believed that while silicons can be released from the crystal lattice thereby resulting in precipitation as a cementing agent, aluminium is largely maintained in the smectite crystal lattice in electrolyte-poor water. Aluminium will migrate to form beidellite and further to illite in electrolyte-rich water through uptake of K from the porewater. The K-source can be either the electrolyte content of the ambient groundwater or feldspars in the backfills or near-field rock.

Both conversion to illite and precipitation of silica have a degrading impact on a smectite buffer: first, by reducing the hydration capacity and expandability and, second, by welding the particle stacks together so that plasticity and expandability can be lost. Although illitization is commonly regarded as the most unwanted reaction, it is in fact not disastrous as the bulk hydraulic conductivity of a dense buffer will not be dramatically changed. Silicification, on the other hand, can change the waste-isolating capacity to the extent that it will become unacceptable. We shall therefore look at this phenomenon in some detail, first by considering actual occurrence of silica precipitation in nature, and second by considering theoretical predictions of the extent to which it can affect the buffer clay in a repository.

Silicification, the most threatening buffer degradation process

Mineralogical analyses of hydrothermally treated smectite clays fully support laboratory-based conclusions concerning illitization and release and associated precipitation of silica in the form of quartz or cristobalite. Conversion of smectite minerals to non-expanding illite, chlorite and kaolinite, and dissolution of accessory minerals such as feldspars and micas, are associated with release, migration and precipitation of released silica. These have been documented by natural analogues represented by the Kinnekulle bentonites (Pusch, 2008) and the Libyan bentonite (Pusch and Yong, 2006), as well as by North Sea sediments. The silica released in the smectite–illite conversion process in nature, as well as in hydrothermal laboratory tests, migrates in the form of H_3SiO_4 and precipitates in colder parts of the buffer and in the near-field rock. Laboratory experiments provide conclusive evidence of cementation by silica precipitation.

Comparison of untreated MX-80 clay and clay hydrothermally treated for 1 year with 90°C and 130°C at opposite ends, described earlier as an SKB/Andra project, included creep tests that testified to significant stiffening of the clay. The higher the temperature, the smaller is the shear strain of samples (see Figure 8.25). The stiffening has been interpreted as the effect of microstructural reformation involving contraction of stacks yielding an increased hydraulic conductivity (Figure 8.20) combined with jointing together of the stacks by precipitation of silicious matter, primarily quartz as specified in Table 8.5, or Fe compounds (Pusch et al., forthcoming). Precipitation of sulphates or chlorides could not have contributed significantly to the stiffening, as the artificially prepared water contained no elements of these types.

In a comprehensive project with MX-80 samples conducted in cells that were uniformly heated, i.e. without temperature gradients, to temperatures up to 200°C, experiments were conducted in which hydraulic interaction was allowed with larger vessels containing distilled water, ocean water diluted by 50 per cent, and 3.5 per cent $CaCl_2$ solution in separate experiments. Mineral analyses conducted by XRD showed that feldspars were largely dissolved at 160°C and disappeared at 200°C. However, the heights and areas of the characteristic XRD peaks of montmorillonite were not significantly

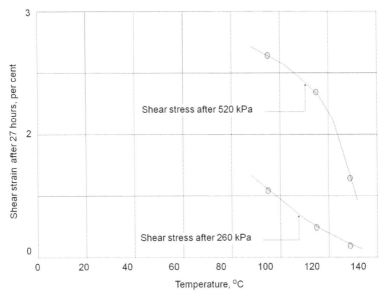

Figure 8.25 Shear box testing of MX-80 clay that had been hydrothermally treated with and without strong γ radiation. Normal stress against shear plane is 6 MPa (adapted from Pusch *et al.*, 1991; Pusch, 2008).

altered, indicating only a slight attack on these minerals. This is surprising, as the pH dropped by about 1 unit from 10 at room temperature for distilled water and from 7–8 for salt solutions, to 4–8 by heating to 130–150°C. This means that dissolution of any carbonate in the buffer must have taken place. Very acid conditions were created by heating to 200°C, which had a naturally strong dissolving impact on all minerals, still without significant impact on the crystal constitution of the smectite. The concentration of Si had increased from around 10 ppm to more than 90 ppm in the pore fluid at the termination of the tests. The components set free by the low pH attack, primarily silica, were precipitated when cooling took place at the termination of the experiments, causing cementation and brittleness as manifested by an increase in undrained shear strength as determined by vane testing. The shear strength doubled or trebled by heating from room temperature to 130–160°C, but dropped at higher temperatures, presumably by dissolution and loss of solid matter.

Kinnekulle – a north European natural analogue

At Kinnekulle, in south-western Sweden, a series of up to 2-m-thick beds that were formed from volcanish ash sedimented in ocean water in Ordovician time were exposed to an ice load of at least 30 MPa of pressure in

Quaternary time. The Ordovician and overlying Silurian sedimentary series, comprising also limestone, shales and sandstones, were penetrated by magma intrusion in Permian time. The temperature evolution of the clay beds has been determined and found to be 120–140°C in the first 500 years after the magma intrusion, followed by a successive drop to about 90°C after 1000 years. Conodont analysis confirmed that the highest temperature was in this order of magnitude (Pusch and Madsen, 1995). The average thermal gradient was 0.02°C/cm in the first hundreds of years

The present montmorillonite content in expandable form (mixed-layer smectite/illite in the bentonite bed is about 25 per cent, whereas illite makes up about 50 per cent. Chlorite/kaolinite, carbonates, quartz and feldspars make up the rest. The water content is 23–35 per cent, indicating that the clay has expanded from a denser state reached during the maximum ice load some tens of thousands of years ago (Pusch, 1983). The evidence of silicification is based on the presence of globular colonies at the rims of many particle stacks and basal surfaces (Mueller-Vonmoos *et al.*, 1990) as seen in scanning electron micrographs in the clay, and the difficulty of dispersion of clay material. Also, geological literature describes that the sediment rock adjacent to the bentonite beds is cemented by precipitated silica (Thorslund, 1945).

Libya – an African natural analogue

In Libya, a 5-m-wide diabase dyke, intersected by up to 5-m-thick sub-horizontal bentonite layers in Tertiary time, provides concrete evidence of cementation by precipitated silica (Pusch and Yong, 2006). The temperature of the bentonite next to the diabase reached a maximum temperature of 500°C after 1 month, and 200°C after 1 year. At about 3 m from the diabase, the maximum temperature had been less than 100°C. The major conclusions from the study were that (a) a silicified brittle zone in which smectite was partly converted to kaolinite was found next to the diabase contact; (b) the specific surface area had dropped close to the diabase; and (c) the grain size distribution indicated coarse, cemented particle agglomerates to within 0.2 m from the diabase contact.

North Sea sediments – deep-sea natural analogues

In the North Sea Basin, the Lower Tertiary sediments are dominated by smectite of mainly volcanic (bentonitic) origin (Roaldset *et al.*, 1998). Burial does not seem to have changed their original composition significantly despite exposure to about 110°C for several million years. Thus, the clay fraction commonly makes up 40–50 per cent of the total mineral substance, and 80–90 per cent of the fraction is smectite. However, there is a major change in dispersibility of the material. It does not readily disperse when submerged in salt water, which can be ascribed to cementation by precipitated silica. It makes the clays slightly brittle and those from the investigated depths, about 2000 m, are therefore termed 'claystone'.

Prediction of silica precipitation in the nearfield of canisters in a repository

In contrast to the conventional Reynold-type models proposed, the Takase and Benbow model of dissolution of smectite and formation of reaction products does not preset smectite to illite as the basic reaction (Grindrod and Takase, 1994). It is in better agreement with the actual smectite–illite (S/I) conversion and precipitation of silica of the Kinnekulle bentonite bed than solid–solution models. The model takes $O_{10}(OH)_2$ as a basic unit and defines the general formulae for smectite and illite as:

$$X_{0.35}Mg_{0.33}Al_{1.65}Si_4O_{10}(OH)_2 \text{ and } K_{0.5-0.75}Al_{2.5-2.75}Si_{3.25-3.5}O_{10}(OH)_2 \quad (8.2)$$

where X is the interlamellar absorbed cation (Na) for sodium montmorillonite. The reactions in the illitization process are:

$$Na_{0.33}Mg_{0.33}Al_{1.67}Si_4O_{10}(OH)_2 + 6H^+ = 0.33Na + 0.33Mg^{2+} + $$
$$1.67Al^{3+} + 4SiO_{2(aq)} + 4H_2O \quad (8.3)$$

causing precipitation of illite and silicious compounds:

$$K\,AlSi_3O_{10}(OH)_2 = K^+ + 3Al^{3+} + 3SiO_{2(aq)} + 6H_2 \quad (8.4)$$

with log K being a function of temperature.
The rate of the reaction R^* used is:

$$R^* = Ae^{-\frac{E_a}{RT}}K^+S^2$$

where A is the coefficient, E_a is the activation energy for S/I conversion, R is the universal gas constant, T is the absolute temperature, K^+ is the potassium concentration in the porewater, and S is the specific surface area for reaction. The practical application of the chemical model to SKB repository conditions requires coupling with the transport model – cast in one-dimensional cylindrical coordinates representing the buffer material. The kinetic reactions are linked with diffusion-dominated transport of the aqueous species, i.e. silica, aluminium, sodium, magnesium and potassium, to form a set of quasi-non-linear partial differential equations for the aqueous species and minerals. They were used to investigate the possibility of silica being precipitated in the direction of the thermal gradient. An important finding was that effectively no illite will be formed despite the conservative assumption of a constant temperature of 90°C at the buffer/canister contact, and 50°C at the buffer/rock contact in the first 500 years. On the other hand, there will be a significant loss of quartz due to dissolution/transport in the inner hot part of the buffer, as illustrated in Figure 8.26. However, assuming the same temperature conditions for 500 years followed by a linear temperature drop with time to 25°C after 10,000 years, precipitation of quartz will take place within about 0.1 m of the rock wall, as shown in the figure (Pusch *et al.*, 1998).

The Grindrod–Takase model appears to adequately describe the chemical evolution with respect to both illitization and silicification. The thermal gradient in the buffer will cause more dissolution in the hot clay at the canister surface than at the clay–rock interface, which generates transport of dissolved species particularly silica towards the cold side. It precipitates as quartz in the buffer (Figure 8.27) and in open fractures in the walls of the deposition holes, which tends to seal them off.

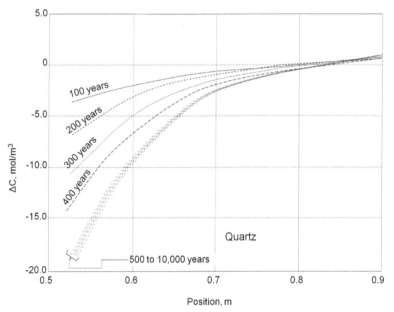

Figure 8.26 Evolution of quartz abundance profiles under conditions of constant temperature of 90°C at the buffer–canister interface and 50°C at the buffer–rock interface in the first 500 years, followed by a linear temperature drop to 25°C at 10,000 years after 500 years. The ordinate ΔC represents the loss of quartz, and the abcissa 'position' expresses the distance from the buffer–canister interface (hot part). Precipitation leading to cementation takes place in the outer and colder 0.1-m part of the buffer (adapted from Pusch *et al.*, 1998; Pusch and Yong, 2006).

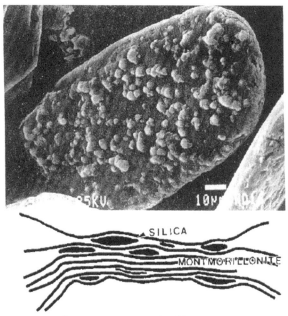

Figure 8.27   Precipitation of quartz in smectite clay. Upper, silt-sized particle in Kinnekulle bentonite with quartz precipitated on the basal surface (adapted from Mueller-Vonmoos *et al.*, 1990); lower, clay microstructure with precipitated cementing silica (Pusch, 1993).

8.3   Concluding remarks

The present ideas of the long-term evolution of the smectite buffer can be summarized as follows:

- In the early stage of evolution desiccation will take place in about 50 per cent of the buffer and frequent fissures and growth of initial voids will appear in the matrix of contracted stacks of lamellae. Precipitation of gypsum and anhydrate will take place, resulting in cementation. These precipitates will dissolve and diffuse out of the buffer when the temperature gradient no longer exists.
- Montmorillonite remains the dominant clay mineral in the buffer even at temperatures of up to 150°C in short-term experiments. However, the fact that natural clays exposed to temperature conditions similar to those of repositories have undergone significant loss of montmorillonite demonstrates that the content of this mineral in the buffer clay will be reduced by dissolution and illitization. The rate of illitization of montmorillonite is not yet known. For the Kinnekulle case, reduction by about 50–75 per cent may have taken place in about 1000 years. However, the higher temperature gradient – about 0.1°C/cm for the

buffer and less than 0.02°C/cm for the Kinnekulle case – may cause much quicker changes in a repository.

- The presence of Fe in contact with buffer, from canisters of iron or steel, will contribute to formation of chlorite and Fe oxyhydroxides. The latter can serve as cementing agent and will reduce or eliminate the expandability of smectite buffer.

- Precipitation of Si in the colder part of the buffer can weld the lamellae and stacks together and eliminate expandability and plasticity. Fe, released from iron-bearing minerals including smectites or from steel canisters, can play the same role. One can conclude that silicification is the most important chemical process in the buffer and near-field rock.

- An increase in shear strength and deformation moduli by around 50 per cent caused primarily by microstructural reorganization can become permanent with Si and Fe precipitation.

Chapter 9

Modelling for prediction and performance assessment

9.1 Modelling and computer codes

9.1.1 Introduction

Assessing and predicting the performance of a system consisting of complex interacting components and subsystems present challenges that require (a) keeping track of a multitude of individual and coupled processes and (b) integrating them in a coherent fashion that reflects how the system will performance under various conditions and scenarios. Mathematical models provide us with the capability to collect and integrate the various mathematical relationships developed to represent the many different processes and their interactions that ultimately result in the portrayal of system performance. Computer simulation of the processes and their interactions that have the benefit of not only using algorithms from mathematical models, but also empirical and general observational relationships in their architecture and computational programme, provide one with the capability for prediction and assessment of system performance.

It is not the purpose of this chapter to enter into a discussion of the development and use of analytical and computer models as such. The reader should consult the many available textbooks that deal directly with development of analytical computer models. The material presented in this chapter is meant to provide an overview of the modelling requirements and challenges as they relate to (a) clay barrier/buffer used in engineered multi-barrier systems for isolation of high-level radioactive wastes (HLWs) in repositories and hazardous solid wastes (HSW) in landfills, and (b) a summary account of some of the models used to address the repository and landfill situations. In the design of engineered barrier systems (EBSs) for deep geological repository containment of HLWs and secure landfills for HSWs, modelling of the integrated performance of all the involved components is required.

At the present time, a number of numerical (computer) codes and analytical methods are available. We have seen in previous chapters that the evolution of the clay buffer/barrier component in the EBS, as a natural consequence of the effects of ageing processes, will significantly change the nature and properties of the clay, with resultant impact on the design performance of the clay as an integral component of the EBS. Although modelling of the evolution of clay buffer/barriers can be reasonably performed for heat transfer, water saturation, and contaminant transport under isothermal and geometrically simple conditions, much remains to be learnt and incorporated in these models when the effects of age-related chemical, biological and radiation-induced processes are involved.

9.1.2   Models and computer codes

There are different types of computer simulation models (commonly known as computer models) available, depending on the purpose to which the models would be used. What separates them into three broad groups is not only the differences in the details of the input and output modules, but also the degree of complexity used to represent and analyse the various processes involved. These groups are:

• *System performance models (SPMs).* The aim of these models is to provide computational tools (computer models) that will inform on the performance of the system under consideration. Included in the major requirements of these models is (a) one's ability to understand or have full quantitative knowledge of the various physical, chemical, and biogeochemical processes participating in the development of system performance and (b) one's capability to account for these independent and interdependent processes in mathematical form and analytical–computational procedures. Determining how the system under consideration will perform under a variety of circumstances and conditions is perhaps the most sought after aspect of SPMs. As will be discussed later, the responsibility for developers of SPMs is to analytically and computationally mimic the operation of the system as a whole. Users of these models include designers, regulators, investigators and researchers.

• *Performance assessment models (PAMs).* These models are generally used as monitoring and assessment tools by regulatory and enforcement agencies responsible for the industry represented by the system under consideration. It is not unusual to see SPMs adapted for use as PAMs. Under such circumstances, changes to the nature of inputs used and output requirements are most frequently the main modifications made. For complex systems in which many interacting processes and couplings are present, and where initial and boundary conditions are not fully understood or identified, the accuracy of analyses or results produced from application of both SIMs and PAMs can be suspect. PAMs, together with SPMs, are useful in that they can help one to identify shortcomings in knowledge of processes and issues governing system performance.

- *Forensic-investigative models (FIMs).* These are commonly used by regulatory and enforcement agencies to determine the source of the problem being investigated, and to assign responsibility for that particular problem. It is not unusual for FIMs to use less complex models to answer specific questions. A good example of this would be the determination of source of contamination of a particular piece of ground.

This chapter will focus first on the general processes-related issues of system conceptualization for model development – including ageing processes – and the requirements for addressing these issues. Second, some commonly available analytical computer models and their computer codes will be examined. Interest is directed towards their capability in analysis and prediction of the status of the clay buffer that is designed and constructed to isolate the canisters containing HLWs. Application of the codes to predict the performance of prototype tests, such as those conducted in the underground research laboratory of the Swedish Nuclear Fuel and Waste Management Company (SKB) at Aespoe, Sweden, provides an insight into code validation.

9.2   System performance conceptualization

9.2.1   System and problem delineation

Developing models that reflect the performance of clay buffer/barrier components in engineered barrier systems requires knowledge of (a) the particular system to be analysed; (b) the various forces/fluxes acting on the system; (c) the external and internal processes involved; and (d) how the outcome of these processes (i.e. the performance of the system) depends on how these processes interact with each other and with the clay. As a start, one needs to have knowledge of the initial and boundary conditions. These are required pieces of information that help (a) to define the system setting (as illustrated in Figure 9.1) and (b) to provide the elements of the scenario for model application.

The question that needs to be asked is: 'What is the system to be analysed?' The answer to this question should lead directly to a delineation of the problem. Take, for example, the clay buffer and clay barrier shown in Figure 9.1 for the HLW canister-repository isolation scheme and the HSW landfill containment respectively. They are part of larger systems that are designed to contain HLWs and HSWs. For problem delineation, we need to consider the clay buffer and the clay barrier (left- and right-hand drawings in Figure 9.1 respectively) as individual systems in themselves. One could say that these individual systems are microsystems within their respective macrosystems. For example, the macrosystem of relevance in the HLW canister-repository scenario shown in Figure 9.1 includes the HLW canister and the repository

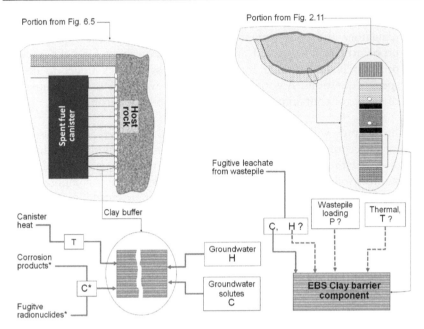

Figure 9.1 HLW canister-repository scenario (left, portion of drawing from Figure 6.5) and HSW landfill with bottom EBS shown on the right-hand side. The EBS drawing is a portion of the drawing from Figure 2.11. The bottom drawings under each isolation–disposal scenario represent the external forces/fluxes imposed on the clay buffer (left) and clay barrier component of the EBS (right). The asterisk shown in the left bottom for corrosion products and fugitive radionuclides refer to age-related (time-related) forces/fluxes, i.e. events that are not expected to occur until much later in the life of the repository. The dashed arrows in the right-hand side of the drawing are meant to depict situations that may or may not occur, depending on the magnitude and influence of the H, P and T factors.

host rock. Groundwater entry from the host rock provides hydraulic forces (hydraulic head) and chemical forces (solutes in groundwater) as external input to the clay buffer. Knowledge of the nature of the external forces and processes acting on these buffer/barrier microsystems is one of the main requirements for realistic prediction of microsystem response. Other requirements include knowledge of the processes generated, material properties, and response performance of the material in interactions with externally and internally generated processes. The outcome of these processes is seen with respect to fluxes generated and manifestations of these fluxes in terms of reactions, responses, and even transformations, as seen, for example, in Figure 9.2, which illustrates the macrosystem–microsystem relationship for HLW problem conceptualization.

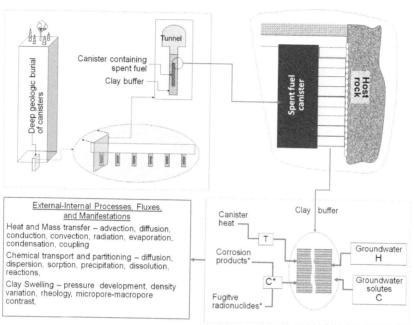

Figure 9.2 System delineation for problem definition for deep geological disposal of HLW using the repository concept, showing where the microsystem fits within the overall scheme of things.

Problem definition–delineation for the clay barrier component of EBSs for HSW or even municipal solid waste (MSW) landfill containment is, in a sense, much simpler than the HLW repository situation. In the 'garbage bag' scheme for land disposal, the clay barrier problem that needs to be analysed is, in essence, a problem that arises when failure of the overlying geomembrane–leachate collection system(s) occurs. In most instances, the clay barrier component of the EBS for HSW landfill containment is considered to be the third or fourth line of defence. Assuming that leachate from the waste-pile does escape from (i.e. penetrate) the overlying EBS membranes, leachate attack on the clay barrier provides the source for hydraulic and chemical transport. Although heat and overburden pressures on the clay barrier component of the EBS are measurable, these are not considered to be particularly significant in so far as their effects on clay performance is concerned. A simple set of measurements and calculations should confirm that these (heat and overburden pressure) can be safely neglected. The primary concern is with respect to the ability of the clay barrier to perform as such, i.e. as a barrier to attenuate and, if possible, to curtail the transport of contaminants into the underlying soil/geological environment.

9.2.2 Processes and interaction mechanisms

It is well understood that the validity of analytical computer model predictions of system performance depends on the accuracy of conceptual portrayal of the various processes and interaction mechanisms involved in production of system performance. In this section, we consider some of the more prominent elements and aspects that impact directly on these processes and interactions, bearing in mind that most of the developed models are essentially deterministic in nature.

HLW repository containment

We can group the events occurring in the clay buffer in HLW repository containment into two separate, but connected, groups. The first group of events deals with the initial phase of the buffer where it is placed in a semi-dry (partly saturated) state, and the second group deals with events that occur under a fully saturated buffer state. With respect to the HLW and HSW containment schemes discussed in previous chapters and represented in Figure 9.1, the main externally generated processes responsible for fluxes include (a) hydraulic forces (fluid flux, H); (b) heat sources generating thermal forces (thermal flux, T); (c) solutes in the groundwater and porewater from the host rock; (d) contaminants from corrosion products, leachates and damaged canisters (chemical flux, C); and (e) overburden/canister pressure (mechanical flux, M). Internally generated processes are often considered to be responsible for sources and sinks in clays, especially when we have a variety of microstructural units constituting the macrostructure of clays. We are reminded from the previous chapters that microstructural units (peds, domains, aggregate groups, etc.) are the major sources of sinks and sources. By and large, the importance and effect of sources and sinks in clay buffer/barriers with respect to response performance are related to (a) the nature and distribution of microstructural units constituting the macroscopic structure of the clay used for the clay buffer/barrier and (b) microporosity–macroporosity contrast of these clays.

The internally generated processes result, for example, in development of swelling pressures (mechanical flux, M), chemical and biologically mediated chemical reactions (chemical fluxes, C), and vaporization and condensation (fluid flux, H) – to name a few. Interactions between externally generated THMC fluxes and internally generated THMC fluxes lead not only to re-enforcement and competition of fluxes, but also to dependencies and interdependencies between some of the fluxes. A simple example of this is coupled heat and mass transfer and developed swelling pressures in a smectitic clay buffer with external heat and water input. Figure 9.2 shows the overall 'picture' for HLW deep geological disposal, and where system-problem delineation fits for conceptualization for modelling. A knowledge of the overall picture is necessary for one to obtain a proper perspective of the system (i.e. microsystem) to be modelled.

Determination of the various external and internal processes that are active in the microsystem isolated for analysis requires consideration not only of the microsystem itself, but also of the overall macrosystem. We have pointed out previously that the externally generated processes are in essence the output from the portion of the macrosystem from which the microsystem – which constitutes the system for performance conceptualization – is 'detached' for analysis. What we need to also point out is that the microsystem itself does not perform its function in a total vacuum. This means to say that the various outputs from the microsystem interact and react with the macrosystem, which in turn will impact on the performance of the microsystem. These have a direct impact on the performance life of the buffer. For the HLW canister-repository scenario, problem conceptualization includes all the different conditions between short- and long-term status of the buffer (Figure 9.3).

In the short-term scenario, the clay buffer (Figure 9.3) is in the partly saturated state. The mechanisms of interaction resulting from the T and H processes are shown in Figure 9.4. In both figures, the impacts of solutes

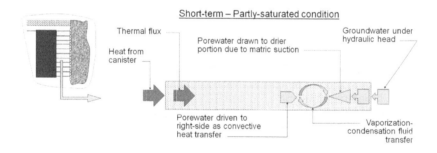

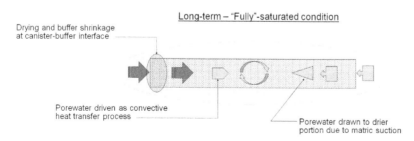

Figure 9.3 Short- and long-term depiction of effect of coupled T–H processes in clay buffer surrounding HLW canister. For simplicity in illustration, the presence of groundwater solutes, corrosion products and fugitive radionuclides are not considered in both partly and 'fully saturated' states.

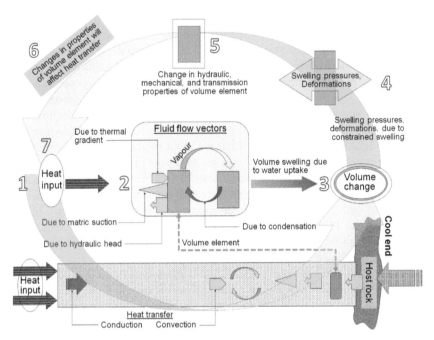

Figure 9.4   Dependent and interdependent mechanisms of interactions resulting from T and H processes. The M process (developed swelling pressures) is a dependent process, dependent on fluid availability and flow. These in turn depend on nature and rate of heat transfer. The cycle starts from position 1 and continues onwards to the heating process, which is now position 7, as a result of the change in properties of the material due to the various interaction mechanisms. The cycle continues until internal equilibrium is achieved, consistent with the boundary conditions.

carried in from the groundwater have not been incorporated in the illustrations – for simplicity in schematic presentation. It is important to recognize the influence of heat on the physical, chemical and biogeochemical processes in the clay buffer/barrier system. These have been discussed in the previous chapters. In the partly saturated short-term state (a) matric suction draws porewater from the portion wetted by available groundwater from the host rock on the right in Figures 9.3 and 9.4 towards the drier end, which is adjacent to the hot HLW canister on the left; (b) the porewater in the wetted portion is transported to the right, against the porewater drawn by the matric suction, because of the thermal gradient, resulting in convective heat transfer; (c) evaporation and condensation contribute to the dynamics of fluid transfer in the buffer; (d) constrained swelling of the wetted and wetting clay buffer results in development of swelling pressures; and (e) the presence of microstructural units and their distribution, coupled with some degree of non-uniformity in proportioning of macro- and microporosities throughout the buffer, will give rise to density and pressure gradients in the buffer. These actions are shown as 'steps' in the bold numerals in Figure 9.4. Changes in the transmittance characteristics and constitutive properties of the buffer mean that heat transmission in the buffer is a continuously changing process.

At least five reasons for water movement in the liquid phase in a partly saturated soil under a thermal gradient are possible, and these are summarized below:

- *Surface tension gradients.* As the surface tension of water against air decreases as the temperature increases, liquid water will flow from the warm regions to the colder or less warm regions because of the influence of surface tension gradients. This has sometimes been referred to as the thermocapillary movement, and is akin to the Gibbs Marangoni effect.
- *Temperature influence on soil water potential* ψ. The soil water potential ψ is a function of both temperature and water content, and can be expressed in a generalized way as:

$$\frac{\partial \psi}{\partial x} = \frac{\partial \psi}{\partial T} \frac{\partial T}{\partial x} + \frac{\partial \psi}{\partial \theta} \frac{\partial \theta}{\partial x}$$

where x, T and θ denote spatial coordinate, temperature, and volumetric water content respectively. For the same volumetric water content, the soil–water potential increases (becomes more negative) as the temperature increases.
- *Flow of water from the less warm regions to the warmer regions.* This occurs on account of differences in the respective heat contents between the liquid water layers sorbed onto the solid particle surfaces and the bulk liquid in the pore spaces. In a sense, this is sometimes interpreted as thermo-osmotic movement.
- *Water movement.* This is generated by the changes in the random kinetic energy of the hydrogen bonds.
- *Ludwig–Soret effect.* This cross-effect between temperature and concentration results in the phenomenon of thermal diffusion, as, for example, in the diffusion of solutes in the porewater from the warm regions to the less warm regions.

Although diffusion of vapour in the clay buffer mass has been generally viewed as a Fickian process, measurements of the circulation of liquid and vapour in closed systems of porous materials show that vapour movement is from four to five times greater than that expected by diffusion predicted from a Fickian model (Gurr *et al.*, 1952; Rollins *et al.*, 1954; Kuzmak and Serada, 1957). This is significant because of the transport of latent heat by water vapour. In essence, if one accounts for only conduction as the means of transport of heat in a partly saturated clay buffer, serious errors can arise because of the underestimated contribution of this latent heat from water vapour in the air-filled pores.

Impact of solutes and microstructure on buffer/barrier performance

The impact of solutes, which are sourced in the groundwater, on short- and long-term performance of the clay buffer in HLW repository containment is felt strongly in changes in the transmittance, swelling and rheological properties of the buffer. The types of solutes in the groundwater range from Na^+, K^+, Ca^{2+} and Mg^{2+} to Cl, HCO_3^- and SO_4^{2-}. We would expect that their individual concentrations and proportions in the groundwater in host rocks would be highly variable and site specific. A useful procedure in determining the impact of solutes is to study the potential energies of interactions between the clay particles in relation to changes in their sodium adsorption ratio (SAR). The SAR measures the milliequivalent weight of sodium divided by the square root of one-half of the sum of the milliequivalent weight of calcium plus the milliequivalent weight of magnesium, i.e.:

$$SAR = \frac{Na^+}{\sqrt{\frac{1}{2}\left(Ca^{2+} + Mg^{2+}\right)}}$$

In Chapter 3 (section 3.7.3), we discussed the DLVO model, which is an interaction energy model that takes into account the nature of the charged (clay) particle surfaces, the chemical composition of the clay–water system and the arrangement of clay particles (and their separation distance) in calculating the magnitude of their interparticle forces. Using this model, we can calculate the energies of interaction between the particles in relation to changes in solute concentrations, for example changes in their SAR. Figure 9.5 is a graphical plot of results reported by Sethi et al. (1980) for the total potential energy of interaction E_T between montmorillonite particles in various modes of interaction (face–face, face–edge and edge–edge), using montmorillonite suspended in 0.3–300 mEq/L solutions with a SAR of 50. The calculations were made using the DLVO model and actual zeta potential ζ measurements, together with the assumptions for particle shapes described in section 3.7.3 and equations 3.14–3.19.

From the results of the calculations shown in graphical form in Figure 9.5, Sethi et al. (1980) indicate that, if one assumes interlayer stacking of particles, there is a high energy barrier of $0.4 \times 10^{12} \kappa T$ against flocculation for face–face particle configuration at low salt concentrations. This accords with the observations reported by Yong and Warkentin (1965, 1975) concerning the proportionality of the shear strength with swelling pressure or

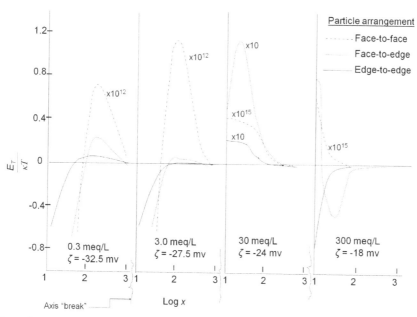

Figure 9.5 Total potential energies of interaction for face–face, face–edge and edge–edge particle configurations, in relation to interparticle distance. Note the scale multipliers associated with face–face curves in all the plots, and for all the curves in the third plot. E_T refers to the total potential energy, κ is the Boltzmann constant, T is the absolute temperature, x is the effective particle separation distance and ζ is the zeta potential.

diffuse double-layer (DDL) repulsion of sodium montmorillonite at low salt concentrations. Increasing salt concentration from 0.3 to 30 mEq/L not only reduces the energy barrier, but also compresses the double layers on both the edges and faces of the clay particles. This allows particle flocculation to occur – a result of the delicate balance between van der Waals attraction and double-layer repulsion. The calculations and test results indicate that formation and reinforcement of existing microstructural units – as in the case of clay buffers embedding HLW canisters or as clay barrier components in engineered barrier systems – are very likely as we increase the salt concentration even higher in the porewater. Particle configurations of edge–edge and face–edge dominate at higher salt concentrations, with occasional face–face arrangements – typical of microstructural units. Differences in morphology of microstructural units are with respect to the gathering of the various particle configurations. These are expressed in terms of the microporosity of the units or microstructural unit density. Variation of solute concentrations in the porewater will lead to changes in the potential energy barrier between particles – with consequent changes in the accumulative, transmission and rheological properties of the clay buffer/barrier.

The preceding discussion shows that we can provide theoretical calculations that point us in the direction of likely configuration of particles in microstructural units. Theoretically, one could include temperature effects and even advective and diffusive forces in the calculations. The addition of these other effects, however, will not change the basic observation with regard to (a) potential energy barriers between particles and their control on particle configuration and (b) development of microstructural units and their individual and collective stabilities. Knowledge of the relationship between the activities and energies in the micropores (pores within a microstructural unit) and the macropores (pores between microstructural units) is necessary if we are to reconcile reality with conceptualization of transmissibility and rheological performance of a smectitic clay buffer. The impact of the nature and distribution of microstructural units in the macrostructure of clays used for construction of buffers and barriers can be significant, especially with respect to macroscopic or *bulk* transmissivity and mechanical (swelling) properties of the clay buffer/barrier system.

In the longer term, although the buffer will attain full water saturation, prolonged exposure to the heated HLW canister will result in drying of the buffer at and near the canister–buffer interface. Prototype repository experiments have shown this to be a prominent feature of long-term buffer–canister interaction (bottom illustration in Figure 9.3). The same features illustrated in Figure 9.3 as the cycle of events for a representative volume element will remain, albeit with different intensities and importance. Although the water content and density gradients may be markedly different in nature and magnitudes, they will still provide the mechanisms for porewater movement to the drier end of the clay buffer. The heat that is transmitted through the clay buffer will be transported to the host rock. What this does to the host rock, and how it reacts with the buffer, will need consideration and analysis, particularly with respect to its role as a thermal boundary and to the groundwater supply to the clay buffer.

Maturation and ageing processes (discussed in Chapter 7) result in changes in the nature of the clay and its properties, to an extent that requires proper accounting of these changes in buffer performance. As the effects of these processes are time dependent, the changes in buffer performance are transitory. A combination of (a) laboratory experiments simulating specific accelerated and/or magnified pseudo long-term conditions, (b) ideal theories of particle-interaction relative to thermal, chemical and biologically associated conditions and (c) full-scale prototype repository experiments, is required to obtain a proper appreciation of the transitory performance characteristics of the clay buffer embedding HLW canisters under the kinds of repository containment environment.

HSW clay barrier – liner and cover

The situation in regard to a clay barrier (as a liner component of EBS) will, by and large, be a case of containment of contaminants escaping from the overlying membrane barrier components of the EBS. Published literature shows that temperature and hydraulic effects are generally not considered to be sufficiently significant for the underlying clay barrier component to warrant incorporating these in analyses involving fate and transport processes. The exception to this will be the top clay barrier component of the surface cover in regions where ground temperatures reach below the freezing point of water – as is the case for many regions of the world. Figure 9.6 shows typical winter and summer ground subsurface temperature profiles. We expect the porewater in most clay soils to freeze at temperatures slightly below 0°C or somewhat lower, depending on the thermal properties of the clay–water system and on the chemistry of the porewater. In the left-hand drawing, the active layer is the layer of ground that freezes in winter and thaws in the summer period. The thickness of this layer, i.e. the depth of frost penetration or depth to which the freezing front will penetrate into the ground subsurface, depends in part on the type of ground cover and the heat capacity of the soil (surface and subsoil). In regions closer to the 'far north' and 'far south', it is possible to have summer temperatures that do not completely thaw the frozen ground. The right-hand diagram shows

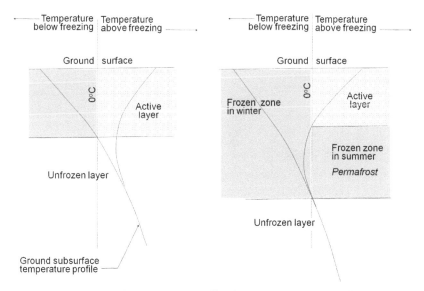

Figure 9.6   Typical ground temperature profiles for winter and summer. The active layer in both diagrams is the layer that freezes in winter and thaws in summer. The frozen zone in the summer, shown in the right-hand drawing, is generally defined as the permafrost (permanently frozen) layer.

such a case. The frozen zone or layer that remains in summer constitutes the permafrost (permanently frozen) layer.

Frost penetration in the active layer is a problem of note for soils that are 'frost susceptible'. These are soils that have a tendency to heave in winter because of the formation of ice lenses in the soil that literally push up the soil, thus causing ground heave. Low-lying buildings, runways, highways and shallow-buried pipelines have been known to suffer considerable distress because underlying frost-susceptible soils developed ice lenses resulting in heaving of the layer in the vicinity of the ice lens. In addition to concerns about the frost susceptibility of clay covers of landfills, there may also be some concern over the clay barrier in the sides of EBS near ground surface in regions where freezing temperatures and frost penetration are significant.

The depth of frost penetration, which is generally taken to be the bottom limit of the active layer, can be determined for non-permafrost regions using an idealized representation of the ground subsurface temperature profile such as that shown in Figure 9.7. This is a highly simplistic representation of the temperature profile and will lead to an overestimation of the depth of frost penetration. Nevertheless, it is useful for scoping calculations and allows one to arrive at a quick appraisal of the likelihood of frost damage to the clay buffer/barrier system. Proceeding with this simple view of the temperature profile, we note that at point Θ, the continuity condition that needs to be satisfied states that the latent heat released as the porewater freezes to a depth Δx in time, Δt must be equal to the rate of heat removed and conducted to the ground surface. Thus, for small values of Δx and Δt, we have:

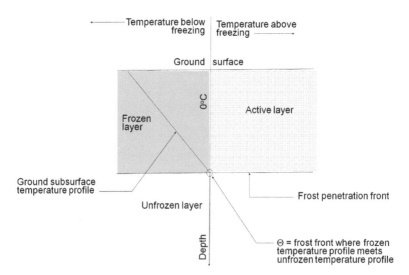

Figure 9.7 Idealized ground subsurface temperature profile for calculation of frost penetration (freezing front) depth.

$$k_f \frac{T_f}{x} = L \frac{dx}{dt} \tag{9.1}$$

where T_f is the temperature below the freezing point and L is the latent heat of the porewater. The solution to equation 9.1 should give us:

$$x = \sqrt{\frac{2k_f \int T_f \, dt}{L}} \tag{9.2}$$

where $\int T_f dt$ is the freezing index F, in degree-days, computed generally as the area of the curve under the freezing line shown in Figure 9.8.

For a more realistic determination of the depth of frost penetration, we can use the diffusion equation to characterize the subsurface ground temperature as follows:

$$\frac{\partial T_f}{\partial t} = a_f \frac{\partial^2 T_f}{\partial x^2} \qquad \text{for frozen layer}$$

$$\frac{\partial T_u}{\partial t} = a_u \frac{\partial^2 T_u}{\partial x^2} \qquad \text{for unfrozen layer} \tag{9.3}$$

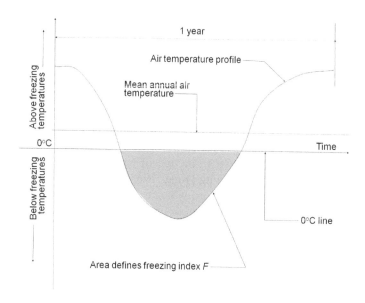

Figure 9.8   Idealized annual air temperature profile. Area under the 0°C line is the freezing index *F*.

where x is the spatial coordinate, T is the temperature, a is the diffusivity coefficient, and the associated subscripts f and u represent frozen and unfrozen respectively. The diffusivity coefficient a is defined as the ratio of the thermal conductivity k of the soil to its volumetric heat C, i.e. $a = k/C$.

The continuity condition at the point Θ requires that the net rate of heat flow from the frost (freezing) front must be equal to the latent heat supplied by the subsurface soil water as it freezes to a depth Δx in time Δt. For vanishingly small values of Δx and Δt, we will have:

$$L\frac{dx}{dt} = \Delta q \tag{9.4}$$

where Δq is the net rate of heat flow at the freezing front. It follows that:

$$k_f \frac{\partial T_f}{\partial x} - k_u \frac{\partial T_u}{\partial x} = L\frac{dx}{dt} \tag{9.5}$$

The solution of the relationship for x, the depth of frost penetration, will include the term $\int T_f dt$.

Frost heaving and formation of ice lenses occur in soils if they are frost susceptible and if water for ice lens growth is available. The Casagrande classification of frost-susceptible soils considers (a) well-graded soils, in which 3 per cent or more of the soil particles are < 0.02 mm in particle size, and (b) uniformly graded soils, in which 10 per cent or more of the soil particles are < 0.02 mm in particle size, to be highly susceptible to frost heave and ice lens formation, provided that water is available for lens formation and that the appropriate freezing index exists. A certain amount of local consolidation or compression of the unfrozen material ahead of the freezing front occurs, the extent of which is a function of the water available for ice lens growth and the degree of heave experienced in the layer behind the ice lens. As Figure 9.9 shows, ice lenses will form in different portions and different positions in the soil as the freezing front penetrates into the soil. The positions and extent of ice lens formation are a function of water uptake at the ice lens as it grows, and the rate of freezing (Yong, 1965). Too rapid a freezing rate will limit the growth of the ice lens, whereas too slow a freezing rate will only create a frozen soil layer, i.e. frozen porewater in the void spaces. The continuity condition that must be satisfied at the freezing front says that the rate of change in thermal energy must be balanced by the spatial rate of change of the heat flux. In this case, the heat flux includes the heat brought in as mass transfer. In all instances, there is an infinitesimal layer of unfrozen water separating the ice lens from the particles and microstructural units. This water layer is the hydrate layer (or layers, depending upon the activity of the soil particle surfaces) and is often called the adsorbed water layer.

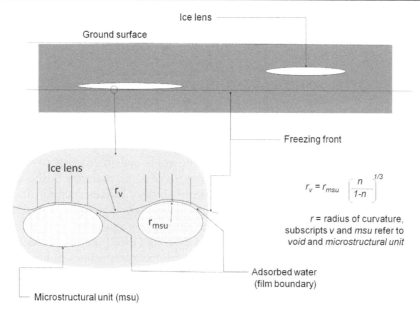

Figure 9.9   Local regime at the freezing front, showing the curvatures at front surface of an ice lens. The relationship between r_v and r_{msu} assumes a cubicle packing of microstructural units. n, porosity of the subsoil.

Numerous studies have reported on this unfrozen water phenomenon in freezing soils (Yong, 1963a,b).

The frost heaving pressures created at the ice lens interface with the overlying soil layer can be theoretically calculated. This requires one to determine the frost heaving pressures as a function of the equivalent radii of representative pore (void) and particle or microstructural unit, beginning with the Maxwell thermodynamic relations written for a curved ice–water interface immediately above the ice lens shown in Figure 9.9. The free energy of the ice must equal that of the water at equilibrium, and so the reversed curvature of the ice front between pore spaces must be accounted for. As the minimum potential energy configuration for the ice lens front must be a flat plane, both the ice radii in the pore r_v and around the microstructural unit r_{msu} must be considered. Taking a local region that encompasses a microstructural unit and a void space, the relationship obtained (Yong, 1967) shows that the pressure generated between the ice and soil particles or microstructural units ΔP can be expressed by the Clapeyron relation as:

$$\Delta P = \frac{L \Delta T}{T \left(V_w - V_i \right)} \tag{9.6}$$

where:

$$\Delta T = \frac{2\sigma T}{\rho_i L}\left[\frac{1}{r_{msu} + r_v}\right]$$ (9.7)

and ΔT is freezing point change at the curved surface, r is the radius of curvature, subscripts v and msu referring to void and microstructural unit, respectively, L is the latent heat of fusion, V is volume, with subscripts w and i referring to water and ice, respectively, T is temperature, ρ_i is density of ice and σ is ice–water interfacial energy.

Representative radii of curvature for the voids and for the microstructural units can be obtained from a study of electron micrographs. Although it would be normal to see a distribution of sizes of microstructural units and macrovoids (voids between microstructural units), it should not be difficult to make a judicious choice of a representative microstructural unit. Experimental frost heaving tests reported by Yong (1967) on fine inorganic silts show that the predictions accord well with measured values of heaving pressures. It is noted that if volume change under frost heaving pressures is allowed, stress relief occurs, resulting thereby in a lower measured heaving pressure. This means that the theoretically calculated values that are based on a no-volume-change condition are the conservative maximum values.

Both freezing and thawing are problems that need to be examined and evaluated for clay liner covers of landfills. In many instances, significant changes in the microstructure of the clay in the clay cover can occur as a result of cyclic freezing and thawing of the clay cover. Figure 9.10 shows the change in microstructural units from the first freeze–thaw period to the 32nd freeze–thaw cycle of a natural high water content clay from northern Québec, Canada. Note the scale for all three pictures is the same, as shown by the 1-nm scale bar (black bar) near the centre-bottom of the images. The restructuring of the microstructural units into more stable units as the number of freeze–thaw cycles is increased is evident – for both open and closed systems. The microstructural units obtained in the open system are less compact than those obtained in the closed system, i.e. the microporosities of the open system microstructural units are larger than those units resulting from the closed system. These impact directly on the production of larger macropores, resulting in greater hydraulic transmission characteristics.

Observations reported by Yong et al. (1984, 1986a) for studies on cyclic free–thaw effects on soil microstructure and properties of clays include the following:

- The formation of numerous ice lenses complicates the prediction of depth of frost penetration and frost heaving pressures.
- Repetitive freezing and thawing causes significant variation in the thermal properties of the clays. The production of greater macropores as a result of the cyclical freeze–thaw reorganization of the microstructures seems to be one of the contributing factors.

Mineralogy in order of
abundance.

Quartz
Illite/mica
Feldspar
Kaolinite
Chlorite
Amphibole

1 cycle – Open

32 cycles - Open

32 cycles - Closed

Figure 9.10 SEM pictures showing change in size and shape of microstructural units in high water content clay (92 per cent natural water content) following one cycle and 32 cycles of freeze–thaw in open and closed systems. The open system allows water to be drawn into the clay or to escape from the clay as it freezes and thaws. The closed system does not allow any water to enter or escape from the clay during freezing and thawing. Note that the scale of all the pictures is the same, as shown by the scale bar at the right-hand corner for all three pictures.

- The thermal properties of the clays depend substantially on the boundary conditions imposed, particularly on the amount of constraint applied against frost heaving.

Interlamellar water in smectite does not freeze until the temperature is significantly below 0°C. Water in external voids (macropores) begins to freeze at temperatures below 0°C, i.e. the critical temperature is below that of ordinary bulk water. As the temperature decreases below critical, some freezing of the water in the micropores will occur. This will continue as the temperature is further lowered, and at some stage, interlamellar water will begin to freeze. At very high densities, the amount of hydrated layers and interlamellar water is significantly larger than water in the macropores, and a large portion of the porewater may remain unfrozen to temperatures below –20°C and even –30°C (Yong, 1965; Anderson *et al.*, 1978).

9.2.3   *Transitory clay properties – conceptualization and specification*

Chapter 7 has discussed many of the significant issues and events relating to evolution of clays under conditions associated with HLW and HSW containment and isolation. One of the most important points that should be noted is that the nature, characteristics and properties of clays will change because of maturation or ageing processes. We consider the changing nature, characteristics and properties of clays as transitory phenomena. Many different ageing processes – biogeochemical, chemical, thermal, radiological, etc. – participate in the development of these phenomena. The end-point of all of these transitory phenomena, which can be considered to be when the nature, characteristics and properties of the clay in question will be 'permanently fixed', will be when all the effects from each of the ageing processes has been exhausted.

To deal with system performance conceptualization to account for the transitory nature and properties of clays used for buffer/barrier systems, we need to see if we can find some characteristic indicators of the changed properties of the clay in question. This can be done by considering what dominant processes, actions, interactions, etc. are imposed or involved in system performance, and what kinds of clay performance properties and/or characteristics are involved. We can group most of the system processes, actions, interactions and/or reactions into three broad categories as follows:

- *Flow processes.* These include processes that involve clay transmittance properties such as heat and fluid transfer, and contaminant transport. Mechanisms that need to be considered include conduction, convection, advection and diffusion.
- *Mechanical processes.* These refer to actions involving rheological properties of the clay and include such actions provoked by constraints against swelling pressure, and pressures developed as a result of canister and overburden loading. Mechanisms to be considered include compression, consolidation, shear resistance and creep.
- *Chemical and biogeochemical processes.* These involve chemical interactions and reactions. Reactions and interactions include contaminant reactions and partitioning, hydrolysis, oxidation–reduction, bacterial degradation and biologically mediated chemical reactions, etc.

The microstructure of the clay is a common clay property that features prominently as a control property involved in all the three categories. Why? Because microstructural units and their distribution control (a) the proportion of micropores to macropores (for example, see Figure 3.23); (b) the transmission characteristics and properties of the clay (for example, see Figure 4.13); (c) the accumulative properties and characteristics of the clay; (d) the nature of sources and sinks in the clay; and (e) the rheological behaviour of the clay. Furthermore, microstructural units vary considerably with respect to their sizes, shapes, and packing density, as seen for example in the bottom two images in Figure 9.10.

Material properties are critical issues in describing buffer/barrier performance. As these will change as a result of a combination of the processes initiated by the HLW/HSW containment scenarios and material maturation/ageing, the importance of proper specification of these properties cannot be overstated. The type of clay and the manner in which it is used or placed to function as a clay buffer or clay liner/barrier component of an EBS will determine what and how the microstructural units will be obtained and distributed. Mineral transformations, together with the physical and chemical composition of clay as a result of the action of ageing processes, will change the nature and distribution of the microstructural units in the clay. Tracking these changes in controlled laboratory tests provides one with a means to determine what material properties to use in seeking solutions to model predictions of buffer/barrier performance. Laboratory tests with accelerated or intensified experimental conditions can be conducted to provide a means for tracking, for example in the cyclical freeze–thaw tests shown in Figure 9.10, which were conducted over a period of a few months.

9.2.4   Elements in problem conceptualization and solution

The previous chapters have described the HLW canister-repository and HSW landfill disposal scenarios. Included in the previous chapters and the preceding sections for this chapter are descriptions of the processes initiated by THMCB (thermal, hydraulic, mechanical, chemical and biological) factors or 'drivers' relative to canister isolation and landfill disposal. For this section, we will define and examine the various elements of the problem as they pertain to conceptualization and model requirements for solution of the problem. The basic elements involved in problem conceptualization and solution fall into two convenient groups as follows: (a) elements for consideration in *problem initialization* and (b) elements and processes for consideration in *problem solution*.

The basic elements of *problem initialization* for HLW canister isolation in the repository scenario include the following:

- long-lasting canister heat source;
- smectite-based buffer barrier or containment system;
- surrounding host rock at a temperatures much lower than the heat source;
- access to water containing dissolved solutes and delivered via fractures and fissures in the host rock to the cooler extremities of the buffer;

- interface conditions at the heat source end (source–buffer interface), at the rock end (buffer–rock interface), and at the top (buffer–backfill interface);
- thermodynamic potentials.

Not all of the elements and processes that need to be considered to arrive at *problem solution* for the HLW canister isolation problem are well identified or fully understood. The ones that have been identified include the following:

- heat and mass transfer;
- vapourization and condensation;
- vapour and liquid transport;
- accessibility and distribution of water intake;
- transport and transmissibility coefficients;
- swelling of buffer material on water uptake;
- swelling and compression pressures;
- solute transport;
- chemical reactions and transformations;
- biotic reactions and transformations;
- transformed mechanical, transmission and transport coefficients;
- couplings between heat, mass, chemical and biological phenomena.

The illustrations shown in Figures 9.1 and 9.2 highlight some of the elements described in *problem initialization* and *problem solution*. The types and nature of the physical contacts between buffer material and host rock, and also with canister and upper backfill material, are particularly important. These contacts impact directly on both the mechanical performance aspects of the buffer and the distribution of water access to the buffer material. The various processes associated with thermal, fluid, vapour, solute and gas transport, geochemical reactions, and abiotic and biotic reactions all lead to maturation and ageing of the clay buffer.

In the case of the HSW landfill problem, the various elements involved in problem conceptualization and solution fall into the same groupings. Considering the top surface cover and the clay liner/barrier component of the multi-barrier EBS, and especially paying attention to the highly variable geological, hydrogeological, and subsurface settings, the basic elements involved in consideration for *problem initialization* include:

- geological, hydrogeological and subsurface settings;
- freezing index and frost penetration;
- water table, accessibility and dissolved solutes;
- competence of overlying multi-barrier components;
- clay liner/barrier and clay cover.

The elements and processes that require consideration for problem solution in the HSW landfill containment–isolation scheme include:

- contaminant transport, partitioning and fate;
- contaminant source input and distribution;
- accumulative and buffering capacities;
- swelling and compression pressures;
- chemical reactions and transformations;
- biological activity and reactions.

Couplings, dependencies and interdependencies

In addition to the coupled relationships between heat and fluid fluxes in a smectite buffer mass, one needs to be aware of the dependent and interdependent relationships between them (heat and fluid fluxes) and the various internal sets of thermodynamic and mechanical forces. An example of interdependencies can be seen in the effect of solute concentration in the porewater of a clay buffer. Solute concentration at any one point has an effect on the osmotic (swelling) pressure at that particular point. In addition it (solute concentration) has an impact on the vapour pressure of the porewater, and on the reactions with other solutes in the porewater and the reactive clay particles' surfaces (see Chapters 3–5). The results of experiments conducted by Gurr *et al.* (1952) on the contributions from vapour and thermocapillary transfer of liquid water under a temperature gradient for a closed soil–column system showed that although total transfer of liquid water was towards the cold end, movement of dissolved solutes was in the opposite direction. Vapour movement was found to be four to five times greater than that predicted by Fickian diffusion.

9.3   Analytical computer modelling

9.3.1   Basic requirements

For analytical computer models to accurately represent, analyse, and/or predict system performance, scientific integrity must be maintained. These models are, in essence, a collection of mathematical expressions that are supposed to represent the results of interactions of complex processes and phenomena. Several points of note are essential:

- The requirement of a clear specification of the objectives for which model development is required is a very necessary condition. The degree or level of complexity of models developed to address the objectives is a function not only of the degree of complexity of the system performance being modelled, but also of the competence/expertise of the model developer and the types and ranges of data and inputs to be considered or available.

The objectives and requirements of regulators, for example, differ from those involved in the design of the constructed facilities.

- The viability of models depends on their comprehension of 'what constitutes the complete system' to be modelled, the various parameters, processes, etc. that must be included in the structuring of the analyses that are central to the final model product.

An example of the level of comprehension needed in reaching the level of 'what constitutes the complete system' is the description of the various stages of water intake and flow in the clay buffer – beginning with hydration, and going onward to redistribution and saturated flow, all of which are mechanisms of water movement that are different and require different mathematical relationships and parameters to express their respective phenomena.

- Complex models are not always the answer. Although complex and complicated numerical codes may be developed to address such system issues as coupled transport processes and other internal dependent–interdependent process-generated phenomena in heterogeneous systems, there are times when simple analytical models that address a particular set of issues would be a better model choice. This means to say that given that physical, physico-chemical, chemical, biological and radiological processes may be operative in the system under consideration, there are occasions when answers can be obtained from simple analytical models using an 'everything-else-being-equal' condition to address a particular problem/question.

A simple example would be a comparison of transport of non-reactive contaminants in a partly saturated buffer with transport of non-reactive contaminants when the buffer has reached full saturation under the same sets of conditions – in which case, one could choose to use the unsaturated transport relationship for the partly saturated case and the often-used diffusion–dispersion relationship for the fully saturated case.

- Although increased model complexity results from the need or desire to address more and different situations of the problem being considered, we need to be aware of the associated requirements of (a) more input data of high quality and relevance and (b) complete accounting, knowledge and conceptualization of all the processes acting on and within the system. Failure to account for all the processes involved and failure to provide proper and complete relevant quality data sets will result in failure of the model to provide accurate outputs.

To maintain scientific integrity, it is necessary to account for all the processes active in the system – coupled, uncoupled etc. – and to supply all the relevant input data. Is it possible that some parameters cannot be accurately described or properly measured/quantified, as for example the question of water or leachate entry points as input to the clay buffer or clay barrier, and variability of water or leachate input.

- High-quality and relevant data need to be site specific, material specific and operation specific. The last set of specific conditions is significant, as the results obtained in laboratory and prototype testing of clays are more often than not operationally defined, i.e. dependent on the conditions imposed as part of the laboratory or prototype test procedure.

  It is necessary to precisely describe the variability of model parameters in light of test methods and constraints.

- As most of the basic laws most generally invoked in progressing from conceptualization to analytical computer model development assume ideal material performance, it is necessary to determine the range of applicability of these laws for the clays and situations being examined. For example, application of Darcy's law developed initially to describe fluid flow in relation to a 'permeability constant' for a fixed porous medium to the clay buffer/barrier system requires careful and informed consideration. Volume change, presence of sources and sinks, local non-laminar flow, partly saturated conditions, and reactions between dissolved solutes and reactive surfaces of clay particles are some of the fluid flow issues not initially considered in the Darcy law formulation.

  The assumptions, idealization and limitations incorporated in the model (from the basic laws) should be recognized and declared.

- Documentation of models should provide complete sets of test problems and input/output samples to allow for model testing, verification and quality assurance.

  In the final analysis, prototype experiments should be designed and constructed to allow for validation of models. This is crucial for complex systems such as deep geological containment of HLW (i.e. HLW canister-repository system). The natural heterogeneity of the geological medium, coupled with complex water chemistry, availability and entry into the clay buffer are issues that can best be addressed by prototype testing for model validation.

9.3.2   Governing relationships

The processes and interactions–reactions in a clay buffer embedding HLW canister in the underground repository (such as those illustrated in Figures 9.1–9.4) and in a clay barrier/liner component of a multi-barrier EBS used in containment of HSW, which need to be described by governing relationship, include the following:

- transport of contaminants (in the clay buffer/barrier);
- partitioning of contaminants;
- heat and mass transport;

- gas and vapour transport;
- porewater pressures and water (hydraulic) pressures in the near-field, and water uptake and redistribution of porewater pressure;
- development of swelling pressure;
- expansion of buffer/barrier from thermal effects and swelling pressures, stresses and displacement;
- dissolution of minerals and precipitation of chemical compounds in the buffer/barrier;
- changes in porewater chemistry.

In this recounting of the governing relationships, the commonly used ones cited in the literature will be presented. It is not the purpose of this section to repeat the background development of many of the relationships presented in previous chapters, except for the purposes of clarity and continuity in presentation. Attention is directed towards the governing relationships used to describe the various processes operative in the clay buffer/barrier systems as they perform their respective roles. In all cases, a continuum approach is used and the clay is assumed to be homogeneous and uniform. This does not mean to say that heterogeneous and probabilistic approaches have not been used. There have been attempts made to account for the heterogeneity of the system, and there have been relationships developed accounting for the probabilistic nature of the problem. However, these appear to have been used for special cases, and more often than not, for simpler situations.

The governing relationships satisfy conservation requirements, which are expressed specifically as (a) mass balance of solids; (b) mass balance of water (liquid); (c) mass balance of air; (d) momentum balance of the medium; and (e) balance of internal energy. The reader will recognize many of these relationships, as they have been developed or discussed in the previous chapters. These are restated in the following subsections to provide continuity in the discussion on governing relationships.

Contaminant transport

The governing relationships used in the development of non-reactive contaminant transport equations include Fickian diffusion (Fick's second law) for solute transport and applicability of Darcy's law for fluid flow.

$$\text{Fick: } \frac{\partial c}{\partial t} = D_s \frac{\partial^2 c}{\partial x^2} \text{ and Darcy: } v = -k\frac{\partial \psi}{\partial x} \tag{9.8}$$

where c is the concentration of solutes (contaminants), t is the diffusion time, D_s is the diffusion coefficient for the solutes, v is the velocity, k is the Darcy permeability coefficient, ψ is the total potential, and x is the spatial coordinate. The temperature dependence of D_s is given as follows:

$$D_s = D_o e^{\frac{-a_d}{RT}}$$
(9.9)

where D_o is the diffusion coefficient at infinite dilution, i.e. maximum diffusion coefficient, a_d is the activation energy, R is the gas constant and T is the absolute temperature.

The governing relationships accounting for partitioning of contaminants take their cue from adsorption isotherms obtained from batch equilibrium tests. The simpler transport models appear to favour a type of Freundlich isotherm: $c^* = k_F c^n$ for inorganic contaminants and $c_{oc} = k_L c_w^m$ for organic chemicals, where c^* is the concentration of inorganic solutes (contaminants) adsorbed, c_{oc} represents concentration of organic chemicals sorbed by the clay solids, c_w is the concentration of organic chemicals remaining in the aqueous phase, k_F and k_L are the partition coefficients, and n and m are Freundlich constants.

Heat transfer

The Fourier relationships for heat conduction are used as follows.
The first law of heat conduction is given as:

$$q_c = -k_c \frac{\partial T}{\partial x}$$

where q_c is the thermal flux and k_c is the thermal conductivity.
The second law of heat conduction is:

$$C \frac{\partial T}{\partial t} = \frac{\partial}{\partial x}\left(k_c \frac{\partial T}{\partial x} \right)$$

where C is the volumetric heat capacity.
The *latent heat transfer* can be incorporated in the second law for heat conduction as follows:

$$C \frac{\partial T}{\partial t} = \frac{\partial}{\partial x}\left(k_c \frac{\partial T}{\partial x} \right) \pm L_T(x,t)$$
(9.10)

where L_T refers to the latent heat transfer due to evaporation or condensation, which explains why we have the $\pm$ sign associated with L_T.

Liquid (water) movement

For liquid movement in partly saturated clays, diffusive movement of water is assumed. The diffusion coefficient for water D_θ is used in the Fickian

relationship and Darcy's law is assumed to be valid for flow in partly saturated clays. Assuming a no-volume condition:

$$\frac{\partial \theta}{\partial t} = \frac{\partial}{\partial x}\left(D_\theta \frac{\partial \theta}{\partial x}\right) \text{ and } D_\theta(\theta) = k\frac{\partial \psi}{\partial \theta} \tag{9.11}$$

where θ represents the volumetric water content. For volume-changing
clays, i.e. clays that change volume because of swelling upon water uptake,
we can consider water movement as motion of water relative to the clay
particles. The flux of clay particles must satisfy three physical conditions of
(a) continuity, (b) Newton's law of motion and (c) rheological equation of
state. For slow fluid flow, the acceleration term implicit in Newton's law can
be safely relegated to zero. The relationship used to describe fluid movement
where small displacements (i.e. clay particles displaced in volume change)
occur has been given as follows (Yong and Warkentin, 1975):

$$\frac{\partial \theta}{\partial t} = \frac{\partial}{\partial x}\left(D_\theta \frac{\partial \theta}{\partial x}\right) - \theta\frac{\partial v}{\partial t} \tag{9.12}$$

where

$$\frac{\partial v}{\partial t} = \frac{\partial V_{sx}}{\partial x}$$

and where V_{sx} is the velocity of clay particles in the x direction and v is the
volumetric strain.

Air (gaseous phase) and vapour

The equilibrium concentration of dissolved gas in the porewater is determined according to Henry's law. The common relationship used is $C_g = k_H P_g$,
where C_g is the concentration of the gas in the porewater, k_H is the Henry's
law constant and P_g is the partial pressure of the gas in the ambient air phase.
The assumption is made that there is no reaction between the gas (solute)
and the solvent (porewater). Generally speaking, Henry's law constant is different for each solute–solvent pair (i.e. gas–porewater in the case of the clay
buffer/barrier system). For dilute solutions of non-reacting gases in water,
the k_H values for dissolution in water at 299 K are $O_2 = 4.34 \times 10^4$ atoms,
$CO_2 = 1.64 \times 10^3$ atoms and $H_2 = 7.04 \times 10^4$ atoms.
   Air–water vapour mixtures and equilibrium states are determined through
psychrometric calculations.

Mechanical behaviour

There are several choices for constitutive relationships describing the response behaviour of clays used for clay buffer/barrier systems, ranging from plasticity concepts to cap models, strain-hardening models, critical state models, etc. The choice of rheological and/or yield-failure model depends on observed behaviour of the clay in laboratory strength tests and as buffer/barrier under 'field-loading' conditions, i.e. 'what happens in the field containment situation'. Examples of these can be seen in section 9.5 in the discussions of model applications.

9.4   Models for special purposes

9.4.1   *Swelling pressures – sequential segment marching model*

The pressures developed in the clay buffer/barrier system are dependent on several factors, for example water availability, hydraulic gradient, osmotic potential ('soil suction'), temperature, initial density, microstructural units and distribution, and constraints. To determine the pressures acting on a representative volume element in an initially partly saturated clay buffer, it is useful to solve the total clay–water potential relationship, as it is equivalent to the total pressure that includes the swelling pressure. We can construct a simple model, which expresses equation 9.12 using:

$$D_\theta(\theta) = k(\theta)\frac{\partial \psi_T}{\partial \theta}$$

to obtain

$$C_\theta \frac{\partial \psi_T}{\partial t} = \frac{\partial}{\partial x}\left(k_\theta(\theta)\frac{\partial \psi_T}{\partial x}\right) - \theta\frac{\partial v}{\partial t} \qquad (9.13)$$

where:

$$C_\theta(\theta) = \frac{\partial \theta}{\partial \psi_T},$$

C_θ is the water capacity and ψ_T is the total soil–water potential. We assume that Darcy's law:

$$v = -k(\theta)\frac{\partial \psi_T}{\partial x}$$

is applicable for v in equation 9.12. If water content profiles and k-θ relationships are given separately, we can obtain the solution for the total potential ψ_T using equation 9.13. To determine the swelling pressure developed in relation to both time of exposure to water and hydraulic gradient, we can use a marching technique that analyses successive segments of clay in sequence to determine the swelling pressures developed as a result of diffusive flow into the swelling clay. Figure 9.11 shows a clay sample that is segmented into n equal layers. The small segment permits one to perform the analysis for each segment as a small increment in the wetting front, coming in from the left in the figure.

Assuming a no-volume-change condition, calculations can be made for the corresponding time of flow across the thickness Δx of one segment, from *segment face A–A* to *segment face B–B*, and the corresponding swelling pressure developed. In so far as the boundary conditions are concerned, at time t_n, the wetting front is at *face A–A*, and the time taken for the front to move the distance of Δx is Δt_{n+1}. This develops a reaction pressure $(p_s)_n$ which acts on *face B–B* to counter the swelling pressure developed because of the wetting of the segment defined by faces *A–A* and *B–B*. As a result, we expect a reduction in the flow of water into the next segment. With the clay being segmented into n equal segments, the swelling pressures developed in the clay buffer/barrier as water enters each succeeding segment can be calculated

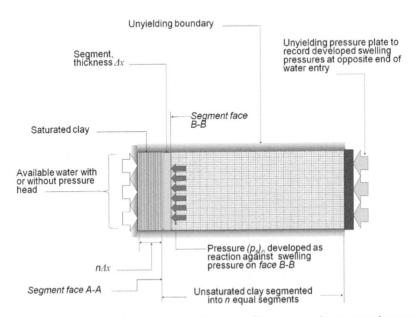

Figure 9.11   Conceptual model of a swelling clay buffer segmented into n equal segments, with water availability at one end for analytical procedure using a marching technique, i.e. marching from one segment to another.

by considering the total soil–water potential ψ_T as the total pressure p_T responsible for water entry and movement in the clay from one segment to another. This total pressure consists of the sum of the matric potential ψ_m expressed as the matric suction p_m, the hydraulic pressure p_h, and the reaction pressure $(p_s)_n$, i.e. $p_T = p_m + p_h - (p_s)_n$. The resultant swelling pressures in the new segment can be used to define the stress boundary condition for the remaining unwetted segments of the clay buffer/barrier.

$$v = -k(\theta)\frac{\partial \psi_T}{\partial x} = k(\theta)\frac{\partial p_T}{\partial x}$$

$$v_{A-A} = k(\theta)\frac{p_m + p_h - (p_s)_n}{n\Delta x}$$

$$\text{(9.14)}$$

The results of experiments conducted to measure axial swelling pressures developed in a swelling clay confined within unyielding boundaries have been compared with calculations by Yong *et al.* (1986b) using this simple marching model, as shown in Figure 9.12 for several hydraulic gradients.

The parameters and conditions used for prediction with the simple marching model were as follows: (a) clay dry density = 2.14 mg/m³, initial weight water content = 7 per cent, full saturated water content = 9.6 per cent and porosity = 0.208; (b) $k(\theta)$ is independent of θ and = 10^{-10} m/s; and (c) at initial water content, $p_m = 55$ kPa.

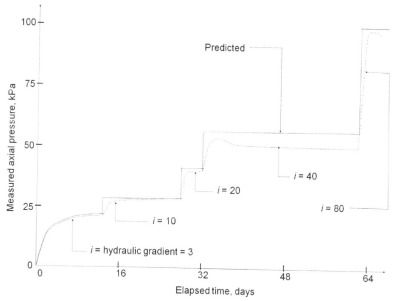

Figure 9.12   Comparison of measured swelling pressure under various hydraulic gradients with predicted values.

9.4.2   Ion diffusion

Porewater chemistry, ion migration and sorption

As we have seen in the previous discussions in the earlier chapters, water uptake and transport of inorganic and organic chemical contaminants are of fundamental importance for a number of microstructurally related processes, such as hydration, cation exchange and chemical reaction rates in clay buffer/barrier systems. The following is a discussion of recent models (Grindrod and Takase, 1994; Pusch *et al.*, 1999) for describing and predicting ionic transport in clays used for buffer/barrier and liner systems.

The porewater chemistry of a clay (or any soil) is the result of complex interactions between the solution and the minerals constituting the clay. Equilibrium modelling is used to explain the basic chemical reactions such as dissolution, precipitation and ion exchange. The small interparticle spaces between mineral surfaces in dense smectite clay means that ion distribution and mobility adjacent to these surfaces play a major role in modelling ion diffusion. MDDL (modified diffuse double layer) theory, which overcomes one of the major drawbacks of unreasonably high cation concentration next to particle surfaces of conventional DDL theory, can be used to calculate the negative potential (Pusch *et al.*, 1999). The standard basic expression for solute diffusion flux J is given as follows:

$$J = -nD_s \frac{\partial c}{\partial x} \tag{9.15}$$

where n is the porosity, D_s is the solute diffusion coefficient, and c is the concentration of solutes. Recalling the 'sources and sinks' phenomena from previous discussions, and recalling also the discussion concerning microstructural units and the ratio of micropores to macropores, it is recognized that the diffusive flux is in reality an apparent diffusive flux with its associated apparent diffusion coefficient D_a:

$$D_a = \frac{nD_s}{n + k_d \rho_d} \tag{9.16}$$

where k_d is the distribution coefficient (discussed in Chapter 5), and ρ_d is the dry density of the clay. Defining k_d as (Pusch *et al.*, 1999):

$$k_d = \frac{\varphi \Gamma}{(1-\varphi)\rho_d}$$

diffusion through the narrow channels typical of microporosity features, we can express J as (Ledesma and Chen, 2003):

$$J = \frac{\varphi D_s \Gamma}{\tau^2}\left(\frac{1+\varsigma}{\eta}\right)$$ (9.17)

where φ is the surface potential, τ is the tortuosity factor, Γ is the surface excess charge at a distance ς, and η is a measure of viscosity of the shear zone at the particle interface.

9.5 Partly saturated clay liquid transfer

Most of the developed numerical models (known generally as *codes*) rely on established relationships governing liquid, vapour, and heat flow. These are the flux laws, respectively (a) Darcy's law for liquid and gas flow, (b) Fick's law for vapour flow and (c) Fourier's law for heat conduction. Some differences in expression of the constitutive or rheological performance of the clay buffer exist between many of the models, for example (a) effective stress relationship for partly saturated soils; (b) elasto-plastic and visco-plastic relationships; and (c) a state surface such as *critical state*, or bounding surface model.

The basic relationships first proposed by Philip and de Vries (1957) are seen in many of the developed models. These basic formulations for coupled heat and moisture flow through porous media are stated as follows:

$$\frac{\partial \theta}{\partial t} = \nabla\left(D_T \nabla T\right) + \nabla\left(D_\theta \nabla\theta\right) + \frac{\partial k}{\partial z}$$ (9.18)

$$C\frac{\partial T}{\partial t} = \nabla\left(\lambda \nabla T\right) + L\nabla\left(D_{\theta v} \nabla\theta\right)$$ (9.19)

where θ is volumetric water content, C is volumetric heat capacity, T is temperature, t is time, λ is thermal conductivity, L is latent heat of water, D_T is thermal moisture diffusivity, D_θ is isothermal moisture diffusivity and $D_{\theta v}$ is isothermal vapour diffusivity. The thermal moisture diffusivity D_T is made up of the thermal vapour diffusivity D_{TV} and the thermal liquid diffusivity D_{TL},

$$D_{TV} = \frac{D_o}{\rho_w}\tau v\, h\beta\frac{\partial \rho_o}{\partial T}\;;\;\; D_{TL} = k_\theta\frac{\psi}{\sigma}\frac{\partial\sigma}{\partial T}$$ (9.20)

where D_0 is molecular diffusivity of water in air, v is mass flow factor, ρ_0 is density of saturated water vapour, τ is tortuosity factor, β is volumetric air content, σ is surface tension of water, ψ *is* total potential and h is relative humidity.

As indicated in section 4.2.4, the isothermal moisture diffusivity D_θ consists of the isothermal vapour diffusivity $D_{\theta v}$ and the isothermal liquid diffusivity $D_{\theta L}$,

$$D_{\theta v} = \frac{\alpha \omega D_{atm} \gamma g \rho_v}{\rho_L RT} \frac{\partial \psi}{\partial \theta} \; ; \; D_{\theta L} = k_\theta \frac{\partial \psi}{\partial \theta} \tag{9.21}$$

where ψ is clay–water potential, α *is* tortuosity, ω *is* volumetric air content, D_{atm} is molecular diffusion coefficient, γ is mass flow factor, g is gravitational acceleration, ρ_v is density of vapour, ρ_L is density of liquid and R is the gas constant. The dependence of the clay water potential ψ on temperature T has been previously given by Philip and deVries (1957) as:

$$\frac{\partial \psi}{\partial T} = \frac{\psi}{\sigma} \frac{\partial \sigma}{\partial T}$$

When it comes to movement of water and vapour in the clay buffer with a hot canister at one end and a cool rock interface at the other, several interesting issues arise. These centre around the role of water content in the buffer. At what point vis-à-vis water content does the dominance of water movement give way to vapour movement? Or vice versa? Is this important? Observations in moisture distribution in surface layer soils in a field-drying situation can give us some guidance (Figure 9.13).

Figure 9.13 shows an initial volumetric water content θ, which is, for the sake of illustration, taken to be constant from the top surface of the soil to some depth below surface. As evapotranspiration progresses, the moisture profile will show a decreasing value of θ, with a characteristic concave shape. As time progresses, further evapotranspiration will cause further decreases in θ and, at some point in time, the moisture profile will show a shape change from concave to convex. This is shown in the figure as the region of inflection. As drying continues, the degree of convexity will increase. The question that needs to be asked is 'At what point will vapour diffusion be more dominant than liquid water transfer?' Why is this important? Because it affects not only one's choice of the diffusivity coefficients, but also the magnitudes of these coefficients and how they are incorporated into the analyses of total liquid transfer in an initially unsaturated clay buffer.

There is an alternative way of looking at the problem of total liquid transfer. This is to use a combined diffusivity coefficient, i.e. a diffusivity coefficient that includes consideration of both vapour and liquid transfer.

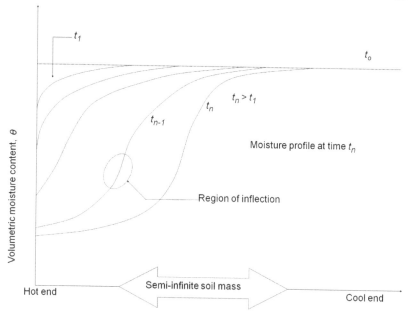

Figure 9.13 Typical field moisture uptake curve, as a function of depth. The 'hot' end is the surface undergoing evapotranspiration and the 'cool' end represents 'depth below ground surface'. The inflection in the moisture characteristic curve develops as the moisture content reaches some minimum threshold, depending on the type of soil.

Figure 9.14 shows this type of coefficient. It is not uncommon to find this type of hooked diffusivity phenomenon in field soils undergoing a surface drying process (Philip, 1974).

9.6   Thermal–hydraulic–mechanical modelling for prototype HLW repository testing

The majority of the models developed to address the transient status of the clay buffer used in deep geological repository containment of HLW have taken a mechanistic approach to problem solution. General broad agreement exists between almost all the models in the structure of the basis functions used to describe the various processes involved in the overall system performance. All of the known processes have been discussed in the previous chapters. Having said this however, there undoubtedly could be processes that have yet to be identified. Many of the known processes are well defined as separate individual processes, for example heat transfer, liquid moisture transfer, vapour transfer, and constitutive performance of the material.

The discussion in this section deals with the analytical computer models developed to address the performance of the clay buffer system in

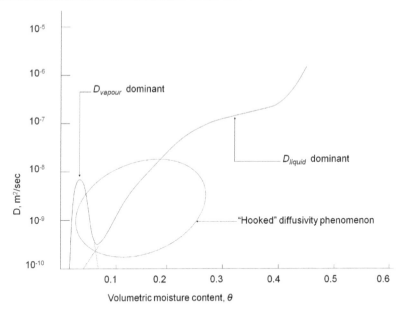

Figure 9.14   In the field soil-drying situation, the combination of vapour and liquid transfer from low θ to field capacity and more involves processes that do not demonstrate a nice smooth *D* coefficient. The hooked diffusivity phenomenon is actually quite common (adapted from Philip, 1974).

a prototype HLW repository founded in crystalline rock. Specifically, the models developed were part of the research programme dealing with model development and verification in the Äspö Hard Rock Laboratory (Äspö HRL) EU-supported Prototype Repository Project. The EU-supported Prototype Repository Project incorporates a full-scale test of the KBS-3V concept at a depth of about 400 m in crystalline rock. The description of the EU-supported Project and the Äspö HRL can be found in Svemar (2002, 2005), SKB (2001a and b) and Svemar and Pusch (2000). The main details of these models have been reported by Pusch (2002) and Pusch and Yong (2006). In this present discussion of the development and use of integrated THM (thermal–hydraulic–mechanical) models, we will use the main elements of the discussion in Pusch and Yong (2006) regarding development and application of THM models. Although the intent of the model development and validation programme was to deal with THMCB factors and processes, most of the models developed and used in the model validation programme did not reach the stage where C (chemistry) and B (biology) reactions, processes and activities were sufficiently verified for application.

The nature of the processes involved in the smectite buffer embedding the long-lived hot HLW canister have been described in the previous chapters. The analytical computer models developed by international research groups working in this particular project for SKB, have been used very recently in the international Prototype Repository Project at Äspö to predict and evaluate major physical and chemical processes. By and large, as most of the models have adopted a mechanistic approach to modelling of system performance there is an implicit requirement for (a) accurate portrayal of the various processes that contribute to the performance of the system; (b) proper delineation and representations of the initial properties of the clay and system; and (c) proper specification of the boundary conditions required for predictions of system performance.

All the pertinent parameters and data relating to clay buffer and the overall EBS performance were made available for THM model validation. The main details relating to the structure and computational techniques used, together with development of the computational codes, are on record in the various reports written for the Prototype Repository project. Prominent amongst these codes are ABAQUS, BRIGHT, COMPASS and THAMES. The details of the codes can be found in the references cited in the course of the discussion of the codes. A summary of the main elements of the codes, together with their application to the HLW repository project, can be found in Pusch and Yong (2006). In repeating many of the main points of the various codes, the terminology, symbols, mathematical expressions, parameters, etc. have been preserved in the forms and format used by the developers of the codes. No attempt has been made to seek uniformity in any of these items amongst the codes.

The various codes use a number of basic physical relationships that are valid for bulk clay – most generally assumed to be homogeneous and uniform. This assumption is common to most of the models developed to deal with transport and fate of contaminants and also with the THM problem. However, there is growing recognition of the heterogeneity of clays, and especially of the role and influence of microstructural units in the control of macroscopic bulk behaviour, leading to the incorporation of probabilistic approaches in the development of THM and transport and fate models. Although water transport in liquid form in unsaturated clays has been treated as unsaturated flow into homogeneous and sometimes rigid porous materials, we need to recognize that this assumption brings with it severe limitations.

9.6.1   Code ABAQUS

The details of ABAQUS, which is a finite element code, can be found in Börgesson (2001), and in the ABAQUS manuals. This discussion focuses its attention on those issues central to system conceptualization and model development. As is common with most complex codes, there are several sub models, each of which deals with specific processes.

Equilibrium, continuity and assumptions

The various processes, behaviour patterns and assumptions considered include:

- Heat conduction is governed by Fourier's law,

$$f = -k \frac{\partial T}{\partial x}$$

  where f is the heat flux, k is the heat conductivity matrix, $k = k(T)$, x is the spatial coordinate, and conductivity can be assumed to be anisotropic or isotropic. The specific heat neglects coupling between mechanical and thermal phenomena.
- Vapour flow is a Fickian phenomenon, and in this case is driven by a temperature gradient.
- Fluid flow is taken to obey Darcy's law and is driven by water pressure differences.
- Mechanical behaviour is assumed to conform to the Drucker–Prager plasticity model, with the effective stresses assumed to be determined by the Bishop unsaturated effective stress model.
- Fluid retention is modelled as a modified van Genuchten model:

$$S_r = S_{r0} + \left(S_{r\max} - S_{r_0}\right) \left[1 + \left(\frac{S}{P_0}\right)^{\frac{1}{1-\lambda}}\right]^{-\lambda} \left[1 - \frac{S}{P_s}\right]^{\lambda_s}$$

- The parameters for the modified van Genuchten model and the simplified effective stress relationship are identified as follows: S_r, S_{r0} and $S_{r\max}$ are degree of saturation, residual degree and maximum degree of saturation, respectively, P_0, P_s, λ and λ_s are material parameters, σ^* and σ are effective and total stress, respectively, and u_w is porewater pressure, χ is parameter dependent on the degree of saturation of the clay, and I is the unitary matrix.
- A simplified effective stress relationship written as:

$$\sigma^* = \sigma + \chi u_w I$$

- The virtual work relationship for equilibrium conditions is written as:

$$\int_V \sigma : \delta \varepsilon \, dV = \int_S t \cdot \delta v \, dS + \int_V F \cdot \delta v \, dV + \int_V S_r n \rho_w g \cdot \delta v \, dV$$

where δv = virtual velocity field, $\delta \varepsilon^{\mathrm{def}} = \mathrm{sym}(\partial \delta v / \delta x)$, $\delta v / \partial x$ = virtual rate of deformation, σ = Cauchy stress, t = surface tractions per unit area, F = body forces per unit volume including the weight of the wetting liquid $f_{w} = S_{r} n \rho_{w} g$, S_{r} = degree of saturation, n = porosity, ρ_{w} = density of the wetting liquid, v = velocity field vector, V = volume of solid material, and g = gravitational acceleration.

- Energy conservation provides the following:

$$\frac{d}{dt} \int_{V} \left(\tfrac{1}{2} \rho v \cdot v + \rho U \right) dV = \int_{S} v \cdot t \, dS + \int_{V} F \cdot v \, dV,$$

where ρ = current density and U = internal energy per unit mass.

- Fluid mass continuity in combination with the divergence theorem implies validity of the pointwise equation as follows:

$$\frac{1}{J} \frac{d}{dt} \left(J \rho_{w} S_{r} n \right) + \frac{\partial}{\partial x} \cdot \left(\rho_{w} S_{r} n v_{w} \right) = 0$$

where J is the determinant of the Jacobian matrix, v_{w} is the fluid velocity, and x is a spatial coordinate.

- For uncoupled heat flow, the basic energy balance is:

$$\int_{V} \rho \dot{U} \, dV = \int_{S} q \, dS + \int_{V} r \, dV$$

where V = is volume of solid material with surface area S; ρ = density of the material; $\dot{U}$ = time rate of the internal energy; q = heat flux per unit area of the body flowing into it; and r = heat supplied externally to the body per unit volume. Thermal and mechanical processes are assumed to be uncoupled in the sense that $U = U(T)$ only, where T = temperature of the material, and q and r do not depend on the strains or displacements of the body,

- Coupled thermal and hydromechanical phenomena are solved through a 'staggered solution technique' as follows:

  - First, a thermal analysis is performed, in which heat conductivity and specific heat are defined as functions of saturation and water content. In the first analysis these parameters are assumed to be constant and in the subsequent analyses they are read from an external file.
  - The hydromechanical model calculates stresses, pore pressures, void ratios, degree of saturation, etc. as functions of time. Saturation and void ratio histories are written onto an external file.
  - The material parameters update module reads the file with saturation and void ratio data and creates a new file containing histories for

saturation and water content. The saturation and water content histories are used by the thermal model in the subsequent analysis.

Constitutive relationships

The constitutive relationships include:

- For the solids: $d\tau^c = H{:}d\varepsilon + g$, where $d\tau^c =$ stress increment, $H =$ material stiffness, $d\varepsilon$ the strain increment, and $g =$ any strain independent contribution (e.g. thermal expansion).
- For the liquid (static) in the porous medium:

$$\frac{\rho_w}{\rho_w^0} \approx 1 + \frac{u_w}{K_w} - \varepsilon_w^{th},$$

where $\rho_w =$ density of the liquid, $\rho_w^0 =$ density in the reference configuration, $K_w(T) =$ bulk modulus of liquid, and $\varepsilon_w^{th} = 3\alpha_w(T-T_w^0) - 3\alpha_{w|T^I}(T^I-T_w^0) =$ volumetric expansion of the liquid caused by temperature change. $\alpha_w(T) =$ liquid thermal expansion coefficient, $T =$ current temperature, $T^I =$ initial temperature at this point in the medium, and $T_w^0 =$ reference temperature for the thermal expansion. Both u_w/K_w and ε_w^{th} are assumed to be small.

- Darcy's law relating the volumetric flow rate of the wetting liquid through a unit area of the medium, $S_r n v_w$, is proportional to the negative of the gradient of the piezometric head:

$$S_r n v_w = -\hat{k}\frac{\partial\phi}{\partial x},$$

where $\hat{k} =$ permeability of the medium, and where the piezometric head ϕ is defined as:

$$\phi \overset{def}{=} z + \frac{u_w}{g\rho_w}$$

- The elevation head at some defined datum is given as z, and g the gravitational acceleration in the direction opposite to z. $\hat{k}$ can be anisotropic and is a function of the saturation and void ratio of the material. For a constant g, we obtain:

$$\frac{\partial\phi}{\partial x} = \frac{1}{g\rho_w}\left(\frac{\partial u_w}{\partial x} - \rho_w g\right)$$

- For unsaturated flow, the hydraulic conductivity of the partly saturated soil is equal to a modified fully saturated hydraulic conductivity. The modification parameter or constant is a function of the degree of saturation of the soil.
- The diffusive vapour flow which is driven by the temperature and vapour pressure gradients is given as: $q_v = -D_{Tv}\nabla T - D_{pv}\nabla p_v$, where q_v = vapour flux, and D_{Tv} = thermal vapour diffusivity, D_{pv} = isothermal vapour flow diffusivity, p_v = vapour pressure, and T = temperature.

Parameters

- Thermal conductivity λ; specific heat c as function of void ratio e, and degree of saturation Sr.
- Hydraulic conductivity of water-saturated material K, as function of void ratio e and temperature T.
- Influence of degree of saturation S_r on the hydraulic conductivity K_p.
- Basic water vapour flow diffusivity D_{vTb} and the parameters a and b.
- Matric suction u_w as a function of the degree of saturation S_r.
- Porous bulk modulus κ and Poisson's ratio ν.
- Drucker–Prager plasticity parameters β, d, ψ, and the yield function.
- Bulk modulus (B_s) and coefficient of thermal expansion of water (B_w, σ_w).
- Bishop's parameter χ (usual assumption $\chi = S_r$).
- Volume change correction ε_v as a function of the degree of saturation S_r (i.e. the 'moisture swelling' factor).
- Initial conditions: void ratio e; degree of saturation S_r; pore pressure u; and average effective stress p.

9.6.2   Code BRIGHT

As with ABAQUS, the details of code BRIGHT, which are given in Ledesma and Chen (2003) and Olivella *et al.* (1996), will not be repeated here. And, as with the discussion of ABAQUS, we will present the main items associated with translation of conceptualization to model development. The reader is advised to consult the code development manuals of this and the other models for details of this and the other codes. Code BRIGHT is a finite element code for the analysis of THM problems in geological media, developed by the Geomechanics group of the Geotechnical Engineering and Geosciences Department, Technical University of Catalunya – Centre for Numerical Methods in Engineering (UPC-CIMNE, Barcelona, Spain).

The modelling approach uses the various constitutive relationships for air, water and heat. In recognition of the unsaturated state of the buffer, the hydraulic conductivity of the material is considered to be dependent on the degree of saturation. Thermal conductivity is also considered to be a function of the hydration state of the material. The effect of vapour diffusion in the buffer material is considered in terms of a coefficient of tortuosity. For the mechanical behaviour of the buffer, a thermoplastic model is used, wherein deformations are a function of net stresses, suction and temperature.

The equilibrium restrictions used in the code include control of (a) mass fraction of water in the vapour phase from psychrometric principles and (b) the amount of air dissolved in water from Henry's law. The code considers all the interacting processes by simultaneous solution of the equations of (a) enthalpy conservation; (b) water mass conservation; (c) air mass conservation; and (d) linear momentum conservation. The geological medium is considered to be a porous medium composed of solid grains, water and gas. Thermal, hydraulic and mechanical aspects are taken into account, including coupling between them in all possible directions. The code deals with chemical processes such as complexation, oxidation–reduction reactions, acid–base reactions, precipitation/dissolution of minerals, cation exchange, sorption and radioactive decay. The total analytical concentrations are adopted as basic transport variables and chemical equilibrium is achieved by minimizing Gibbs free energy. Analysis of the problem is formulated in a multi-phase and multi-species approach as follows:

- Phases: solid phase (s); liquid phase (l), consisting of water and dissolved air; and gas phase (g), consisting of a mixture of dry air and water vapour.
- Species: solid (–); water (w), existing as liquid or evaporated in the gas phase; and air (a), consisting of dry air or gas or dissolved in the liquid phase.

Assumptions, continuity and governing relationships

The assumptions used in problem formulation and the continuity and governing relationships include:

- Dry air is considered as a single species. It is the main component of the gaseous phase. Henry's law is used to express equilibrium of dissolved air.
- Thermal equilibrium between phases is assumed. This means that the three phases are at the same temperature.
- Vapour concentration, determined according to the psychrometric law, is in equilibrium with the liquid phase.
- State variables (also called unknowns) are solid displacements u (three spatial directions), liquid pressure P_l, gas pressure P_g and temperature, T.
- Balance of momentum for the medium, as a whole, is reduced to the equation of stress equilibrium together with a mechanical constitutive model to relate stresses with strains. Strains are defined in terms of displacements.

- Small strains and small strain rates are assumed for solid deformation. Advective terms due to solid displacement are neglected after the formulation is transformed in terms of material derivatives (in fact, material derivatives are approximated as Eulerian time derivatives). In this way, volumetric strain is properly considered.
- The elastoplastic model is used to describe the stress–strain relationship of the bentonite.
- Balance of momentum for dissolved species and for fluid phases are reduced to constitutive equations (Fick's law and Darcy's law).
- Physical parameters in constitutive laws are functions of pressure and temperature. For example: concentration of vapour under planar surface (in psychrometric law), surface tension (in retention curve), and dynamic viscosity (in Darcy's law) are strongly dependent on temperature.
- Mass balances: (a) of solids; (b) of water; and (c) of air.
- Momentum balance for the medium.
- Internal energy balance for the medium.

9.6.3   Code COMPASS

The COMPASS code developed by Thomas and his co-workers (1996) is based on a mechanistic theoretical formulation, by which various aspects of soil behaviour under consideration are included in an additive manner. It adopts and extends the Philip and de Vries formulations (section 9.5) to include the other processes involved in system performance. In addition to the adoption of the flux laws for water and air, heat transfer included the effects of conduction, convection and latent heat of vaporization. Temperature effects on suction are considered in terms of their effect on surface energy. A number of constitutive relationships have been implemented to describe the contributions. In particular for the net stress, temperature and suction contributions, both elastic and elastoplastic formulations are used. To describe the contribution of the chemical solute on the stress–strain behaviour of the soil, as a first approximation, an elastic state surface concept that describes the contribution of the chemical solute through an elastic relationship, based on osmotic potentials, is proposed. The basic unknowns are porewater pressure, pore air pressure, temperature and displacements. As with all the other models/codes, the partly saturated soil is considered as a three-phase porous medium consisting of solid, liquid and gas (air), with the liquid phase (porewater) containing multiple chemical solutes.

Processes and assumptions

- Moisture flow includes both liquid and vapour flow. Liquid flow is assumed to be described by a generalized Darcy's law, whereas vapour transfer is represented by a modified Philip–de Vries approach. The governing equation for moisture transfer is expressed, in primary variable form, as:

$$C_{ll}\frac{\partial u_l}{\partial t}+C_{lT}\frac{\partial T}{\partial t}+C_{la}\frac{\partial u_a}{\partial t}+C_{lu}\frac{\partial \mathbf{u}}{\partial t}=$$
$$\nabla.\left[K_{ll}\nabla u_l\right]+\nabla.\left[K_{lT}\nabla T\right]+\nabla.\left[K_{la}\nabla u_a\right]+\nabla.\left[K_{lc_s}\nabla c_s\right]+J_l$$

where C_{lj}, K_{lj} and J_l are coefficients of the equation ($j=l$, T, a, c_s, u).

- Heat transfer includes conduction, convection and latent heat of vaporisation transfer in the vapour phase. The governing equation for heat transfer is expressed, in primary variable form, as:

$$C_{Tl}\frac{\partial u_l}{\partial t}+C_{TT}\frac{\partial T}{\partial t}+C_{Ta}\frac{\partial u_a}{\partial t}+C_{Tu}\frac{\partial \mathbf{u}}{\partial t}=$$
$$\nabla.\left[K_{Tl}\nabla u_l\right]+\nabla.\left[K_{TT}\nabla T\right]+\nabla.\left[K_{Ta}\nabla u_a\right]+$$
$$V_{Tl}\nabla u_l+V_{TT}\nabla T+V_{Ta}\nabla u_a+V_{Tc_s}\nabla c_s+J_T$$

where C_{Tj}, V_{Tj}, K_{Tj} and J_T are coefficients of the equation ($j=l$, T, a, c_s, u), u_l and u_a are the pore pressures for liquid and air, respectively and $\mathbf{u}$ is a nodal deformation vector.

- Flow of dry air due to the bulk flow of air arising from an air pressure gradient and dissolved air in the liquid phase is considered. The bulk flow of air is again represented by the use of a generalized Darcy's law. Henry's law is used to calculate the quantity of dissolved air and its flow is coupled to the flow of pore liquid. The governing equation for dry air transfer is expressed, in primary variable form, as:

$$C_{al}\frac{\partial u_l}{\partial t}+C_{aT}\frac{\partial T}{\partial t}+C_{aa}\frac{\partial u_a}{\partial t}+C_{au}\frac{\partial \mathbf{u}}{\partial t}=$$
$$\nabla.\left[K_{al}\nabla u_l\right]+\nabla.\left[K_{aa}\nabla u_a\right]+\nabla.\left[K_{ac_s}\nabla c_s\right]+J_a$$

where C_{aj}, K_{aj} and J_a are coefficients of the equation ($j=l$, T, a, c_s, u).

- Deformation effects are included through either a non-linear elastic, state surface approach or an elastoplastic formulation. In both cases deformation is taken to be dependent on suction, stress and temperature changes. The governing equation for stress-strain behaviour is expressed, in primary variable form, as:

$$C_{ul}du_l+C_{uT}dT+C_{ua}du_a+C_{uc_s}dc_s+C_{uu}d\mathbf{u}+db=0$$

where C_{uj} are coefficients of the equation ($j = 1$, T, a, c_s, **u**) and **b** is the vector of body forces.

- Chemical solute transport for multi-chemical species includes diffusion dispersion and accumulation from reactions due to the sorption process. The governing equation for chemical solute transfer is expressed, in primary variable form, as:

$$C_{c,l}\frac{\partial u_l}{\partial t} + C_{c,a}\frac{\partial u_a}{\partial t} + C_{c,c_s}\frac{\partial c_s}{\partial t} + C_{c,u}\frac{\partial \mathbf{u}}{\partial t} = \nabla.\left[K_{c,l}\nabla u_l\right] + \nabla.\left[K_{c,T}\nabla T\right] +$$

$$\nabla.\left[K_{c,a}\nabla u_a\right] + \nabla.\left[K_{c,c_s}\nabla c_s\right] + J_{c_s}$$

where $C_{c,j}$, $K_{c,j}$ and $J_{c,j}$ are coefficients of the equation ($j = 1$, T, a, c_s, **u**).

9.6.4 Code THAMES

The THAMES finite element code developed by Ohnishi *et al.* (1985) extends the Philip and de Vries work for treatment of coupled heat and mass transfer. The model also considers water movement due to osmotic potentials, in addition to the thermally driven fluid flow process. Moisture transfer is considered in terms of the internal driving force expressed by the gradient of the moisture potential ψ, and external driving forces. In addition, moisture transfer in the unsaturated zone is considered to be a diffusion process, and transfer in the saturated zone to be governed by the Darcy relationship.

The mathematical formulation for the THAMES model utilizes Biot's theory, with Duhamel–Neuman's form of Hooke's law, and an energy balance equation. The governing equations are derived with fully coupled thermal, hydraulic and mechanical processes. The assumptions invoked include (a) the medium is poro-elastic; (b) Darcy's law is valid for the flow of water through a saturated–unsaturated medium; (c) heat flow occurs only in solid and liquid phases (the phase change of water from liquid to vapour is not considered); (d) heat transfer among three phases (solid, liquid and gas) is disregarded; (e) Fourier's law holds for heat flux; and (f) water density varies depending upon temperature and the pressure of water.

Accordingly, the continuity condition for moisture transfer is expressed as:

$$\left[\xi\rho_w D_\theta \frac{\partial \theta}{\partial \psi}\left(h_{,j} - z_{,j}\right) + \left(1 - \xi\right)\frac{\rho_w^2 gK}{\mu_w}h_{,j}\right]_{,i} + \left[\rho_w D_T T_{,i}\right]_{,i}$$

$$-\rho_{wo} nS_r \rho_w g\beta_p \frac{\partial h}{\partial t} - \rho_w \frac{\partial \theta}{\partial \psi}\frac{\partial \psi}{\partial t} - \rho_w S_r \frac{\partial u_{i,i}}{\partial t} + \rho_{wo} nS_r \beta_T \frac{\partial T}{\partial t} = 0$$

(9.22)

where ξ is the saturation parameter and is zero in the unsaturated zone and equal to unity in the saturated zone; ψ = soil moisture potential, ρ_w = density of water, S_r = degree of saturation, β_p = compressibility of water, β_T = thermal expansion coefficient of water, μ_w = viscosity of water, K = hydraulic conductivity and subscript o = reference state.

The pressures developed and the equilibrium relationship in terms of the pressures developed from the swelling behaviour of the clay buffer material is given as follows (M. Chijimatsu, T. Fujita, A. Kobayashi and M. Nakano, Valuclay Project, unpublished):

$$\left[\frac{1}{2}C_{ijkl}\left(u_{k,l}+u_{l,k}\right)-F\pi\delta_{ij}-\beta\delta_{ij}\left(T-T_0\right)+\chi\delta_{ij}\rho_w b\right]_{,j}+\rho b_i = 0 \qquad (9.23)$$

where C_{ijkl} = elastic matrix, F = coefficient related to swelling pressure in the buffer, π = swelling pressure and χ = effective stress parameter (which varies from zero for the unsaturated zone and is equal to unity in the saturated zone). The Bishop–Blight extension of the Terzaghi effective stress relationship to include saturated–unsaturated media is used in the following form: $\sigma'_{ij} = \sigma'_{ij} + \chi\delta_{ij}\rho_f g\psi$, where σ'_{ij} is the effective stress, δ_{ij} is the Kronecker delta, ρ_f is the unit weight of water, and g is the acceleration of gravity. The general equilibrium equation for the effective stress is $(\sigma'_{ij}+ \chi\delta_{ij}\rho_f g\psi) + \rho b_i$, where $(\chi\delta_{ij}\rho_f g\psi)$ is a term implying that changes in the pressure head influence the equilibrium equation. The swelling pressure in the buffer π is assumed to be a function of the soil water potential ψ. It is useful and interesting to note that in equation 9.22, the water potential is assumed to be a single-valued function of the volumetric water content θ. The relationship for swelling pressure in the buffer is given as follows:

$$\pi\left(\theta_1\right)=\rho_w g\left(\Delta\psi\right)=\rho_w g\left\{\psi\left(\theta_1\right)-\psi\left(\theta_0\right)\right\}=\rho_w g\int_{\theta_0}^{\theta_1}\frac{\partial\psi}{\partial\theta}d\theta \qquad (9.24)$$

The effects of temperature considered in the constitutive relationship takes its cue from Duhamel–Neuman's relationship for solid media. The stress equilibrium equation, which takes into account the effects of temperature and pore pressure changes, is given as follows:

$$\left[\frac{1}{2}C_{ijkl}\left(u_{k,l}+u_{l,k}\right)-\beta\delta_{ij}\left(T-T_0\right)+\chi\delta_{ij}\rho_f g\psi\right]_{,j}+\rho b_i = 0 \qquad (9.25)$$

where the $(-\beta\delta_{ij}(T-T_0))_{,j}$ term indicates the influence of heat transfer on the equilibrium equation.

9.7   Application of models for prediction

9.7.1   Preliminary information

To demonstrate and illustrate how analytical computer models have been developed for use in analysis and prediction of clay buffer/barrier systems, we will cite some of the results obtained in the model development and verification programme in the Äspö Hard Rock Laboratory (Äspö HRL) EU-supported Prototype Repository Project. The conditions for applying the THM models referred to in the previous include the initial and boundary conditions, the heat source and the changes with time that it may undergo, and whether the material properties and characteristics change because of water uptake and chemical changes in the system, and access to water for saturation of the initially partly saturated clay. A rectangular cut-out of the deep geological prototype system used is shown in Figure 9.15. Although the width (*W*), height (*H*) and depth (*D*) dimensions of the cut-out were geometrically set by the modellers for their three-dimensional finite element calculations, the sizes of the deposition holes, canister (heater) and buffer system were consistent with the concept of the final HLW containment scheme.

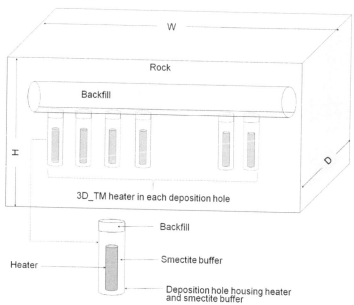

Figure 9.15 View of block cut-out from actual deep geological prototype repository system, with six deposition holes and their respective heater canisters, used to generate information and data for testing of models. For details of the system and tests conducted, refer to Svemar (2005).

A key element in the design of the HLW repository system to protect the canister is the physical–thermal–chemical buffering effect of the smectite buffer embedding the canister. To achieve this buffering effect, complete water saturation and a degree of resultant homogenization of the buffer are needed. In a repository system such as that shown in Figure 9.15, one relies on the water-transport attributes and features of both the host rock and the buffer itself. The process of water uptake from the rock and backfill should ultimately result in complete water saturation and the final density distribution of the smectite buffer. The rock around the deposition holes allows transport of water through discrete hydraulically and mechanically active or activated discontinuities. Hydraulic transport capacity is a function of the nature (frequency and conductive properties) of the discontinuities. The distribution over the periphery of the holes is controlled both by the location of the intersecting fractures and by the conductivity of the shallow boring-disturbed zone. If the host rock permits very little water to be transported to the deposition holes, the backfill may serve as the major water source. The primary agents for water transport from the backfill to the buffer are water pressure and matrix potential of the smectite constituting the buffer, and obviously the degree of water saturation of the backfill. Water pressure in the relatively more permeable backfill, at the contact between buffer and backfill, may rise relatively quickly if access to water from intersected hydraulically active fracture zones is available. If access is denied, the rise in water pressure will be relatively low and will also be slow.

In the cut-out of the repository system shown in Figure 9.15, the average bulk hydraulic conductivity of the near-field rock around the deposition holes was $k = 10^{-12}$ to $k = 4 \times 10^{-10}$ m/s. From information relative to the rock structure in the repository, one would expect that a deposition hole would be intersected by about three steep and four flat-lying, highly water-bearing fractures – a condition that is perhaps somewhat overrated. The average inflow of water in the holes was less than 0.006 L/min in all the holes, except for hole 1, which had an inflow of 0.08 L/min. Water pressure at about a 2-m distance from the tunnel wall was between 100 kPa and 1.5 MPa. The hydraulic conductivity of the backfill varied from 10^{-11} to 10^{-9} m/s. Hydration of the buffer and backfill by water uptake from the rock was associated with expansion of the upper part of the buffer and consolidation of the backfill as well as with axial displacement of the canisters.

9.7.2   Temperatures and pressures in buffer

Actual temperatures obtained in the buffer in relation to time after introduction of the heating source in Figure 9.16. After about 700 days (close to 2 years), the temperature in the clay adjacent to the surface of the canister was about 72°C. The temperature at the rock interface was around 60°C at

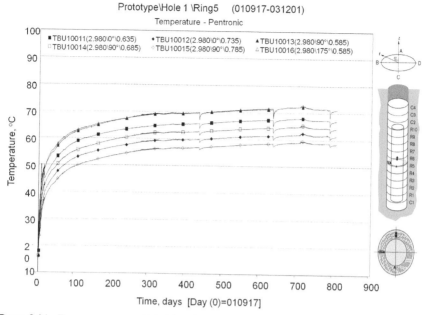

Figure 9.16   Temperature at mid-height of canister in hole 1 (highest *T* close to the rock). Terminology, parameters, etc. refer to notations used in Svemar (2005).

the rock after about 700 days, or around 2 years. The average temperature gradient was about 0.34°C per centimetre radial distance.

The swelling pressures developed in the smectite buffer – in response to water uptake in the initially partly saturated clay – are shown in Figure 9.17. Based on the nature of the curves, we can speculate that equilibrium swelling pressure development had not been attained. The upward trajectories of the pressures recorded by the various pressure sensors, in relation to time, suggest very strongly that maturation will continue for a longer period of time, i.e. much longer than that obtained in the repository test period. The information presented in Figure 9.17 is consistent with the upward temperature trajectories shown in Figure 9.16, for which equilibrium temperatures do not appear to have been attained in the same repository test period.

9.7.3   Comparison – model and actual

Computational methods used and detailed comparisons between calculated and measured values for the sets of experiments conducted in the Prototype Repository project can be found in Svemar (2005). By and large, models/ codes BRIGHT, COMPASS and THAMES showed satisfactory agreement between computed temperatures and different radial positions in the

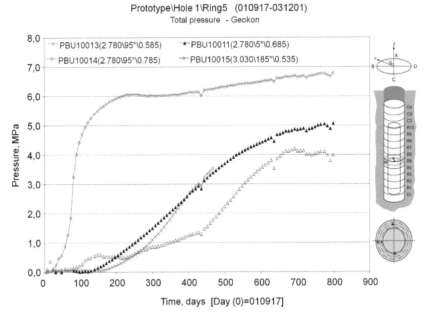

Prototype\Hole 1\Ring5   (010917-031201)
Total pressure - Geokon

Figure 9.17   Development of swelling pressure at mid-height in the smectite buffer in
the wettest deposition hole. The highest pressure (~6.7 MPa) was reached,
at a position close to the host rock after about 2 years. The lowest pressure
(4 MPa) was obtained at a position close to the canister. Terminology,
parameters, etc. refer to notations used in Svemar (2005).

borehole with recorded temperature readings. The following observations
have been reported in Pusch and Yong (2006).

- The models provided calculated results that were of the same order of
  magnitude as the measurements, and can be used for rough prediction
  of the temperature, hydration and pressure build-up in buffer of the type
  used in the Prototype Repository project.
- Good agreement between predictions and measurements were obtained
  for the evolution of temperature with time. Some over-prediction of
  temperature occurred with some models for the first 2 years, whereas
  others showed good agreement between predicted (calculated) and
  measured values. This however can be considered to be a plus, as this
  offers a safe, conservative prediction. There were indications that the
  thermal conductivity of the buffer was higher than assumed and that the
  heat transfer could have been reinforced by some undefined mechanism,
  as, for instance, convection through vapour flow.
- Almost all of the models provide calculated results that were fairly well
  in agreement with the recordings. They provided information on the
  wetting rate for deposition holes with 'unlimited' access to water from

the rock that can be useful for scoping purposes. The predicted rate of saturation was in general too high, indicating thereby that all the processes associated with water uptake were not fully understood and therefore not included in the models.

- Prediction of wetting phenomena for those deposition holes with limited access to water for hydration showed more uncertainty in the comparisons. One of the models provided fairly accurate predictions of hydration in a 'dry' hole by basing its calculations on measured inflow before placement of the buffer in the hole (deposition hole #3). The comparison of calculated and observed results indicated that it was more difficult to predict the wetting process for a repository with much less access to water.

- Providing theoretical calculations relating to the evolution of pressure and mechanical response of the buffer presents the biggest challenge to the models. This is owing not only to the requirement for specification of fracturing and displacements in the buffer for inclusion in the models, but also to the need for specification or prescription of the inter-relationships between both hydration and dehydration, and swelling and drying. Prediction of the hydration rate has been shown to be less than accurate. Accordingly, forecasting of the mechanical response, which is a function of the hydration phenomenon, becomes more uncertain. Nevertheless, one can argue that models can provide calculated results that can be usefully used as scoping data.

- It needs to be stated that all the predictions and comparisons were made for the short-term performance of the Prototype Repository and, in comparison with the long-term period of 100,000 years of expected operational time for the real system, this is a very short time period. How well the models will perform will depend on how well they can represent the results of evolution of the buffer and how well one can describe the boundary conditions. Both of these issues are highly significant, as they represent knowledge requirements that demand considerable appreciation of maturation processes both in the buffer clay and at the interface between canister and buffer and between buffer and host rock.

9.8   Concluding remarks

The importance of developing mathematical computer models that reflect the performance of clay buffer/barrier systems, either as stand-alone clay buffers as in the case of the smectite buffers used to embed HLW canisters in repositories, or as liner–barrier components in engineered barrier systems, cannot be overstated. The often-repeated statement of 'garbage in, garbage out' has been well proven historically in test comparisons between predictions made by models in comparison with actual field results. There is no substitute for obtaining complete and thorough knowledge and understanding of the particular system to be analysed and the various processes involved in the performance of the system.

In the comparison of model predictions and actual performance of the clay buffer in the Prototype Repository project, one accepts that questions should be raised with respect to the capability of both models and field measurement devices/techniques. Lack of correspondence between calculated/predicted and measured values can be due to (a) model inadequacies; (b) inadequate and/or improper input and material properties/parameters; (c) inadequacies and improper measurements of field/laboratory experiments used for comparison with model predictions; and (d) all of the preceding. The case of hydration of the smectite buffer in the Prototype Repository project is a good example of 'what might have happened in the field experiment'. One could speculate that some pressure gauges may have reacted too soon because of water migration along cables and this may imply that the maturation of the buffer is in fact even slower than indicated by the recordings.

It bears repeating that almost all, if not all, comparisons or validations of models with actual field performance data have been carried out with short-term tests. These have shown that they can perform well for situations in which proper appreciation of actual system performance and processes exists. However, in the face of operating lives of greater than tens of thousands of years, and armed with the knowledge that ageing processes can significantly alter the characteristics and properties of buffer clay and also severely impact on the nature of the boundary conditions, one can foresee considerable difficulties ahead. Much work remains in developing proper knowledge of the impact of ageing processes on material properties and performance, and on the nature of boundary conditions. Much work also remains in seeking ways to provide the means for model validation for long-term performance prediction. What this tells us is that there is an absolute requirement for model developers, researchers and stakeholders to continue research and development work on long-term performance assurance and prediction.

9.8.1   Points to ponder

The following are some points to consider in the performance of clay buffers and clay barriers in EBS.

- Wetting of the clay buffer in the HLW repository experiment, or in the clay barrier of an HSW-EBS, could result in precipitation of Ca^{2+}, Mg^{2+}, SO_4^{2-} and Cl^- as minerals with reversed solubility, such as sulphates and carbonates, at the wetting front. In the case of a HLW repository, these will move towards the hot canister and will inevitably interact with the smectite minerals and the canister. In the case of the clay barrier in the HSW-EBS, this could degrade the permeability and accumulative properties of the clay.

• In a HLW repository, a temperature gradient can eventually result in the dissolution of silicate minerals, including the smectite component. The extent of the dissolution is greater at locations near the heater, and lesser at regions close to the rock. This differential will result in the migration of the released silica towards the host rock boundary, and accumulation of precipitates in the outer part of the bentonite. Silicification causes cementation and brittleness of the clay. Regarding the clay barrier in the HSW-EBS, temperature gradients with respect to colder temperatures in winter conditions are more of a concern, and the problem of silicification from temperature gradients is not expected to be a problem. That being said, one should not discount silicification from chemically related events.

• The major effects and consequences of the water uptake by the smectite in the HLW repository are:

  – Water uptake provides the desired condition of the canisters of complete embedment by a practically impermeable clay medium.

  – The canister may undergo displacement in the course of the wetting of the buffer and subsequently.

  – Expansion of the upper part of the buffer and compression of the contacting backfill will change their hydraulic conductivities. Thermally or tectonically induced compression and shearing of hydraulically active rock discontinuities may have an impact on the ability of the rock to transmit water to the buffer in the deposition holes.

  – If the rate of water uptake is slow, salt precipitation in the buffer close to the canisters is enhanced. This also enhances corrosion of the copper canisters. Exposure to vapour given off from the wetting front will affect the clay minerals and cause dissolution and precipitation of silica.

Chapter 10

Safety assessment and performance determination

10.1 Safety, reliability and risks

10.1.1 What is meant by 'safe'

To close the discussions in this book on the fundamental aspects of clay properties and their requirements and performance as engineered clay buffers and barriers in multi-barrier systems, we have chosen to focus on *safety assessment* in this chapter instead of the more traditional *risk management* treatment. The issues of concern in any design, construction and performance venture must always be directed towards the safe performance of the particular venture. The same holds true for engineered clay buffers and barriers in multi-barrier systems. These clay buffers and barriers, which are only one part of a multi-barrier system that is designed to contain high-level radioactive waste (HLW) and/or hazardous solid waste (HSW), must perform according to their designed purpose, as the overall safety of the multi-barrier requires the safe performance of each of its components.

It is not realistic or even possible to conduct a safety assessment of a particular system without having a framework that defines the safe standard, or criterion, or level, etc. What is meant by *safe*? In the context of the topic of this book, the question becomes more specific: 'What is meant by *safe performance?*' Obviously, the definition or specification or criteria needed to establish 'safe performance' will be directly contingent on (a) the system, facility, venture, operation, etc. being examined and (b) the susceptibility and tolerance of the targets and/or receptors affected by the detrimental outcome resulting from 'unsafe performance'. By all accounts, most *safety* considerations pay initial attention to protection of human health and the environment.

For the HLW deep geological repository and HSW landfill scenarios considered in this book, the principal sets of concern relate to exposure of biotic receptors and the environment to fugitive radionuclides from the HLW repository, and contaminants escaping from the HSW landfill (Figure 10.1). That being said, we cannot overlook the hazards posed by HLW and

HSW during operations leading to their final containment in the repository or landfill, respectively. Although, strictly speaking, these are not direct clay buffer/barrier performance hazards, they fall within the purview of hazard awareness, as the containment operations relate directly to the transport or transfer of (a) the HLW canister from above ground to the final borehole or tunnel repository containment scheme and (b) the discharge or placement of HSW in the landfill – including daily cover operations.

For the two situations illustrated in Figure 10.1, we recognize that it is the failure of the overall containment system to fulfil its design function that has allowed the escape of radionuclides and contaminants. The primary concern with regard to *safety* for the two situations shown in the figure is (a) to prevent the fugitive radionuclides from coming near to or reaching the ground surface, or (b) failing that, to make sure that the half-lives of the radionuclides reaching ground surface are below the threshold limits; and (c) to ensure that the concentrations of contaminants reaching the control monitoring stations register well below threshold concentrations. It is not within the purview of this book to enter into a discussion of health and environment threats and sustainability, nor is it the intent of this discussion to examine the possible scenarios leading to failure of total system and/or individual components in the HLW and HSW multi-component containment systems.

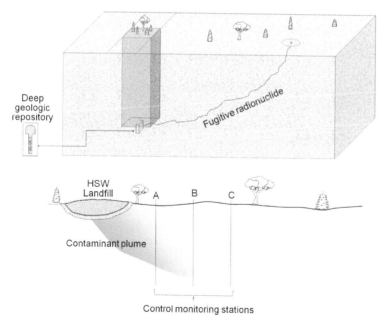

Figure 10.1   Threats to biotic receptors and environment from fugitive radionuclides from deep geological HLW repository (top illustration) and contaminants in contaminant plume emanating from HSW landfill (bottom illustration).

Safety is a direct measure of the *reliability* of a system to perform according to design expectations in its design lifetime and under design conditions. The probability of success for a system to perform according to design expectations constitutes a measure of reliability. For the HLW/HSW containment systems, we consider an *adequate* system performance to be one that avoids the failure scenarios depicted in Figure 10.1, meaning that the system is *safe*. That being said, a system performance above *adequate* is generally taken to mean that the system is *safer*, whereas less-than-adequate performance means failure of the system and hence an unsafe system.

There are many different techniques and methodologies that can be used to determine how safe a system is. All of these techniques and methodologies agree that *safety* and/or *reliability* are knowledge-based subjective concepts. They rely on knowledge on the probability of occurrence of events acting in series and/or in parallel or various combinations, the result of which would lead to success of operation of the system under consideration.

Complete and absolute *safety* in any course of action, venture or operational system (facility) requires that the course of action, venture or system performance must be absolutely *risk free,* i.e. absent of any risk of failure to perform according to design expectations and standards (Figure 10.2). One assumes that the design of the system itself is also absent of flaws and

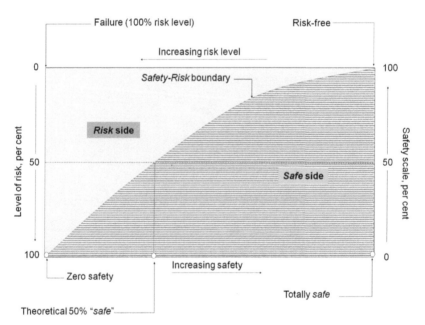

Figure 10.2 Relationship between safety and risk. Increasing safety is obtained by decreasing the level of risk.

imperfections. Experience tells us that the chances of obtaining a perfect flawless operating system are next to zero. In the case of the clays used for engineered clay buffers and barriers, a *risk-free* scenario with respect to its application does not exist. There are many steps along the way, between 'selection of candidate clay' for the buffer or barrier to 'final placement and function', wherein *events* such as unexpected inadequacies and 'failure to perform or conform' can arise from various sources of uncertainties, etc. In the final analysis, failure to perform up to design standards is the central and over-riding issue that dominates safety assessment.

The diagram shown in Figure 10.2 illustrates the intimate linking between *safety* and *risk*. The greater is the risk of failure in any single subset of activities or performance of the clay, the less safe the system will be. Increasing the safety of the system, i.e. making the system safer, requires a reduction in the level of risk. Probably one of the best demonstrations of the linkage between safety and risk with respect to the design function of engineered clay buffers and barriers is to ask the question: 'How safe is *safe?*' Technically, or analytically, the answer to this question would generally be 'What level of risk is one willing to accept?' At this juncture, one will embark on studies dealing with *risk assessment and analysis* – the end result of which should provide one with the information needed to determine an acceptable level of risk required to articulate an *acceptance criterion* consistent with the requirements of a safety assessment.

10.1.2 Safety assessment contents

To determine if a system under consideration or any component of a multi-component system can function according to design-constructed expectations, a performance assessment is generally undertaken. For completed facilities, the goal of a successful performance assessment process is to meet the requirements set forward in the *acceptance criterion* which is the core of a *safety assessment*. To facilitate discussions, we will use the acronyms when we refer to these regulatory instruments as such. To meet the requirements of an acceptance criterion, one must demonstrate that the facility or operation in question has met all the standards and performance specifications. In multi-component systems, this means that all the individual and interacting components that contribute to the successful operation of the system must meet their individual and respective specifications.

Detailing of requirements, criteria, etc. in an acceptance criterion for a safety assessment of a facility or an operating system is the responsibility of the appropriate governmental regulatory agency or authority. That being said, it is not uncommon for stakeholders to conduct their own safety assessments throughout the design, construction and operation of their systems – in all probability using sets of criteria and requirements that are similar to those set forward by the regulatory agency or authority. Most jurisdictions (countries, states, etc.) have set forward the contents of a safety assessment for containment or management of HSWs, in all probability through the regulatory or safety branches of their respective environmental protection agencies. Not all jurisdictions however have reached the stage of complete articulation of safety assessment contents for permanent isolation of HLW. To a large extent, this is because the scenarios for permanent isolation of HLW generated within those jurisdictions have not been totally determined.

The basic elements in the list of contents for a safety assessment include the following:

- *Description of system or component system* – e.g. a clay buffer in the HLW repository containment scenario and/or the clay barrier in an engineered multi-barrier system for HSW landfills.
- *Specification or description of the objectives of the system under consideration* – e.g. to provide a competent isolation barrier against deleterious actions initiated by the contained HLW or HSW.
- *Specification of the requirements needed to meet the objectives* – e.g. contaminant accumulation and mitigation, heat dissipation barrier, etc.
- *Specification of criterion or criteria that must be met if safety is to be assured* – it is at this stage that determination of acceptance criterion as a measure of the level of risk is required. As stated previously, the underlying question that must be answered when one seeks to specify an acceptance criterion is 'What level of risk are we willing to accept?'

By and large, articulation of the various specifications that constitute an acceptance criterion requires one to undertake *probabilistic risk assessment*. There are some who will argue that it is *probabilistic safety assessment (PSA)* that should be required, as opposed to safety assessment, to encourage the use of the tools of *reliability* and *risk analysis* in the assessment process. That being said, it is useful to recall that at the outset of this discussion it was pointed out that the contents or details of a safety assessment vary considerably depending on the nature of the system, activity, facility, etc. that is being scrutinized. Thus, the term *safety assessment* is perhaps more widely used, simply because it is understood that when required, analyses of reliability and risk are employed in determining 'what is *safe* and what is not *safe*'. Although our discussions will utilize some of the concepts and principles of reliability and risk analysis, they will not develop the basic elements and details of the subjects of *reliability* and *risk analysis* that underpin PSA. These (reliability and risk analyses) are important and wide-ranging subjects beyond the scope of this book and therefore the reader is advised to consult specialized textbooks for further information.

10.2 Probabilistic safety assessment

10.2.1 Elements in a probabilistic safety assessment process

Probabilistic safety assessment instead of regular safety assessment

As stated at the outset, there is no standard procedure or specific set of rules that governs safety assessment contents and procedures. This is understandable as safety assessments cover everything ranging from industrial activities to anthropogenic activities, to agro- and forest industries, to physical performances, etc. Adopting a probabilistic approach to safety assessment provides one with a systematic methodology to determine and evaluate the risks of *failure to perform* according to design requirements and expectations. When applied to complex multi-component systems such as HLW repositories in deep geological formations and HSW landfills, implementation of a PSA process requires one to have access to a breadth of information that may not be readily available. Specifically, one needs to have sufficient information of prior events if the relative frequency approach of probability analysis is to be undertaken with a high degree of confidence.

As deep geological repositories to house HLW canisters have yet to be built, and as information on the performance of multi-component barriers for HSW landfills is readily available, we can see that this probabilistic approach is not appropriate. Instead, the axiomatic approach to probability may be more appropriate for the PSA process. That being said, it should be recognized that there may be some considerable difficulty in assigning probabilities that will strictly obey the axioms and laws of probability. This will be true for the third axiom for clay buffers used as HLW canister embedment material in repositories, particularly with respect to the treatment of intersection of events $(A_1 \cap A_2)$ and union of events $(A_1 \cup A_2)$. In this discussion of the implementation of probabilistic safety assessment procedures, we will use the spirit of the axiomatic interpretation of probability, i.e. an adaptation of the conventional approach. Once again, the scarcity of performance information for constructed structures in this field of interest combined with a paucity of research in critical issues constituting clay buffer/barrier performance for these structures forces one to use the subjective approach in determination of uncertainty of events.

Probabilistic safety assessment contents

Applying a PSA process to only one component of a multi-component system, such as the clay buffer surrounding a HLW canister or the clay barrier of an engineered multi-barrier system of a HSW landfill, requires one to take into account interacting relationships with contiguous components (see

Figure 10.1). Figure 10.3 shows some of the typical basic elements embodied in the assessment process. Implementing the process requires (a) exercise in judgement, particularly with respect to specification of events and treatment of $A_1 \cap A_2$ and $A_1 \cup A_2$ and (b) performance of several types of tasks, such as those shown in the rectangular and square toolboxes in Figure 10.3. The principal elements of the process include:

- *Element 1.* Specification of the system together with a description of the system and its function. This could be interpreted as a detailing of the objectives of the system being assessed. For clay buffers and barriers, one would need to elaborate on the isolation–protection role of these buffers and barriers.
  Subelement. (a) Specification or articulation of performance requirements and (b) specification of acceptance criterion or criteria.
- *Element 2.* Scenario specification – the objectives of the specified scenario or scenarios must be provided, together with the acceptance criterion or criteria. This is not to be confused with *element 1.* Specific situations such as exposure to high pollutant concentrations and/or high pore pressures are examples of scenarios facing clay buffers and barriers.
- *Element 3.* Specification of failure mode and determination of failure impacts and analysis of effects of failure. One could say that this element is the heart of the PSA.

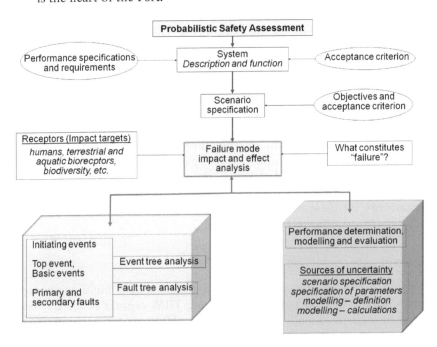

Figure 10.3 Typical principal elements constituting a PSA.

Subelement. (a) Information on the nature of the *receptors* (targets of the failure impacts); (b) determination of impact tolerance levels and limits of receptors; and (c) defining 'what constitutes failure'.

We call the rectangular and square boxes in the lower portion of Figure 10.3 *toolboxes* because they include the evaluation and analytical tools that can be used to (a) determine and analyse possible and potential mechanisms leading to distress or failure of the system under investigation; (b) determine the probability of distress or failure under the specified scenarios provided in *element 2*; and (c) determine the risk level of distress or failure for different acceptance criteria. There are some that would argue that if one uses inductive safety analysis for *failure mode impact and effect* studies, there will not be any need to conduct fault tree analysis. However, for complex multi-component systems, these analyses (fault and event tree analyses) provide one with more necessary information to render judgements on level of risks associated with the various criteria articulated in the acceptance criterion.

10.2.2 Failure mode – impacts and effects

Failure events

To define a failure mode, we need to define what we mean by *failure* – in the context of a clay buffer and clay barrier. *Failure to perform its design function* – consistent with design performance standards, and/or criteria – has often been used as an indicator of failure. Clay buffers/barriers are often viewed in the macroscale as single physical components in multi-component engineered barrier systems (EBSs) (see top illustration in Figure 10.6). However, in the meso- and microscales, they are multi-component systems consisting of solid, fluid and gaseous phases interacting between themselves and with the immediate environment and external forces/fluxes. Failure of the clay buffer/barrier system can be considered to be a *top failure event*, i.e. macrosystem failure (hereafter *failure events* will be referred to as *events*). In the microsystem, we recognize that there are other failures that occur in the clay that may be *top events* in themselves or *subevents,* i.e. *events* that are subordinate to the clay microsystem *top event*. The subject of events will be examined in greater detail in a later section when we discuss *fault tree analysis.*

Defining the broad objective of clay buffers and barriers to be isolation barriers is insufficient. One needs to elaborate on their performance requirements, and to specify 'what will constitute failure as an isolation barrier'. The following is an example of some of the major performance requirements that might be included in 'what constitutes failure': using a form of *fault tree analysis*, these may be considered as *top failure events* or as *subordinate events* depending on the *scenario specification.*

a *Failure to impede the flow of leachate from wastepile and/or groundwater flow into repository from surrounding host rock.* If one specifies a limiting maximum permeability coefficient for the clay buffer or barrier, for example $k < 10^{-9}$ m/s, any value of k higher than 10^{-9} m/s would constitute failure. By allowing leachate (with its constituent contaminants) to pass through the barrier with little hindrance, the local subsurface environment will receive the contents of the leachate. Transport of contaminants in the leachate to receiving waters and to other regions in the subsurface and even the surface will pose potential health and environmental threats.

b *Significant cautionary issues surrounding specification of limiting k values.* These revolve around (a) the fact that the k values are 'average' (bulk) values and therefore do not differentiate between flow through macro- and micropores, and in particular, channelized flow – issues that impact directly on the accumulative ability of the clay and (b) measurements designed to obtain k values are generally conducted on laboratory samples that are confined. Field bulk samples do not have the same form of constraint – a feature that allows for the development of considerable differences in characterization of flow paths and tortuosity. Field experience teaches us that field measurements of k do not often accord with laboratory-measured values.

c *Failure to provide a stable cushion support for canister and/or wastepile.* One could specify a minimum shear strength and a maximum coefficient of consolidation of the clay as the threshold values. The clay densities of engineered clay buffers/barriers when constructed are not uniform, despite strict controls on placement techniques. Added to this fact are non-homogeneity and non-uniformity in distribution of clay minerals and other minerals. These all result in local differences and variations in pore spaces and strengths of the in-place clay buffers and barriers. These are critical issues for HLW canisters (see section 3.10).

d *Failure to provide a barrier between host environment and canister and/or wastepile.* This could include failure to provide a physical buffer against the host environment. The sequence of actions and interactions shown in Figure 6.6 in Chapter 6 demonstrate the importance of creation of the buffer as a cushion barrier hosting the HLW canister while isolating it from the host rock.

e *Failure to attenuate contaminants.* This *failure* is not to be confused with the failure described in (a). Although the coefficient of permeability of the clay constituting the buffer/barrier may satisfy the limiting maximum value, it does not follow that the buffer/barrier will attenuate contaminants. As has been discussed in Chapter 5, this is a function of the accumulative and buffering properties of the clay by adsorption and precipitation of radioactive and hazardous solids. Improper selection of clay material and placement can result in failure of the buffer/barrier to attenuate contaminants in transport through the buffer/barrier. Contaminant concentrations in excess of defined threshold contaminant concentrations at specific spatial distances from contaminant source, and at specific elapsed time periods, will be considered as *failure to attenuate*.

Figure 10.4 shows an example of monitoring of spread of contaminants in the ground from a contaminant source. Samples obtained from the monitoring wells are analysed to determine concentrations of contaminants, and to compare against published threshold values to determine if these values have been exceeded. Establishing threshold values will be discussed when we address the subject of receptors and impact targets.

f   *Failure to properly locate HLW canisters and/or HSW in landfills from transfer stations.* The problems arise with respect to escape of radionuclides and/or contaminants if HLW canisters and/or HSW are damaged or mishandled, thereby allowing for unplanned discharge of these hazardous contaminants. Although the transport of these hazardous wastes is regulated under the Transport of Dangerous Goods Act (TDGA), or its equivalent in all jurisdictions, control over the transfer of these hazardous radioactive and solid wastes when they arrive at the

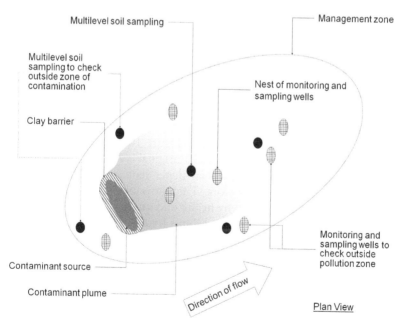

Figure 10.4   Use of a management zone to monitor and determine if concentration of contaminants emanating from contaminant source exceed threshold values. Determination of threshold values depend on the type of performance requirements and restrictions placed on clay barrier.

containment/disposal sites is generally under the jurisdiction of the stakeholder. Handling of these hazardous radioactive and solid wastes at the site includes transfer and placement into the repository environment or landfill.

Receptors, impact targets and limiting (threshold) values

In general, criteria for *failure* are defined in relation to (a) impacts of individual and collective 'failures' on receptors and other impact targets and (b) a measure of the risk that one is willing to accept, vis-à-vis individual and collective impacts. Setting impact levels, limits and criteria requires knowledge of (a) receptors and other targets impacted by 'failure' and (b) what constitutes a distress to the receptor and other targets. The primary groups of receptors and targets include:

- *Biotic receptors.* These include humans, terrestrial and aquatic animals. Birds and other airborne species are also considered as biotic receptors at risk because their habitats are land based and their food sources can be both land and aquatic based.
- *Environment.* The spatial extent and the nature of the environment included for consideration will vary depending on the type of impact involved. The concern for the environment is directed towards (a) degradation of environmental resources such as biodiversity, renewable and non-renewable resources and (b) contamination of the environment with pollutants, thereby creating threats to biotic receptors.

Specifying threshold or limit dosage values for humans – dosages that are deemed to cause distress to biotic receptors of concern – requires knowledge of exposure pathways. This will indicate the likely manner of contact with contaminants as shown, for example, in Figure 10.5 for the case of a piece of contaminated ground as the source of original contamination. One could investigate the probability of pathway-dependent exposures to assess the health risk of the human receptors. The allowable values such as acceptable daily intakes (ADIs) issued by health and safety agencies can be used to determine the appropriate threshold limits for the various pollutants under investigation. By and large, most jurisdictions adopt their interpretation of ADIs as threshold limits. The same methodology for specification can be said to apply to other biotic receptors.

Different jurisdictions and their respective health and safety units may have different interpretations and measurements on what constitutes acceptable daily intakes (ADIs) for humans. Dose–response assessments and exposure assessments are essential tools used in conjunction with analyses for ADIs (Purchase, 2000). Variable factors include body size, weight, sex, age, diet, etc. Determination of ADIs generally requires information on 'no observed

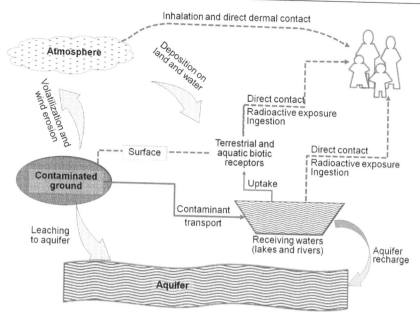

Figure 10.5 Demonstration of the various possible pathways from contaminant source to humans as the biotic receptors.

adverse effects level' (NOAELs) and 'low observed adverse effects level' (LOAELs). As one will conclude from this 'effect' information, considerable historical information is necessary to set the ADIs. In all instances, one deals with probabilities of occurrence, exposure, type of contact (pathway), and type and concentration of pollutant or contaminant.

Setting ADIs and threshold limits is a complicated exercise, not only because of the various issues related to finding a 'representative human', but also because of the complex issue related to 'type of health distress'. General toxicological testing (National Academy of Science, 1982) includes three groups of tests: (a) mutagenic activity; (b) cumulative toxicity; and (c) chronic carcinogenic and non-carcinogenic bioeffects. Threshold limits will therefore vary according to the type of health distress caused, and by all the other factors mentioned previously. Furthermore, many types of pollutants can cause distress through more than one pathway of exposure, for example inhalation, ingestion and contact. This means that one needs to know about *relative source concentration* when decisions are made in regard to threshold limits.

An obvious example of exposure and pathways can be seen in Figure 10.5. The majority of inorganic and organic contaminants offer threats to biotic receptors by pathways of ingestion, inhalation and in some instances, direct contact. Take manganese, for example, a metal that occurs naturally and is found in the air and on the ground. Exposure pathways include inhalation and ingestion from food and water. Although it is considered to be an essential element for humans, records of suggested ADIs for this element vary considerably depending on whether the manganese is ingested in food or water or whether it is through inhalation – with the higher limit being assigned to food sources. This example shows that control of contaminant transport in the ground is essential, not only because of the likelihood of pollution of aquifers and receiving waters, but also because of their capability to become airborne when they reach the ground surface.

The case of fugitive radionuclides (see section 2.3 for details of types) offers another pathway dimension – direct exposure. Fugitive radionuclides finding their way to the surface and the biosphere in general pose threats to biotic receptors. To appreciate threat dosages and their implication with respect to when a fugitive radionuclide finds its way to the biosphere, we can compare dose exposures as follows. An unprotected human near an unshielded fuel can suffer a fatal dose (5000 mSv) exposure in less than 5 min. That same human, 100 m removed from the unshielded fuel, will be exposed to a dose rate of about 6.5 mSv/h, meaning of course that distance and dosage are important considerations in consideration of threat impact. With respect to what it means for the clay buffer and repository containment of HLW canisters, distance away from the radiation source is an important factor – a fact that is well known. If we factor in another protective measure such as *time*, we will add another level of protection to humans and other biotic receptors. To continue the example, consider that after 1000 years of containment, the dose rate exposure of fugitive radionuclides will have dropped to about 50 mSv/h. Extending this time period to 10,000 years or 100,000 years will bring down this dose rate to about 5 mSv/h and to less than 1 mSv/h respectively. Recognizing this important aspect of protection for biotic receptors, the Swedish Radiation Protection Institute (SSI), which uses the sets of rules articulated by the International Atomic Energy Agency (IAEA), requires that the annual risk of exposure be less than 10^{-6} for the first 100,000 years after closure.

Considerable attention has been focussed on ingestion as a direct health hazard – more specifically towards ingestion of water. Contaminant transport in the subsurface, as shown in Figure 10.5, can have significant impact on drinking water, i.e. receiving waters and aquifers. These contaminants can be in the form of leachates emanating from HSW landfills or in the form of fugitive radionuclides escaping from the underground repositories. Whatever the type or source of these dangerous contaminants, they have the potential for polluting drinking water sources. The subjects of bioaccumulation and bioconcentration of pollutants in receiving waters and the effect of these on the food chain – where humans are at the top – are important subjects that cannot be overlooked. However, they are not within the purview of this discussion.

Specification of impact threshold levels for the environment is in a sense more complex or more difficult, depending on one's understanding of the problem at hand. This is owing to a combination of several factors, not the least of which include (a) incomplete understanding of what constitutes the 'environment' – from the different compartments to the various elements and parts within those compartments; (b) under-appreciation of what a particular impact on the environment means; (c) lack of adequate baseline data and lack of research efforts directed towards determination of distress levels; and (d) lack of defined protocols and structures for determination of *no impact, low impact* and *unacceptable impact*. One of the central issues concerning evaluation of impact to the environment is the selection of markers and criteria used as indicators of distress.

10.2.3 Events and fault tree analysis

Subsystem and total system failure

As stated before, to define what is meant by 'failure' in clay buffers and barriers, it is necessary to establish the objectives and/or specific design performance criteria, i.e. performance requirements of the system. This applies not only to multi-component systems, but especially to any of the individual components within the system. Why *especially*? Because failure in any of the individual component (as a subsystem) could lead to failure of the total system. The failure of an individual component is considered an *initiation event* in the total system context. Application of *event tree analysis* will inform us of the various possible outcomes. In this instance, we will consider only the outcomes that are dependent on technology and science, as opposed to outcomes that are a function of human intervention (driven by economics, politics, etc.).

Going back to the top illustration in Figure 10.1, total system failure is when the radiotoxicity (on ingestion) of fugitive radionuclides reaching ground surface are deemed to be in excess of permitted levels. This means that the subsystem clay buffer in the multi-component barrier (canister–clay, buffer–host rock) system has failed to retard or mitigate the transport of the fugitive radionuclides. Accounting for decay of the radionuclides, radiotoxicity in the long term (thousands of years) is dominated by plutonium-239 and plutonium-240 (Hedin, 1997).

For the bottom illustration in Figure 10.1, one could consider failure of the total multi-component barrier system to occur when concentrations of target contaminants in monitoring position A or B or C at specified time limits reaches LOAEL levels or exceeds ADIs. As in the case of the HLW repository, we consider the clay barrier (as the underlying component of the multi-component barrier system) to have failed to attenuate the transport of contaminants to positions A or B or C.

Fault tree analysis

Construction of fault trees to assist in quantitative determination of the probability of fault occurrences under various performance scenarios is not an exact science. One needs a combination of experience and an in-depth knowledge of how the system functions (Vesely *et al.*, 1981). We will use the spirit of conventional fault tree analysis in our discussion in this section. To illustrate the primary failure events and the relationship of subevents to the primary event, we use a scenario setting of contaminant transport in a clay buffer/barrier. The many levels of contributing and basic events are shown in Figure 10.6. These are the events that will populate the fault tree. Beginning from the top event ('Contaminant attenuation *failed*'), the failure of any of the three principal functions directly involved in contaminant attenuation is the direct result of failures of some of the events in the group of contributing events shown in the bottom of Figure 10.6. In the hierarchy of events, considering the clay buffer/barrier to be the total system (total microsystem in this case), failure of the clay buffer/barrier is a *top event*. We depart somewhat from conventional *fault analysis* by allowing that there may be more than one top event for our system – as discussed, for example, in section 10.2.2. Another example of a top event for the clay buffer/barrier could be failure of the material to support the HLW canister or failure of the material against shearing forces as shown in Figure 6.16 in Chapter 6.

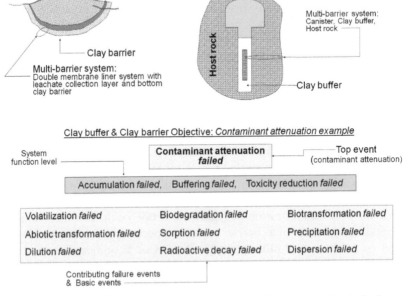

Figure 10.6 Example of failed contaminant attenuation objective in a clay buffer/barrier together with the many levels of contributing failure events.

The second level shown in Figure 10.6 is the *system functioning level*. This level describes the primary mechanisms that establish the system capability, which, in this case, is its ability to attenuate contaminants. Failures in this level can be considered as *contributing events*. These are failures in (a) accumulation of contaminants by chemical and physical means; (b) physical and chemical buffering; and (c) reduction in toxicity by dilution, degradation, transformation, etc. Failure of contributing events may be due to other events contributing at a lower level and finally to a set of *basic events*, i.e. events that can no longer be subdivided. Taking 'dilution' (of contaminants) as an example, we could argue that 'dilution' fails when we do not have sufficient water for dilution. In this instance, we consider 'dilution' fails as a *contributing event* and insufficient water as a *basic event*.

The formation of a tight clay buffer embedding the HLW canister in a repository situation is used to illustrate a means for quantitative determination of the risk of failure of a particular element or system. Figure 6.6 in Chapter 6 is a scenario that shows the required stages of formation of a tight clay buffer under initial conditions as indicated in the top left-hand sketch in the figure. A sample of fault tree analysis for that scenario is shown in Figure 10.7. The probability of occurrence of Stage 4 failure event can be determined in relation to the probability of occurrence of the contributing failure subevents (Stages 1– 3). By using Boolean algebraic equations for each gate, analysis of the fault tree will lead to solution for the top event (Stage 4).

10.3   Performance determination and evaluation

10.3.1 Uncertainty analysis

Performance determination and evaluation in a safety assessment process are required procedures leading to a *permit to operate*, assuming a positive safety assessment. As the system has yet to be operational, performance determination and evaluation must be conducted as a paper exercise, using information obtained from laboratory tests, prototype testing and mathematical modelling of system performance. For this to be successful, proper understanding of the operation of the system is needed, together with proper conceptualization for development of the appropriate and relevant mathematical tools. As has been pointed out repeatedly in this and previous chapters, this is not a simple task in complex multi-component systems. As stated previously, and often, lack of historical performance data impedes one from relying on 'experience' as the foundation for mechanistic and conceptual modelling.

The bottom right-hand box in Figure 10.3 gives us an idea of the sources of uncertainty that can impact severely on safety performance determination and evaluation – key elements in a safety assessment process. The tools for uncertainty analysis should be utilized to sort out errors and uncertainties

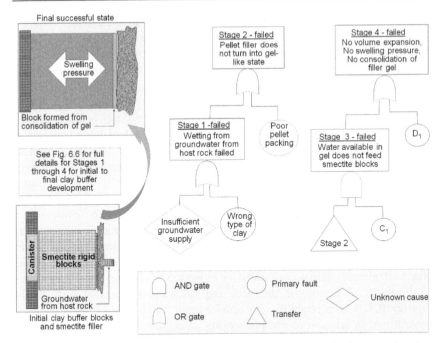

Figure 10.7   Example of fault tree for development of a competent clay buffer embedding a HLW canister in an underground repository (see Figure 6.6 for sequence of stages). C_i and D_i represent events that have yet to be elaborated. 'AND' gate, operations involving intersection of events; 'OR' gate, operations involving union of events. The 'unknown cause' (diamond symbol) can be considered to be a primary fault.

and to allow one to gain more confidence in performance assessment. The reader is advised to consult the standard texts on *uncertainty analysis* for more information on this subject.

Amongst a host of uncertainties, three groups stand out prominently, which include, for example, site characterization, material properties and characteristics determination, and inputs. The problem of clay buffer/barrier containment of the escape and/or transport of errant radionuclides or contaminants will be used as the focus of this discussion.

10.3.2 Sources of uncertainty

Scenario specification

There are at least two *uncertainty* groups in *scenario specification* for a safety assessment process: (a) *ab initio* conditions and (b) processes and events. The example of a scenario requiring information and performance of clay buffer/barrier contaminant attenuation provides the focus for this discussion:

- *Ab initio conditions.* Scenario description includes the physical setting of the system being studied, such as HLW canister embedment in the clay buffer in a repository and/or HSW landfill sited in a particular landform. For the problem under study, one presumes that the scenario will provide elaboration of contaminant and radionuclide contact or ingress into the clay buffer/barrier, such as escape of corrosion products and radionuclides from a HLW canister, and/or escape for toxic contaminants from the overlying double-membrane HSW landfill liner system.

  Uncertainties in the physical features with regard to the specified scenario include knowledge of species, concentrations and distribution of contaminants (including radionuclides), hydrogeological and geological settings, physical interfaces between buffer/barrier and host material (rock or subsurface), construction artefacts, etc.

- *Properties, processes, mechanisms and events.* The bottom half of Figure 10.6 shows the various processes, interactions and mechanisms required to provide the functional capability of the clay buffer/barrier. The questions concerning relevance, accuracy and, especially, validity apply to all of the preceding. Have we identified all the processes? Have we been able to determine all the relevant properties? How realistic or how applicable are these (properties and processes)? And, most importantly, what is the effect of maturation processes on interactions, properties, processes, etc.? All of these impact directly on one's capability to determine the probability of failure of subevents and top events.

Specification of parameters

Included in this group are: material characterization, information on laboratory-derived material property and performance, and operational parameters (for mathematical/computer modelling). The question of pertinence is added to the list of uncertainties articulated previously (relevance, accuracy, validity). What can go wrong and what are the uncertainties? The accuracy of results obtained from analytical tools and laboratory testing of materials is a subject that has occupied researchers time and again. The basic underlying fact that continues to elude many is that many types of test results are operationally defined – that is to say, the test results are conditioned by the test technique, sample preparation and handling, equipment constraints, and data reduction model used. These are often lumped into the category of experimental errors and inaccuracies. Relevance, reliability and pertinence are significant issues that point towards a myriad of uncertainties and fault events.

Model definition and calculations

Pertinent and comprehensive system conceptualization is key to successful development of analytical/computer models. By virtue of the fact that these are simulation models, one accepts that simplification of some, or even all, of the phenomena and processes represented by the function of the system may be required. These simplifications in conceptualization, coupled with assumptions made, will provide one with the necessary formulation that is the structure of the model.

Questioning the validity and robustness of the model, two of the key issues, is best done by examining the degree of uncertainty in the simplifications and assumptions made, including choice and relevance of operational parameters. The question of accuracy of model calculations is another issue that must be added to the mix of uncertainties. For complex computer codes, there exist many sources of inaccuracies – round-off errors, discretization, mathematical simplications, numerical algorithms, etc.

In addition to the preceding issues of model viability, it is important to note that initial and boundary conditions are significant issues in computational accuracy and pertinence. The pertinence, reliability and validity of the results obtained from model calculations of simulated performances of clay buffer and barriers are dependent on how well the initial and boundary conditions of the problem setting are represented. This means to say that even if the mathematical computer model has been correctly developed, the calculated results obtained from the model will not have credence if the initial and boundary conditions for the system problem setting are not properly posed and not properly represented. This situation is particularly critical in both HLW repository containment and HSW landfill containment. At the boundaries of clay buffers/barriers, the nature of the physical, chemical and biological properties and interactions (with the clay solids) can gradually and sometimes rapidly change with time – because of the interactions between clay and the geological/hydrogeological setting and/or changes in the HLW canister or HSW wastepile, and/or changes in the nature of the clay itself due to ageing processes. These issues will significantly impact and impose limitations on the calculated results and will inevitably increase the uncertainties for performance determination and prediction. What this tells us is that evaluation of the boundary conditions for problem setting must include the information on likely physical, chemical, and biological changes (with time), and their impact on the boundary condition specifications.

Threshold limits

The discussion in section 2.3 concerning threshold limits for hazardous contaminants (radionuclides, inorganic and organic chemical pollutants) has indicated that considerable attention needs to be paid to (a) exposure pathways; (b) manner of exposure; (c) nature of health distress (e.g. chronic, carcinogenic, mutagenic); (d) determination of acceptable levels relative to 'representative' biotic species, diet, environment, etc.; and (e) when onset of irreversible health distress occurs. Establishing threshold limits or acceptable daily intake or exposure to hazardous contaminants is not an easy task. Guidelines have been issued by almost all health and safety agencies in all countries for drinking water, and in many cases for food and even for exposure to 'harmful sun rays'. The difficulty in arriving at specific numbers for any of the types of hazardous contaminants lies not only in determining 'what (contaminant) is harmful', but also in distinguishing between 'how much (concentration, dosage, etc.) is harmful' and 'what does *harmful* mean?' These questions demonstrate that uncertainty abounds when we have to consider threats posed by escaping contaminants and radionuclides.

It is pertinent perhaps, at this juncture, to indicate that one of the greatest problems or barriers towards issuing absolute threshold or acceptable limits for exposure to hazardous contaminants is that if stakeholders or the general public abide by those limits, the following situation may arise: presuming that a governmental agency is responsible for issuing the limits, one could ask 'If a health mishap or a disaster occurs in circumstances that have not contravened any of the specified limits, is the governmental agency therefore liable?' Given the wide range of uncertainties associated with specification of any kind of exposure value, it is small wonder that most reputable agencies shy away from issuing anything more than 'guidelines'.

10.4   Concluding remarks

Safety assessments are conducted for many situations that are not necessarily connected to engineering structures or facilities or operational systems. For example, situations involving activities associated with operating a vehicle (terrestrial, airborne, aquatic), mountain climbing, firefighting, logging, etc. constantly undergo safety assessments. Given the very wide range of situations, activities, facilities, systems, etc. it will not be surprising to learn that the 'rules of safety assessment' will vary, not only in terms of the scope, but also in the nature of the assessment – according to the particular situation under consideration.

We design and build structures to operate safely. In the main, our concern is protection of public health and the environment. This chapter has focussed on the need to build safe structures and, in this case, the use of clay as a buffer or barrier material for containment of some very dangerous wastes – HLW and HSW. Although we have not directly focussed our attention on the risks associated with the use of the material, we have nevertheless been cognizant of such risks. In actual fact, the material developed in the various chapters has done nothing but focus on the many details and requirements for the proper use of clays for clay barriers, with particular attention to the performance requirements in the short- and long-term service lives. By undertaking a safety assessment of our system, we focus attention not only on 'what might go wrong', but also on 'what is required to allow the system to function safely throughout its design service life'. We are concerned with developing a safe containment structure, using clay as one of the containment barriers in a multi-component system. Hence, *safety assessment* constitutes a hallmark of such a venture. This is a very necessary procedure for *permitting* (i.e. permit to operate).

References

Chapter 1

Pusch, R., and Yong, R.N., 2006, *Microstructure of Smectite Clays and Engineering Performance*, Taylor and Francis, London.

Whittaker, R.H., 1969, 'New concepts of kingdoms or organisms: Evolutionary relations are better represented by new classifications than by the traditional two kingdoms', *Science*, 163(863): 150–60.

Yong, R.N., 2001, *Geoenvironmental Engineering: Contaminated Soils, Pollutant Fate, and Mitigation*, CRC Press, Boca Raton, FL.

Yong, R.N., and Mulligan, C.N., 2004, *Natural Attenuation of Contaminant in Soils*, Lewis Publishers, CRC Press, Boca Raton, FL.

Chapter 2

Department for Environment, Food and Rural Affairs (DEFRA), 2001, Managing Radioactive Waste Safely: Proposals for Developing a Policy for Managing Solid Radioactive Waste in the UK, DEFRA, Product Code PB 5957.

Environment Agency (UK), 2003, Hazardous Waste: Interpretation of the Definition and Classification of Hazardous Waste, Technical Guidance WM2, Version 1.

Hedin, A., 1997, Spent Nuclear Fuel – How Dangerous Is It? SKB Technical Report, 97–13.

International Atomic Energy Agency (IAEA) Waste Technology Section, 2007, Categorizing Operational Nuclear Waste, IAEA-TECDOC-1538.

Pusch, R., 1994. *Waste Disposal in Rock*, Developments in Geotechnical Engineering, 76, Elsevier, Amsterdam.

Swedish Nuclear Fuel and Waste Management Company (SKB), 1999, Deep Repository for Spent Nuclear Fuel: SR97 – Post-closure safety, SKB Report, TR-99–06, Main Report, Volume 1.

US Nuclear Regulatory Commission (US NRC), 2002, Radioactive Waste: Production, Storage, Disposal, NUREG/BR-0216 Rev. 2.

Yong, R.N., 2001, *Geoenvironmental Engineering: Contaminated Soils, Pollutant Fate and Mitigation*, CRC Press, Boca Raton, FL.

Yong, R.N., 2004, 'On engineered soil landfill barrier-liner systems', Proceedings of Malaysian Geotechnical Conference, pp. 109–118.

Chapter 3

Alammawi, A.M., 1988, 'Some aspects of hydration and interaction energies of montmorillonite', PhD thesis, McGill University.

Bockris, J.O'M. and Reddy, A.K.N., 1970, *Modern Electrochemistry*, Plenum Press, New York, Volumes 1 and 2.

Bolt, G.H., 1955, 'Analysis of the validity of the Gouy–Chapman theory of the electric double layer', *Journal of Colloid Science* 10: 206–18.

Bolt, G.H., 1956, 'Physico-chemical analysis of compressibility of pure clays', *Geotechnique* 6: 86–93.

Börgesson, L., Hökmark, H., and Karnland, O., 1988, Rheological Properties of Sodium Smectite Clay, SKB Technical Report TR 88–30, SKB, Stockholm.

Bowden, J.W., Oisner, A.M., and Quirk, J.R., 1980, 'Adsorption and charging phenomena in variable charge soils', in B.K. Theng (ed.) *Soils with Variable Charge*, New Zealand Society of Soil Science, Lower Hutt.

Brunauer, S., Emmett, P.H., and Teller, E., 1938, 'Adsorption of gases in multimolecular layers', *Journal of the American Chemistry Society* 60: 309–19.

Buckingham, E., 1907, 'Studies on the movement of soil moisture', US Department of Agriculture, Bureau of Soils, Bulletin 38, p. 61.

Carter, D.L., Mortland, M.M., and Kemper, W.D., 1986, 'Specific surface', in A. Klute (ed.) *Methods of Soil Analysis, Part 1: Physical and Mineralogical Methods*, Monograph 9, American Society of Agronomy, Madison, WI, pp. 413–23.

Cloos, P., Leonard, A.J., Herbillon, A., and Fripiat, J.J., 1969, 'Structural organization in amorphous silico-aluminas', *Clays and Clay Minerals* 17: 279–85.

Darcy, H., 1856, 'Les fontaines publiques de la ville de Dijon; Exposition et application des principles à suivre et des formules à employer', in V. Dalmont (ed.) *Histoire les Fontaines Publiques de Dijon*, Librarie des Corps Imperiaux des Ponts et Chaussees et des Mines, Paris, pp. 305–311.

Deresiewicz, H., 1958, 'Mechanics of granular materials', *Advances in Applied Mechanics* 5: 233–306.

Farmer, V.C., 1978, 'Water on particle surfaces', in D.J. Greenland and M.H.B. Hayes (eds) *The Chemistry of Soil Constituents*, John Wiley and Sons, New York, Chapter 6.

Flaig, W., Beutelspacher, H., and Reitz, E., 1975, 'Chemical composition and physical properties of humic substances', in J.E. Gieseking (ed.) *Soil Composition*, Springer-Verlag, Berlin, pp. 1–219.

Flegmann, A.W., Goodwin, J.W., and Ottewill, R.H., 1969, 'Rheological studies on kaolinite suspensions', *Proceedings of the British Ceramic Society* pp. 31–44.

Gast R.G., 1977, 'Surface and colloid chemistry', in R.C. Dinauer (ed.) *Minerals in Soil Environment*, Soil Science Society of America, Madison, WI, Chapter 2.

Grahame, D.C., 1947, 'The electrical double layer and the theory of electocapillarity', *Chemistry Review* 41: 441–501.

Greenland, D.J., and Mott, C.J.B., 1985, 'Surfaces of soil particles', in D.J. Greenland and M.H.B. Hayes (eds) *The Chemistry of Soil Constituents*, John Wiley and Sons, New York, pp. 321–54.

Hansbo, S., 1960, 'Consolidation of clay, with special reference to influence of vertical sand drains', Swedish Geotechnical Institute, Proceedings No. 18, pp. 41–61.

Hayes, M.H.B., and Swift, R.S., 1985, 'The chemistry of soil organic colloids', in D.J. Greenland and M.H.B. Hayes (eds) *The Chemistry of Soil Constituents*, John Wiley and Sons, New York, pp. 179–320.

Hogg, R., Healy, T.W., and Fuerstenay, D.W., 1966, 'Mutual coagulation of colloidal dispersions', *Transactions of the Faraday Society* 62: 1638–51.

Horseman, S.T., Harrington, J.F., and Sellon, P., 1999, 'Gas migration in clay barriers', *Engineering Geology* 54: 139–49.

Ichikawa,Y., Kawamura, K., Nakano, M., Kitayama, K., and Kawamura, H., 1999, 'Unified molecular dynamics and homogenization analysis for bentonite behaviour: current results and future possibilities', *Engineering Geology* 54: 21–31.

Janbu, N., 1967, 'Settlement calculation based on the tangent modulus concept', Guest lectures at Moscow University, Geoteknikk Medd 2, Norwegian Institute of Technology, Trondheim.

Kawamura, K., 1992, 'Interaction potential models for molecular dynamics simulations of multi-component oxides', in F. Yonezawa (ed.) *Molecular Dynamics Simulations*, Springer Series in Solid State Sciences, Berlin, Volume 103, pp. 88–97.

Kruyt, H.R., 1952, *Colloid Science*, Volume I, Elsevier, Amsterdam.

Kumagai, N., Kawamura, K., and Yokokawa, T., 1994, 'An interatomic potential model for H_2O systems and the molecular dynamics applications to water and ice polymorphs', *Molecular Simulations* 12: 177–86.

Lambe, T.W., 1953, 'The structure of inorganic soil', *Proceedings of the American Society of Civil Engineers*, No. 315.

Lambe, T.W., 1958, 'The structure of compacted clay', *Journal of Soil Mechanics and Foundations Division*, ASCE 84: SM2, 34.

Lutz, J.F., and Kemper, W.D., 1959, 'Intrinsic permeability of clay as affected by clay–water interactions', *Soil Science* 88: 83–90.

Mooney, R.W., Keenan, A.C., and Wood, L.A., 1952, 'Adsorption of water vapour by montmorillonite, II: Effect of exchangeable ions and lattice swelling as measured by x-ray diffraction', *Journal of the American Chemistry Society* 74: 1371–4.

Mortland, M.M., and Kemper, W.D., 1965, 'Specific surface', in C.A. Black (ed.) *Methods of Soil Analysis: Part 1*, American Society of Agronomy, Madison, WI, pp. 532–44.

Muurinen, A., 2006, Ion Concentration Caused by an External Solution into the Porewater of Compacted Bentonite, Working Report 2006–96, Posiva, Eurajoki, Finland.

Nakano, M., and Kawamura, K., 2006, 'Adsorption sites of Cs on smectite by EXAFS analyses and molecular dynamics simularions', *Clay Science* 12 (Suppl. 2): 76–81.

Newman, A.C.D., and Brown, G., 1987, 'The chemical constitution of clays', in A.C.D. Newman (ed.) *Chemistry of Clays and Clay Minerals*, Mineralogical Society Monograph No. 6, John Wiley and Sons, New York, pp. 1–128.

van Olphen, H., 1977, *An Introduction to Clay Colloid Chemistry*, 2nd edn, Wiley, New York.

Ouhadi, V.R., Yong, R.N., and Sedighi, M., 2006, 'Influence of heavy metal contaminants at variable pH regimes on rheological behaviour of bentonite', *Journal of Applied Clay Science* 32: 217–31.

Pearson, R.G., 1963, 'Hard and soft acids and bases', *Journal of the American Chemical Society* 85: 3533–9.

Philip, J.R., 1957, The physical principles of soil water movement during the irrigation cycle', Congress of International Commission on Irrigation and Drainage, 8: 125–54.

Pusch, R., 1966, 'Quick clay microstructure', *Journal of Engineering and Geology* 3: 433–43.

Pusch, R., 1993, Evolution of Models for Conversion of Smectite to Non-expandable Minerals, SKB Technical Report TR 93–33, SKB, Stockholm.

Pusch, R., 1994, *Waste Disposal in Rock*, Developments in Geotechnical Engineering, 76, Elsevier, Amsterdam.

Pusch, R., 2002, The Buffer and Backfill Handbook. Part 1: Definitions, basic relationships and laboratory methods, SKB Technical report TR-02-20, SKB, Stockholm.

Pusch, R., and Karnland, O., 1988, Hydrothermal Effects on Montmorillonite. A Preliminary Study, SKB Technical Report TR 88–15, SKB, Stockholm.

Pusch, R., and Yong, R.N., 2006, *Microstructure of Smectite Clays and Engineering Performance*, Taylor & Francis, London.

Quigley, R.M., Sethi, A.J., Boonsinsuk, P., Sheeran, D.E., and Yong, R.N., 1985, 'Geologic control on soil composition and properties, Lake Ojibway clay plain, Matagami, Quebec', *Canadian Geotechnical Journal* 22: 491–500.

Quirk, J.P., 1968, 'Particle interaction and soil swelling', *Israel Journal of Chemistry* 6: 213–34.

Ritchie, G.S.P., and Sposito, G., 2002, 'Speciation in soils', in A.M. Ure and C.M. Davidson (eds) *Chemical Speciation in the Environment*, Wiley, Malden, MA.

Singh, A. and Mitchell, J.K., 1968, 'General stress-strain-time functions for soils', *Proceedings of the American Society of Civil Engineers*, Volume 94, No. SM 1.

Singh, U., and Uehara, G., 1986, 'Electrochemistry of the double layer principles and applications to soils', in D.L. Sparks (ed.) *Soil Physical Chemistry*, CRC Press, Boca Raton, FL, pp.1–38.

Sposito, G., 1981, *The Thermodynamics of Soil Solutions*, Oxford University Press, New York.

Sposito, G., 1984, *The Surface Chemistry of Soils*, Oxford University Press, New York.

Stern, O., 1924, 'Theorie der electrolytischen Doppelschicht', Z. *Electrochemistry*, 30: 508–16.

Suquet, H., de la Calle, C., and Pezerat, H., 1975, 'Swelling and structural organization of saponite', *Clays and Clay Minerals*, 23: 1–9.

Tanai, K., and Yamamoto, M., 2003, Experimental and Modelling Studies on Gas Migration in Kunigel V1 Bentonite, JNC TH8400, 2003-024, Japan Atomic Energy Agency.

Terzaghi, K., and Peck, R.B., 1948, *Soil Mechanics in Engineering Practice*, John Wiley and Sons, New York.

Warkentin, B.P., and Schofield, R.K., 1962, 'Swelling pressure of sodium montmorillonite in NaCl solutions', *Journal of Soil Science* 13: 98.

Yong, R.N., 2001, *Geoenvironmental Engineering: Contaminated Soils, Pollutant Fate and Mitigation*, CRC Press, Boca Raton, FL.

Yong, R.N., and Mourato, D., 1988, 'Extraction and characterization of organics from two Champlain Sea subsurface soils', *Canadian Geotechnical Journal* 25: 599–607.

Yong, R.N., and Mourato, D., 1990, 'Influence of polysaccharides on kaolinite structure and properties in a kaolinite-water system', *Canadian Geotechnical Journal* 27: 774–8.

Yong, R.N., and Mulligan, C.N., 2003, 'The impact of clay microstructural features on the natural attenuation of contaminants', *Applied Clay Science* 23: 179–86.

Yong, R.N., and Mulligan, C.N., 2004, *Natural Attenuation of Contaminants in Soils*, Lewis Publishers, Boca Raton, FL.

Yong, R.N., and Ohtsubo, M., 1987, 'Interparticle action and rheology of kaolinite-amorphous iron hydroxide complexes', *Applied Clay Science* 2: 63–81.

Yong, R.N., and Warkentin, B.P., 1966, *Introduction to Soil Behaviour*, Macmillan, New York.

Yong, R.N., and Warkentin, B.P., 1975, *Soil Properties and Behaviour*, Elsevier, Amsterdam.

Yong, R.N., Sethi, A.J., Ludwig, H.P., and Jorgensen, M.A., 1979, 'Interparticle action and rheology of dispersive clays'. ACSE, Geotechnical Engineering Division 105: 1193–1209.

Yong, R.N., Boonsinsuk, P., and Yiotis, D., 1985, 'Creep behaviour of a buffer material for nuclear fuel waste vault', *Canadian Geotechnical Journal* 22: 541–50.

Yoshimi, Y., and Osterberg, J.O., 1963, 'Compression of partially saturated cohesive soils', *Journal of Soil Mechanics*, ASCE, SM, 1–24.

Chapter 4

Albrigh, J.N., 1972, 'X-ray diffraction studies of alkaline earth chloride solutions', *Journal of Chemical Physics* 56: 3783–6.

Arrhenius, S., 1887, 'On the dissociation of substances dissolved in water', *Zeitschrift fur physikalische Chemie* 1: 631.

Bresler, E., McNeal, B.L. and Carter, D.L., 1982, 'Saline and sodic soils', *Principles–Dynamics–Modeling Advanced Series in Agricultural Sciences*, New York: 85–101.

Brønsted, J., 1923, 'Some remarks on the concept of acids and bases', *Recueil des Travaux Chimiques des Pays-Bas* 42: 718–28.

Chijimatsu, M., Fujita, T., Kobayashi, A., and Nakano, M., 2000, 'Experiment and validation of numerical simulation of coupled thermal, hydraulic and mechanical behaviour in engineered buffer materials', *International Journal for Numerical and Analytical Methods in Geomechanics* 24: 403–24.

de Groot, S.R., 1961, *Thermodynamics of Irreversible Processes*, Interscience Publishers, New York.

Einstein, A., 1905, 'Uber die von der molekularkinetischen theorie der warme geforderte bewegung von in ruhenden flussigkeiten suspendierten teilchen', *Annalen der Physick* 4: 549–660.

Elzahabi, M., and Yong, R.N., 1997, 'Vadose zone transport of heavy metals', in R.N. Yong, and H.R. Thomas (eds) *Geoenvironmental Engineering – Contaminated Ground: Fate of Pollutants and Remediation*, Thomas Telford, London, pp. 73–180.

Hillel, D., 1998, *Environmental Soil Physics*, Academic Press, San Diego.

Jost, W., 1960, *Diffusion in Solids, Liquids, Gases*, Academic Press, New York.

Lerman, A., 1979, *Geochemical Processes: Water and Sediment Environments*, John Wiley and Sons, New York.

Lewis, G.N., 1923, *Valence and the Structure of Atoms and Molecules*, The Chemical Catalogue, New York.

Li, Y.H., and Gregory, S., 1974, 'Diffusion of ions in sea water and in deep-sea sediments', *Geochimica et Cosmochimica Acta* 38: 603–714.

Manheim, F.T., and Waterman, L.S., 1974, Diffusimetry (diffusion constant estimation) on Sediment Cores by Resistivity Probe, Initial Report of the Deep Sea Drilling Project, Volume 22, US Government. Printing Office, pp. 663–70.

Nakano, M. and Miyazaki, T., 1979, 'The diffusion and non-equilibrium, thermodynamic equations of water vapour in soils under temperature gradients', *Soil Science* 128: 184–8.

Nakano, M., Amemiya Y. and Fujii K., 1986, 'Saturated and unsaturated hydraulic conductivity of swelling clays', *Soil Science* 1410: 1–6.

Nakano, M., Kawamura, K., Emura, S., 2004, 'Local structural information from EXAFS analyses and adsorption mode of strontium on smectite', *Clay Science* 12, 311–19.

Neretnieks, I., and Moreno, L., 1993, 'Fluid flow and solute transport in a network of channels', *Journal of Contaminant Hydrology* 14: 163–92.

Nernst, W., 1888, 'Zur knetik der in losung befinlichen korper', *Zeitschrift für Physikalische Chemie* 2: 613–37.

Ohtomo, N., Arakawa, K., 1979, 'Neutron diffraction study of aqueous ionic solutions. 1. Aqueous solutions of lithium chloride and cesium chloride', *Bulletin of the Chemistry Society of Japan* 52: 2744–59.

Paissioura, J.B., 1971, 'Hydrodynamic dispersion in aggregated media: I. Theory', *Soil Science* 111: 339–44.

Pearson, R.G., 1963, 'Hard and soft acids and bases', *Journal of the American Chemical Society* 85: 3533–9.

Perkins, T.K., and Johnston, O.C., 1963, 'A review of diffusion and dispersion in porous media', *Journal of the Society of Petroleum Engineers* 17: 70–84.

Philip, J.R., and de Vries, D.A., 1957, 'Moisture movement in porous materials under temperature gradients', *Transactions American Geophysics Union* 38: 222–32.

Philip, J.R., 1968, 'Diffusion dead-end pores, and linearized adsorption in aggregated media', *Australian Journal of Soil Research* 6: 31–9.

Philip, J.R., and Smiles, D.E., 1969, 'Kinetics of sorption and volume change in three-component systems', *Australian Journal of Soil Research* 7: 1–19.

Pusch, R., Moreno, L., and Neretnieks, I., 2001, 'Microstructural modelling of transport in smectite clay buffer', in K. Adachi and M. Fukue (eds) *Clay Science for Engineering*, Balkema, Rotterdam, pp. 47–54.

Pusch, R., and Yong, R.N., 2006, *Microstructure of Smectite Clays and Engineering Performance*, Spon Research, Taylor and Francis.

Renkin, E.M., 1954, 'Filtration, diffusion, and molecular sieving through porous cellulose membranes', *Journal of General Physiology* 38: 225–43.

Rao, P.S., Rolston, D.E., Jessup, R.E., and Davidson, J.M., 1980, 'Solute transport in aggregated porous media: Theoretical and experimental evaluation', *Soil Science Society of America Journal* 44: 1139–46.

Robinson, R.A., and Stokes, R.H., 1959, *Electrolyte Solutions*, 2nd edn, Butterworths, London.

Rollins, F.L., Spangler, M.G., and Kirkham, D., 1954, 'Movement of soil moisture under a thermal gradient', *Proceedings of the Highway Research Board* 33: 492–508.

Skopp, J., and Warrick, A.W., 1974, 'A two-phase model for the miscible displacement of reactive solutes through soils', *Soil Science Society of America Proceedings* 38: 545–50.

Sörenson, S.P.L., 1909, 'Enzyme studies II: The measurement and meaning of hydrogen ion concentration in enzymatic processes', *Biochemische Zeitschrift* 21: 131–200.

Wagenet, R.J., 1983, 'Principles of salt movement in soils: Chemical mobility and reactivity in soil systems', *Soil Science Society of America (Special Publication)* 11: 123–40.

Yong, R.N., 2001, *Geoenvironmental Engineering: Contaminated Soils, Pollutant Fate and Mitigation*, CRC Press, Boca Raton, FL.

Yong, R.N., and Xu, D.M., 1988, 'An identification technique for evaluation of phenomenological coefficients in unsaturated flow in soils', *International Journal for Numerical and Analytical Methods in Geomechanics* 12: 283–99.

Chapter 5

Arnold, P.W., 1978, 'Surface–electrolyte interactions', in D.J. Greenland and M.H.B. Hayes (eds) *The Chemistry of Soil Constituents*, John Wiley and Sons, New York, pp. 355–401.

Bezile, N., Lecomte, P., and Tessier, A., 1989, 'Testing readsorption of trace elements during partial chemical extractions of bottom sediments', *Environmental Science and Technology* 23: 1015–20.

Bohn, H.L., 1979, *Soil Chemistry*, John Wiley and Sons, New York.

Bolt, G.H., 1979, *Soil chemistry, Part II: Physico-chemical Models*, Elsevier Scientific, New York.

Burchill, S. and Hayes, M.H.B., 1980, 'Adsorption of polyvinyl alcohol by clay minerals', Proceedings of the International Soil Science Society Symposium 'Organic Chemicals in the Soil Environment', Jerusalem, 1976, pp. 109–21.

Chang, A.C., Page, A.L., Warneke, J.E. and Grgurevic, E., 1984, 'Sequential extraction of soils heavy metals following a sludge applications', *Journal of Environmental Quality* 13: 33–8.

Chang, F.C., Skipper, N.T. and Sposito, G., 1995, 'Computer simulation of interlayer molecular structure in sodium montmorillonite hydrate', *Langmuir* 11: 2734–41.

Chester, R., and Hughes, R.M., 1967, 'A chemical technique for the separation of ferromanganese minerals, carbonate minerals and adsorbed trace elements from Pelagic sediments', *Chemical Geology* 2: 249–62.

Chiou, C.T., Schmedding, D.W., and Manes., 1982, 'Partitioning of organic compounds on octanol–water system', *Environmental Science and Technology* 16: 4–10.

Clevenger, T.E., 1990, 'Use of sequential extraction to evaluate the heavy metals in mining waste', *Water, Air and Soil Pollution* 50: 241–54.

Elliott, H.A., Liberati, M.R., and Huang, C.P., 1986, 'Competitive adsorption of heavy metals by soils', *Journal of Environmental Quality* 15: 214–19.

El-Rahman, K.M.A., El-Sourougy, M.R., Abdel-Monem, N.M., and Ismail, I.M., 2006, 'Modelling the sorption kinetics of cesium and strontium ions on zeolite A', *Journal of Nuclear and Radiochemical Sciences* 7: 21–7.

Emmerich, W.E., Lund, L.J., Page, A.L. and Chang, A.C., 1982, 'Solid phase forms of heavy metals in sewage sludge-treated soils', *Journal of Environmental Quality* 11: 178–81.

Engler, R.N., Brannon, J.M., Rose, J. and Bigham, G., 1977, 'A practical selective extraction procedure for sediment characterization', in T.F. Yen (ed.) *Chemistry of Marine Sediments*, Ann Arbor Science Publishers, Ann Arbor, MI.

Farrah, H., and Pickering, W.F., 1977, 'The sorption of lead and cadmium species by clay minerals', *Australian Journal of Chemistry* 30: 1417–22.

Frenkel, M., 1974, 'Surface acidity of montmorillonites', *Clays and Clay Minerals* 22: 435–441.

Gibson, M.J., and Farmer, H.G, 1986, 'Multi-step sequential chemical extraction of heavy metals from urban soils', *Environmental Pollution Bulletin* 11: 117–35.

Guy, R.D., Chakrabarti, C.L., and McBain, D.C., 1978, 'An evaluation of extraction of copper and lead in model sediments', *Water Resources* 12: 21–24.

Jones, G., and Dole, M., 1929, 'The viscosity of aqueous solutions of strong electrolytes with special reference to barium chloride', *Journal of the American Chemical Society* 52: 29–50.

Karickhoff, S.W., Brown, D.S., and Scott, T.A., 1979, 'Sorption of hydrophobic pollutants on natural sediments, *Water Research* 13: 241–8.

Kenaga, E.E., and Goring, C.A.I., 1980, Relationship between Water Solubility, Soil Sorption, Octanol-water Partitioning and Concentration of Chemicals in Biota, ASTM-STP 707, pp. 78–115.

Kibe, K., Takahashi, M., Kameya, T., and Urano, K., 2000, 'Adsorption equilibriums of principal herbicides on paddy soils in Japan', *Science of the Environment* 263: 115–25.

McCarty, P.L., and Semprini, L., 1994, 'Groundwater treatment for chlorinated solvents', in Bioremediation of Ground Water and Geologic Material: A Review of in situ Technologies, Section 5, Government Institutes, Lanham, MD.

McCarty, P.L., Reinhard, M., and Rittman, B.E., 1981, 'Trace organics in groundwater', *Environmental Science & Technology* 15: 40–51.

MacDonald, E., 1994, 'Aspects of competitive adsorption and precipitation of heavy metals by a clay soil', MEng Thesis, McGill University, Montreal, Quebec.

Montgomery, J.H., and Welkom, L.M., 1991, *Groundwater Chemicals Desk Reference*, Lewis Publishers, Chelsea, MI.

Morrill, L., Mahilum, B., and Mohiuddin, 1982, *Organic Compounds in Soils: Sorption, Degradation, and Persistence*, Ann Arbor Science Pubishers, Ann Arbor, MI.

Mortland, M.M., 1970, 'Clay-organics complexes and interactions', *Advances in Agronomy* 22: 75–117.

Nakano, M. and Kawamura, K., 2006, 'Adsorption sites of Cs on smectite by EXAFS analyses and molecular dynamics simulations', *Clay Science* 12(Suppl. 2): 76–81.

Olsen, R.I., and Davis, A., 1990, 'Predicting the fate and transport of organic compounds in groundwater, Part 1', *Hazardous Materials Control* 3: 38–64.

Onsager, L., 1931, 'Reciprocal relation in irreversible processes, II', *Physical Review* 38: 2265–79.

Ouhadi, V.R., Yong, R.N., and Sedighi, M., 2006, 'Influence of heavy metal contaminants at variable pH regimes on rheological behavior of bentonite', *Journal of Applied Clay Science* 32: 217–31.

Puls, R.W., and Bohn, H.L., 1988, 'Sorption of cadmium, nickel, and zinc by kaolinite and montmorillonite suspensions, *Soil Science Society of America Journal* 52: 1289–92.

Pusch, R., Kasbohm, J., and Thao, H.M., 2007, 'Evolution of clay buffer under repository-like conditions', Proceedings of the Workshop on Long-Term Performance of Smectite Clays Embedding Canisters with Highly Radioactive Waste, Lund, pp. 1–12.

Rao, P.S.C., and Davidson, J.M., 1980, 'Estimation of pesticide retention and transformation parameters required in nonpoint source pollution models', in M.R. Overcash and J.M. Davidson (eds) Environmental Impact of Nonpoint Source Pollution, Ann Arbor Science Publishers, Ann Arbor, MI, pp. 23–7.

Schwarzenbach, R.J., and Westall, J., 1981, 'Transport of non-polar organic compounds from surface water to groundwater: laboratory sorption studies', Environental Science & Technology 15: 1360–7.

Schwarzenbach, R.J., Gschwend, P.M., and Imboden, D.M., 1993, Environmental Organic Chemistry, Wiley and Sons, New York.

Smith, D.E., 1998, 'Molecular computer simulations of the swelling properties and interlayer structure of cesium montmorillonite', Langmuir 14: 5959–68.

Solomon, D.H., and Murray, H.H., 1972, 'Acid-base interactions and properties of kaolinite in non-aqueous media, Clays and Clay Minerals 20: 135–41.

Sposito, G., 1984, The Surface Chemistry of Soils, Oxford University Press, New York.

Tessier, A., Campbell, P.G.C. and Bisson, M., 1979, 'Sequential extraction procedure for the speciation of particulate trace metals, Analytical Chemistry 51: 844–51.

US EPA Science Advisory Board, 2001, Monitored Natural Attenuation; USEPA Research Programme: An EPA Science Advisory Board Review, Science Advisory Board (1400A), Washington, DC.

Verscheuren, K., 1983, Handbook of Environmental Data on Organic Chemicals, 2nd edn, van Nostrand Reinhold, New York.

Xu, D., Zhou, X., and Wang, X., 2008, 'Adsorption and desorption of Ni^{2+} on Na-montmorillonite: Effect of pH, ionic strength, fulvic acid, humic acid and addition sequences', Journal of Applied Clay Science 39: 133–41.

Yanful, E.K., Quigley, R.M., and Nesbitt, H.W., 1988, 'Heavy metal migration at a landfill site, Ontario, Canada – II: Metal partitioning and geotechnical implications', Applied Geochemistry 3:623–9.

Yong, R.N., 2001, Geoenvironmental Engineering: Contaminated Soils, Pollutant Fate and Mitigation, CRC Press, Boca Raton, FL.

Yong, R.N., and Mulligan, C.N., 2003, Natural Attenuation of Contaminants in Soils, Lewis Publishers, Boca Raton, FL.

Yong, R.N., and Phadungchewit, Y., 1993, 'pH influence on selectivity and retention of heavy metals in some clay soils', Canadian Geotechnical Journal 30: 821–33.

Yong, R.N., and Samani, H.M.V., 1987, 'Modelling of contaminant transport in clay via irreversible thermodynamics', Proceedings of the Geotechnical Practice for Waste Disposal '87, ASCE, Ann Arbor, MI, pp. 846–60.

Yong, R.N., Galvez-Cloutier, R., and Phadungchewit, Y., 1993, 'Selective sequential extraction analysis of heavy metal retention in soil', Canadian Geotechnical Journal 30: 834–47

Yong, R.N., Desjardins, S., Farant, J.P., and Simons, P., 1997, 'Influence of pH and exchangeable cation on oxidation of methylphenols by a montmorillonite clay', Journal of Applied Clay Science 12: 93–110.

Chapter 6

Börgesson L., 1990, Interim Report on the Laboratory and Theoretical Work in Modeling the Drained and Undrained Behavior of Buffer Materials, SKB Technical Report TR 90-45, SKB, Stockholm.

Campbell, G.S. and Norman, J.M., 1988, *An Introduction to Environmental Biophysics*, 2nd edn, Springer-Verlag, Berlin.

Feltham P., 1979, *The Inter-relationship: Earth Science—Material Sciences. Mechanisms of Deformation and Fracture*, Pergamon Press, London, Volume 29, No. 42.

Grindrod, P., and Takase, H., 1994, 'Reactive chemical transport within engineered barriers', in Proceedings of the 4th International Conference on the Chemistry and Migration Behaviour of Actinides and Fission Products in the Geosphere, Oldenburg Verlag, Munich, pp. 773–9.

Hedin, A., 1997, Spent Nuclear Fuel – How Dangerous Is It? SKB Technical Report 97–13.

Huang, P.M., and Schnitzer, M., 1986, Interactions of Soil Minerals with Natural Organics and Microbes, 3rd print, SSSA Special Publication 17.

Huang, P.M., Wang, M.K., and Chiu, C.Y., 2005, 'Soil mineral–organic matter–microbe interactions: Impacts on biogeochemical processes and biodiversity in soils', *Pedobiologia* 49: 609–35.

Ichikawa, Y., Kawamura, K., Nakano, M., Kitayama, K., and Kawamura, H., 1999, 'Unified molecular dynamics and homogenization analysis for bentonite behavior: current results and future possibilities', *Engineering Geology* 54: 21–31.

Jacobsson, A., and Pusch, R., 1972, 'Thixotropic action in remoulded quick clay', *Bulletin of the International Association of Engineering Geology* 5: 105–10.

Kasubuchi, T., 1977, 'Twin transient-state cylindrical-probe method for the determination of thermal conductivity of soil', *Soil Science* 124: 255–8.

Kasubuchi, T., 1984, 'Heat conduction model of saturated soil and estimation of thermal conductivity of soil solid phase', *Soil Science* 138: 240–7.

Katchalsky, A., and Curran, P.F., 1967, *Non-equilibrium Thermodynamics in Biophysics*, Harvard Books in Biophysics, 1, Harvard University Press, Cambridge, MA.

Kosika, J.E., Dtucki, J.W., Nealson, K.H., and Wu, J., 1996, 'Reduction of structural Fe(III) in smectite by a pure culture of the Fe-reducing 'Shewanella putrefaciens' strain MR-1', *Clays and Clay Minerals* 44: 522–9.

Krumbein, W.E. (ed.), 1978, *Environmental Biogeochemistry and Geomicrobiology: Methods, Metals and Assessment*, Ann Arbor Science Publishers, Ann Arbor, MI.

Made, B., Clement, A., and Fritz, B., 1994, 'Modeling mineral/solution interactions: The thermodynamic and kinetic code KINDISP', *Computers and Geosciences* 30: 1347–63.

Manose, T., Sakaguchi, I., and Kasubuchi, T., 2008, 'Development of an apparatus for measuring one-dimensional steady state heat flux of soil under reduced air-pressure', *European Journal of Soil Science* 59: 982–9.

Marty, N,. Montes-Hernandez, G., Fritz, B., Clement, A., and Michau, N., 2007, 'Modelling the long term alteration of the engineered bentonite barrier in an underground radioactive waste repository', Proceedings of the Workshop on Long-Term Performance of Smectitic Clays Embedding Canisters with Highly Radioactive Waste, Lund, Sweden, pp. 222–31.

Mohamed, A.M.O., Yong, R.N., and Cheung, S.C.H., 1992, 'Temperature dependence of soil water potential', *Geotechnical Testing Journal* 15: 330–9.

Motosuke, M., Nagasaka, Y., and Nagashima, A., 2003, 'Measurement of dynamically changing thermal diffusivity by the Forced Rayleigh Scattering method (measurement of the gelation process)', Paper presented to 15th Symposium on Thermophysical Properties, June, Boulder, CO.

Murray, J.D., 1993, *Mathematical Biology*, 2nd edn, Biomathematics 19, Springer-Verlag, Berlin.

Nadeau, P.H., and Bain, D.C., 1986, 'Composition of some smectites and diagenetic illitic clays and implications for their origin', *Clays and Clay Minerals* 14: 455–64.

Nakano, M. and Miyazaki, T., 1979, 'The diffusion and non-equilibrium, thermodynamic equations of water vapour in soils under temperature gradients', *Soil Science* 128: 184–8.

Osako, M., and Ito, E., 1997, Simultaneous Thermal Diffusivity and Thermal Conductivity Measurements of Mantle Materials up to 10 GPa, Technical Report of ISEI (Institute for Study of the Earth's Interior), Ser.A. No. 67, Okayama University, Japan.

Pedersen, K., 2000, 'Microbial Processes in Radioactive Waste Disposal', Technical Report TR-00-04, SKB, Stockholm.

Pusch, R., 1983, Stress/Strain/Time Properties of Highly Compacted Bentonite, SKBF/KBS Technical Report 83-47, SKB, Stockholm.

Pusch, R., 1994, *Waste Disposal in Rock*, Developments in Geotechnical Engineering, 76, Elsevier, Amsterdam.

Pusch, R., 2000, On the Risk of Liquefaction of Buffer and Backfill, Technical Report TR-00-18, SKB, Stockholm.

Pusch, R., and Adey, R., 1986, 'Settlement of clay-enveloped radioactive canisters', *Applied Clay Science* 1: 353–65.

Pusch, R, and Feltham, P, 1981, 'Computer simulation of creep of clay', *Journal of Geotechnical Engineering* (Div, ASCE) 107(GT1): 95–104.

Pusch, R., and Kihl, A., 2004, 'Percolation of clay liners of ash landfills in short and long time perspectives', *Waste Management & Research* 22: 71–7.

Pusch, R., and Yong, R.N., 2006, *Microstructure of Smectite Clays and Engineering Performance*, Taylor & Francis, London.

Pusch, R, Kasbohm, J., and Thao, H.M., 2007, 'Evolution of clay buffer under repository-like conditions', Proceedings of the Workshop on Long-Term Performance of Smectitic Clays Embedding Canisters with Highly Radioactive Waste, Lund, Sweden, pp. 1–12.

Pusch, R., Zhang, L., Adey, R., and Kasbohm J., 2009, 'Rheology of artificially prepared clays', *Journal of Applied Clay Science* (in press).

Pytte, A.M., 1982, 'The kinetics of smectite to illite reactions in contact metamorphic shales', MA Thesis, Dartmouth College, NH.

Robert, M., and Berthelin, J., 1986, *Interactions of Soil Minerals with Natural Organics and Microbes*, 3rd print, SSSA Special Publ. 17.

Seabaugh, J.L., Dong, H., Eberl, D.D., Kim, J., and Kukkadapu, R.K., 2004, 'Reduction of structural Fe(iii) in nontronite by a thermophilic bacterium', Symposium on 'Microbial Impacts on Clay Transformation and Reactivity', Clay Minerals Society 41st Annual Meeting, Richland, WA.

Sims, J.L., Sims, R.C., and Mathews, J.E., 1990, 'Approach to bioremediation of contaminated soil', *Journal of Hazardous Waste and Hazardous Matter* 4: 117–49.

SKB, 1999, *Deep Repository for Spent Nuclear Fuel: SR97 – Post-closure Safety*, SKB Report, TR-99-06, Main Report, Volume 1.

Stroes-Gascoyne, S., and West, J.M., 1996, 'An overview of microbial research related to high-level nuclear waste disposal with emphasis on the Canadian concept for the disposal of nuclear fuel waste', *Canadian Journal of Microbiology* 42: 1133–46.

Stucki, J.W., and Getty, P.J., 1986, 'Microbial reduction of iron in nontronite', *Agronomy Abstracts*, p. 279.

Stucki, J.W., and Roth, C.B., 1976, 'Interpretation of infrared spectra of oxidized and reduced nontronite', *Clays and Clay Minerals* 24: 293–6.

Stucki, J.W., Low, F.F., Roth, C.B., and Golden, D.C., 1984, 'Effects of oxidation state of octahedral iron on clay swelling', *Clays and Clay Minerals* 32: 357–62.

Stucki, J.W., Lee, K., Zhang, L., and Larson, R.A., 2002, 'The effects of iron oxidation state on the surface and structural properties of smectites', *Pure and Applied Chemistry* 74: 2079–92.

Wang, Y., and Francis, A.J., 2005, 'Evaluation of microbial activities for long-term performance assessments of deep geologic nuclear waste repositories', *Journal of Nuclear and Radiochemical Science* 6: 43–50.

Weaver, C.E., 1979, Geothermal Alteration of Clay Minerals and Shales: Diagenesis, Tech. Report, ONWI-21, ET-76-C-06–1830, Battelle Office on Nuclear Waste Isolation, Paris.

West, J.M., Cristofi, N., and McKinley, I.G., 1985, 'An overview of microbiological research relevant to the geologic disposal of nuclear waste', *Radioactive Waste Management Nuclear Fuel Cycle* 6: 79–95.

Wu, J., Low, P.E., and Roth, C.B., 1989, 'Effects of octahedral-iron reduction and swelling pressure of interlayer distances in Na-nontronite', *Clays and Clay Minerals* 37: 211–18.

Yan, L., and Stucki, J.W., 2000, 'Structural perturbations in the solid–water interface of redox transformed nontronite', *Journal of Colloid and Interface Science* 225: 429–39.

Yong, R.N., and Mulligan, C.N., 2004, *Natural Attenuation of Contaminants in Soils*, Lewis Publishers, Boca Raton, FL.

Yong, R.N., and Warkentin, B.P., 1975, *Soil Properties and Behaviour*, Elsevier Scientific, Amsterdam.

Yong, R.N., and Xu, D.M., 1988, 'An identification technique for evaluation of phenomenological coefficients in coupled flow in unsaturated soils', *Journal of Numerical and Analytical Methods in Geomechanics* 12: 283–99.

Yong, R.N., Xu, D.M., Mohamed, A.M.O., and Cheung, S.C.H., 1992, 'An analytical technique for evaluation of coupled heat and mass flow coefficients in unsaturated soil', *International Journal of Numerical and Analytical Methods in Geomechanics* 16: 233–46.

Chapter 7

Eberl, D.D., and Hower, J., 1976, 'Kinetics of illite formation', *Bulletin of the Geological Society of America* 87: 1326–20.

Ehrlich, H.L., 2002, 'Interactions between microorganisms and minerals under anaerobic conditions' in P.M. Huang, J.M. Bollag, and N. Senesi (eds) *Interactions between Soil Particles and Microorganisms: Impact on the Terrestrial Ecosystem*, IUPAC Series on Analytical and Physical Chemistry of Environmental Systems, John Wiley and Sons, Chichester, Volume 8, pp. 439–4.

Grim, R.E., 1953, *Clay Mineralogy*, McGraw-Hill, New York.

Huang, P.M., Wang, M.-K., and Chiu, C.-Y., 2005, 'Soil mineral–organic matter–microbe interactions: Impacts on biogeochemical processes and biodiversity in soils', *Pedobiologia* 49: 609–35.

Hofbauer, H., Sigmund, K., 1988, *The Theory of Evolution and Dynamical Systems*, Cambridge University Press, Cambridge.

Jackson, M.L., 1964, 'Chemical composition of soils', *American Chemistry Society Monograph* 160: 71–141.

Konishi, Y., Asai, S., Katoh, H., 1990, 'Bacterial dissolution of pyrite by *Thiobacillus ferrooxidans*', *Bioprocess Engineering* 5: 231–7.

Konishi, Y., Asai, S., Yoshida, N., 1995, 'Growth kinetics of *Thiobacillus thiooxidans* on the surface of elemental sulfur', *American Society for Microbiology* 61: 3617–22.

Kurek, E., 2002, 'Microbial mobilization of metals from soil minerals under aerobic conditions', in P.M. Huang, J.M. Bollag and N. Senesi (eds) *Interactions between Soil Particles and Microorganisms: Impact on the Terrestrial Ecosystem*, IUPAC Series on Analytical and Physical Chemistry of Environmental Systems, John Wiley and Sons, Chichester, Volume 8, pp. 189–225.

Mann, S., Archibald, A., Didymus, J.M., Douglas, T., Heywood, R.R., Meldrum, F.C., and Reeves, N.J., 1993, 'Crystallization at inorganic–organic interfaces: Biominerals and biomimetic synthesis', *Science* 261: 1286–92.

Marshall, K.C., 1976, *Interfaces in Microbial Ecology*, Harvard University Press, Cambridge, MA.

Molz, F.J., Widdowson, M.A., Benefield, L.D., 1986, 'Simulation of microbial growth dynamics coupled to nutrient and oxygen transport in porous media', *Water Resource Research* 22: 1207–16.

Monod, J., 1949, 'The growth of bacterial culture', *Annual Review of Microbiology* 3: 371.

Moodie A.D., and Ingledew J., 1990, 'Microbial anaerobic respiration', *Advances in Microbiology and Physiology* 31: 225–69.

Murphy, E.M., Ginn, T.R., 2000, 'Modelling microbial processes in porous media', *Hydrogeology Journal* 8: 142–58.

Murray, J.D., 1993. *Mathematical Biology*, 2nd edn, Biomathematics 19, Springer-Verlag, Berlin.

Nakano, M., and Kawamura, K., 2008, 'Estimating the corrosion of compacted bentonite by a conceptual model based on microbial growth dynamics', *Applied Clay Science*, DOI: 10.1016/j.clay.2008.08.009.

Pusch, R., Kasbohn, J., and Thao, H.T.-M., forthcoming, 'Chemical stability of montmorillonite buffer clay under repository-like conditions – A synthesis of relevant experimental data', *Applied Clay Science* (in press).

Pusch, R., and Yong, R.N., 2006, *Microstructure of Smectite Clays and Engineering Performance*, Taylor & Francis, London.

Regnier, P., O'Kane, J.P., Steefel, C.I., and Vanderborght, J.P., 2002, 'Modelling complex multi-component reactive-transport systems: towards a simulation environment based on the concept of a knowledge base', *Applied Mathematical Modelling* 26: 913–27.

Thullner, M., van Cappellen, P., and Regnier, P., 2005, 'Modelling the impact of microbial activity on redox dynamics in porous media', *Geochimica et Cosmochimica Acta* 69: 5005–19.

Chapter 8

Forslind, E., and Jacobsson, A., 1975, *Clay-water Systems. Water, a Comprehensive Treatment*, Volume 5. Plenum Press, New York.

Gray, M.N., 1993, *OECD/NEA* International Stripa Project Overview Volume III: Engineered Barriers, Swedish Nuclear Fuel and Waste Management Co. (SKB), Stockholm.

Grindrod, P., and Takase, H., 1994, 'Reactive chemical transport within engineered barriers', 4th International Conference in Chemistry and Migration Behaviour of Actinides and Fission Products in the Geosphere, R. Oldenburg Verlag, Munich.

Gueven, N., Carney, L.L., and Ridpath, B.E., 1987, Evaluation of Geothermal Drilling Fluids using a Commercial Bentonite and a Bentonite/saponite Mixture. Contractor Report SAND 86-7180, SANDIA Nat. Laboratories, Albuquerque, NM.

Mantovani, M., Escudero, A., Alba, M.D., and Becerro, A.I., 2008, 'Stability of phyllosilicates in $Ca(OH)_2$ solution. Influence of layer nature, octahedral occupation, presence of tetrahedral Al and degree of crystallinity', *Applied Geochemistry* (in press).

Mueller-Vonmoos, M., Kahr, G., Bucher F, and Madsen, F.T., 1990, 'Investigations of the metabentonites aimed at assessing the long-term stability of bentonites under repository conditions. Artificial Clay Barriers for High Level Radioactive Waste Repositories', *Engineering Geology* 28: 269–80.

Pacovsky, J., Svoboda, J., and Zapletal, L., 2005, 'Saturation development in the bentonite barrier of the Mock-up CZ geotechnical experiment. Clay in Natural and Engineered Barriers for Radioactive Waste Confinement – Part 2', *Physics and Chemistry of the Earth* 32: 767–79.

Pusch, R., 1983, Stability of Deep-sited Smectite Minerals in Crystalline Rock – Chemical Aspects, Technical Report 83-16, SKB, Stockholm.

Pusch, R., 1993, Evolution of Models for Conversion of Smectite to Non-expandable Minerals, Technical Report TR-93-33, SKB, Stockholm.

Pusch, R., 1994, 'Waste disposal in rock', *Developments in Geotechnical Engineering*, 76, Elsevier, Amsterdam.

Pusch, R., 2008, *Geological Storage of Highly Radioactive Waste*, Springer, New York.

Pusch, R., and Madsen, F., 1995, 'Aspects on the illitization of the Kinnekulle bentonites', *Clays and Clay Minerals* 43: 261–70.

Pusch, R., and Yong, R.N., 2006, *Microstructure of Smectite Clays and Engineering Performance*, Taylor & Francis, London.

Pusch, R., Börgesson, L., and Ramqvist, G., 1985, Final Report of the Buffer Mass Test – Volume II: Test Results, Technical Report 85-12, SKB, Stockholm.

Pusch, R., Karnland, O., Hökmark, H., Sandén, T., and Börgesson, L., 1991, Final Report of the Rock Sealing Project – Sealing Properties and Longevity of Smectitic Clay Grouts, Stripa Project Technical Report 91-30, SKB, Stockholm.

Pusch, R., Karnland, O., Lajudie, A., and Decarreau, A., 1993, MX-80 Clay Exposed to High Temperatures and Gamma Radiation, Technical Report 93-03, SKB, Stockholm.

Pusch, R., Takase, K., and Benbow, S., 1998, Chemical Processes Causing Cementation in Heat-affected Smectite – the Kinnekulle Bentonite, Technical Report TR-98-25, SKB, Stockholm.

Pusch, R., Kasbohm, J., Pacovsky, J., and Cechova, Z., 2005, 'Are all smectite clays suitable as "buffers"?. Clay in Natural and Engineered Barriers for Radioactive Waste Confinement – Part 1', *Physics and Chemistry of the Earth* 32: 116–22.

Pusch, R., Kasbohm, J., and Thao, H.T.M., forthcoming. 'Chemical stability of montmorillonite buffer clay under repository-like conditions – a synthesis of relevant experimental data', *Applied Clay Science* (in press).

Pusch, R, Kasbohm, J., and Thao, H.M., 2007, 'Evolution of clay buffer under repository-like conditions', Proceedings of the Workshop on Long-Term Performance of Smectitic Clays Embedding Canisters with Highly Radioactive Waste, Lund, Sweden, pp. 1–12.

Roaldset, E., He Wei, and Grimstad, S., 1998, 'Smectite to illite conversion by hydrous pyrolosis', *Clay Minerals* 33: 146–58.

Svemar, C., 2005, Cluster Repository Project (CROP), Final Report of European Commission Contract FIR1-CT-2000-20023, Brussels, Belgium.

Thorslund, P., 1945, 'Om bentonitlager i Sveriges kambrosilur', *Geologiska Föreningens i Stockholm Förhandlingar* 67: 286.

Chapter 9

Anderson, D.M., Pusch, R., and Penner, E., 1978, 'Physical and thermal properties of frozen ground', in B. Andersland and D.M. Anderson (eds) *Geotechnical Engineering for Cold Regions*, McGraw Hill, New York, Chapter 2.

Börgesson, L., 2001, Äspö Hard Rock Laboratory: Selection of THMCB models, SKB International Progress Report IPR-01-66, pp. 46–54.

Grindrod, P., and Takase, H., 1994, 'Reactive chemical transport within engineered barriers', 4th International Conference on the chemistry and migration behaviour of actinides and fission products in the geosphere, Charleston, SC USA, R. Oldenburg Verlag, pp. 773–9.

Gurr, C.G., Marshall, T.J., and Hutton, J.T., 1952, 'Movement of water in soil due to a temperature gradient', *Soil Science* 74: 335–45.

Kuzmak, J.M., and Serada, P.J., 1957, 'The mechanism by which water moves through a porous material subjected to a temperature gradient: 2. Salt tracer and streaming potential to detect flow in the liquid phase', *Soil Science* 84: 419–22.

Ledesma, A., and Chen, G., 2003, 'T-H-M modelling of the Prototype Repository experiment at Äspö HRL, Sweden', Proceedings of the Geoproc Conference, KTH, Stockholm, pp. 370–5.

Ohnishi, Y., Shibata, H., and Kobayashi, A., 1985, 'Development of a finite element code for the analysis of coupled thermo-hydro-mechanical behaviour of a saturated-unsaturated medium', Proceedings of the International Symposium on Coupled Processes Affecting the Performance of a Nuclear Waste Repository, Berkeley, CA, pp. 263–8.

Olivella, S., Carrera, J., Gens, A., and Alonso, E.E., 1996, 'Numerical formulation for a simulator (CODE_BRIGHT) for the coupled analysis of saline media', *Engineering Computations* 13: 87–112.

Philip, J.R., 1974, 'Water movement in soil', in D.A. de Vries and N.H. Afgan (eds) *Heat and Mass Transfer in the Biosphere*, Halsted Press-Wiley, New York, pp. 29–47.

Philip, J.R., and de Vries, D.A., 1957, 'Moisture movement in porous materials under temperature gradients', *Transactions of the American Geophysics Union* 38: 222–32.

Pusch, R., 2002, 'Prototype repository project – selection of THMCB models', Proceedings of the 2nd CLUSTER URL's seminar, Mol, Belgium, EC Report, EUR 19954, pp. 119–33.

Pusch, R., and Yong, R.N., 2006, *Microstructure of Smectite Clays and Engineering Performance*, Taylor & Francis, London.

Pusch, R., Muurinen, A., Lehikoinen, J., Bors, J., and Eriksen, T., 1999, Microstructural and Chemical Parameters of Bentonite as Determinants of Waste Isolation Efficiency, Final Report of European Commission Contract F14W-CT95-0012, Brussels, Belgium.

Rollins, R.L., Spangler, M.G., and Kirkham, D., 1954, 'Movement of soil moisture under a thermal gradient', *Highway Research Board Proceedings (Washington)* 33: 492–508.

Sethi, A.J., Yong, R.N., and Jorgensen, A., 1980, 'Influence of salt concentration on interparticle action and rheology of montmorillonite suspensions', Proceedings of the International Symposium on Salt-Affected Soils, Karmal, India, pp. 169–78.

SKB, 2001a, Äspö Annual Report, Technical Report TR 01-10, SKB, Sweden.

SKB, 2001b, Research, Development AND Demonstration Programme 2001, Technical Report TR 01–30, SKB, Sweden.

Svemar, C., 2002, 'Äspö HRL – In-situ programme and prototype repository', Proceedings of the 2nd CLUSTER URL's seminar, Mol, Belgium, EC Report, EUR 19954, pp. 57–67.

Svemar, C., 2005, Prototype Respository Project, Final report of European Commission Contract FIKW-2000–00055, Brussels, Belgium.

Svemar, C., and Pusch, R, 2000, Prototype Repository – Project Description, FIKW-CT-2000–00055, SKB International Progress Report IPR 00-30, Sweden.

Thomas, H.R., He, Y., Sansom, M.R., and Li, C.L.W., 1996, 'On the development of a model of the thermo-mechanical-hydraulic behaviour of unsaturated soils', *Engineering Geology* 41: 197–218.

Yong, R.N., 1963a, 'Soil freezing considerations in frozen soil strength', Proceedings of the International Conference on Permafrost, NationalAcademy of Science, Washington, DC, Publ. No. 1287, pp. 315–18.

Yong, R.N., 1963b, 'Research on fundamental properties and characteristics of frozen soils', National Research Council of Canada, Tech. Memorandum No. 76, pp. 84–93.

Yong, R.N., 1965, 'Soil suction effects on partial soil freezing', *Highway Research Board Research Record* 68: 31–42.

Yong, R.N., 1967, 'On the relationship between partial soil freezing and surface forces', in H. Oura (ed.) *Physics of Snow and Ice*, Bunyeido Printing Co., Sapporo, Japan, pp. 1375–86.

Yong, R.N., and Warkentin, B.P., 1965, 'Studies of the mechanisms of failure under load in expansive soils', Proceedings of the International Research Engineering Conference on Expansive Soils, Texas A & M, College Station, Texas, pp. 69–77.

Yong, R.N., and Warkentin, B.P., 1975, *Soil Properties and Behaviour*, Elsevier, Amsterdam.

Yong, R.N., Boonsinsuk, P., and Tucker, A.E., 1984, 'A study of frost-heave mechanics of high clay content soils', *Transactions of ASME* 106: 502–8.

Yong, R.N., Boonsinsuk, P., and Tucker, A.E., 1986a, 'Cyclic freeze–thaw influence on frost heaving pressures and thermal conductivities of high water content clays', Proceedings of the 5th International Offshore Mechanics and Arctic Engineering Symposium, Volume 4, pp. 277–84.

Yong, R.N., Xu, D.M., Boonsinsuk, P., and Cheung, S.C.H., 1986b, 'Theoretical analysis of unsaturated flow in a swelling soil', Proceedings of the International Conference of IASTED, Vancouver, Canada.

Chapter 10

Hedin, A., 1997, Spent Nuclear Fuel – How Dangerous Is It?, Report TR-97-13, SKB, Stockholm.

National Academy of Science, 1982, *Quality Criteria for Water Reuse*, National Academy Press, Washington, DC.

Purchase, I.F.H., 2000, 'Risk assessment: Principles and consequences', *Pure Applied Chemistry* 72: 1051–6.

Vesely, W.E., Goldberg, F.F., Roberts, N.H., and Haasl, D.F., 1981, Fault Tree Handbook, Report NUREG-0492, US NRC, Washington DC.

Suggested reading

Bedford, T., and R. Cooke, 2001, *Probabilistic Risk Analysis: Foundations and Methods*, Cambridge University Press, Cambridge.

Polson, N.G., and Tiao, G.C., 1995, *Bayesian inference*, Volumes I and II, Cambridge University Press, Cambridge.

Reed, M.E., and Whiting, W.B., 1993, 'Sensitivity and uncertainty of process designs to thermodynamic model parameters: A Monte Carlo approach', *Chemical Engineering Communications* 124: 39–48.

Singpurwalla, N.D., 2006, *Reliability and Risk: A Bayesian Perspective*, Wiley Series in Probability and Statistics, New York.

Todinov, M., 2005, *Reliability and Risk Models: Setting Reliability Requirements*, John Wiley and Sons, New York.

Whiting, W.B., Tong, T.M., and Reed, M.E., 1993, 'Effect of uncertainties in thermodynamic data and model parameters on calculated process performance', *Industrial and Engineering Chemistry Research* 32: 1367–71.

Index